Learning and Relearning Equipment Complexity

With industrial systems becoming ever more mechanized and reliant on advanced technology, the complexity of equipment, especially in risky industries, is increasing on a daily basis. A thorough understanding of operations and providing safety for these complex systems has become a firm requirement for many. This book offers the knowledge required by safety professionals to provide and maintain the safety of engineering complex systems.

Through a scientific and engineering approach to designing, implementing, operating, and maintaining complex systems, *Learning and Relearning Equipment Complexity: Achieving Safety in Engineering Complex Systems* details the need for more engineering and scientific knowledge to understand and maintain their safety. It gives clear explanations of reasons for a system's complexity, based on control systems and non-linear dynamics. In addition, the book addresses the necessary changes in the approach and the procedures for the safety assessment of engineering complex systems. The reader will develop a thorough understanding of what complex systems are, why they are complex, and how they are utilized.

This book will appeal to any safety professional tasked with complex systems. This extends to professionals in risky industries such as aviation, nuclear power, chemicals, railway and transport, and pharmaceuticals.

Developments in Quality and Safety

Series Editor
Sasho Andonov

Quality-I Is Safety-II
The Integration of Two Management Systems
Sasho Andonov

Bowtie Methodology
A Guide for Practitioners
Sasho Andonov

The Art of Safety Auditing
A Tutorial for Regulators
Sasho Andonov

Safety Accidents in Risky Industries
Black Swans, Gray Rhinos and Other Adverse Events
Sasho Andonov

Learning and Relearning Equipment Complexity
Achieving Safety in Engineering Complex Systems
Sasho Andonov

For more information about this series, please visit: www.routledge.com/Developments-in-Quality-and-Safety/book-series/CRCDQS

Learning and Relearning Equipment Complexity

Achieving Safety in Engineering Complex Systems

Sasho Andonov

CRC Press

Taylor & Francis Group

Boca Raton London New York

CRC Press is an imprint of the
Taylor & Francis Group, an **informa** business

First edition published 2024
by CRC Press
6000 Broken Sound Parkway NW, Suite 300, Boca Raton, FL 33487–2742

and by CRC Press
4 Park Square, Milton Park, Abingdon, Oxon, OX14 4RN

CRC Press is an imprint of Taylor & Francis Group, LLC

© 2024 Sasho Andonov

Library of Congress Cataloging-in-Publication Data
Names: Andonov, Sasho, author.
Title: Learning and relearning equipment complexity : achieving safety in engineering
    complex systems / Sasho Andonov.
Description: First edition. | Boca Raton : CRC Press, [2024] | Series: Developments in
    quality and safety | Includes bibliographical references and index.
Identifiers: LCCN 2023003462 (print) | LCCN 2023003463 (ebook) |
    ISBN 9781032518350 (hbk) | ISBN 9781032520094 (pbk) | ISBN 9781003404811 (ebk)
Subjects: LCSH: System safety. | Safety education. | Industrial equipment—Safety
    measures. | Automation—Safety measures. | Chaotic behavior in systems.
Classification: LCC TA169.7 .A53 2024 (print) | LCC TA169.7 (ebook) |
    DDC 363.11028/4—dc23/eng/20230216
LC record available at https://lccn.loc.gov/2023003462
LC ebook record available at https://lccn.loc.gov/2023003463

ISBN: 978-1-032-51835-0 (hbk)
ISBN: 978-1-032-52009-4 (pbk)
ISBN: 978-1-003-40481-1 (ebk)

DOI: 10.1201/9781003404811

Typeset in Times
by Apex CoVantage, LLC

Contents

PART I Engineering Complex Systems

PART II   Non-linearity in Complex Systems

PART III   Safety of Complex Systems

Preface

There is an interesting story how I started and wrote this book . . .

It all started in Muscat (Oman) in 2016 . . .

I was working as Senior Lecturer at Military Technological College (MTC) and I was alone, without my family. My wife was occasionally visiting me. At afternoons and weekends, after discovering the beauties of Oman, I had plenty of time to dedicate myself to ideas which were coming to my mind very often.

I read a lot there!

I was very often at the MTC Library or I was reading books downloaded from Internet. In all these readings, my attention was drawn by two particular books . . .

The first book was *The Black Swan: The Impact of Highly Improbable* by Nassim Nicholas Taleb[1] and the second book was *Chaos: Making a New Science* by James Gleick. I read these books as a method for relaxation at the evenings, but these books had more profound effect on my future ideas.

Reading about Black Swan events (BSe) and Theory of Chaos (ToC), I realized that these two things are totally neglected in safety community. This was a trigger for the idea for new book on BSe and ToC in the quality and safety areas to be born.

I started to write a new book trying to describe the influence and contribution of Humans to Black Swan and analysis of ToC in the faults of equipment, but I made a mistake . . .

Obviously, there was huge amount of material available on Internet and reading all the books and articles, I realized that I need to split those two areas. I focused myself on Black Swans and in December 2021, my new book *Safety Accidents in Risky Industries: Black Swans, Gray Rhinos and Other Adverse Events* was published by CRC Press.

At that time, I was unemployed and it gave me a chance to dedicate myself on Chaos. I was pretty much present on LinkedIn and I noticed that the area of dealing with Complex System in safety bothers Safety Managers. It was strange to me because, whatever the Complexity of the system is, it must be maintained (implemented, monitored, controlled, etc.) in the same way. The area of Complex System is huge and I found plenty of material dialing with it. The common ground with the Chaos was found everywhere. Depending on how you define Complexity, mostly the uncertainty of operations is caused by the non-linear systems, Humans, and Software and the effect of these uncertainties on the control of operations of Complex Systems are the biggest problem.

I started to gather material regarding the non-linear dynamics, Complex Systems, and I dedicated myself to read it. Reading and studying of these two aspects of Safety took a lot of time, but I needed to go deep into science to get competence for book dedicated to practitioners. I spent a lot of time on non-linear dynamics because

---

[1] In this book, when I would like to refer to Mr. Nassim Nicholas Taleb, I will use acronym NNT. This is done only from practical reasons and I do not like to undermine in any ways the personality of Mr. Taleb!

my knowledge was not on level to deal with all these Complex Systems and Chaos theorems.

Although the non-linearity cannot be split from the Complexity of the systems, so I realized that the uncertainty triggered by the non-linearity is the main aspect of overall Complexity in the engineering systems. Reading the posts regarding quality and safety on LinkedIn, I realized that there is real ignorance about the Nature of determinism, randomness, and uncertainty which affects our daily professional (and private) lives. So, this book is dedicated to making an influence and a contribution toward the missing knowledge about non-linear systems (processes, operations, activities, etc.) as one of the (neglected) aspects of the control of the Complexity of the Complex Systems in industry. This problem of control of the non-linearity very much contributes to the faults of Complex Systems, which causes failures of operations.

As a medical doctor (physician) must know in detail what is normal functioning of the human body to find out what causes abnormal human body functioning (illness), I do believe that this is a way how the engineers must be familiar with the normal functioning of the Complex Systems to understand what would be the abnormal situation with it. The Safety is not about bureaucracy established with the Safety Management System (SMS), but it is about understanding the Complexity of your systems and operations. If you think that you do not need knowledge about that, you are very much wrong!

In June 2019, I attended IMEKO[2] Symposium in St. Petersburg (Russia) and there I was frustrated by the misunderstandings of the scientists regarding "applied science" called *engineering*. During presentations, the presenters (scientists!) used the engineering phrases in totally wrong context. Simply they did not understand that Reliability applies to equipment, not to data (data have integrity), that Quality Control do not improve quality, etc. There was even presentation where one "respected" scientist tried to establish the term "technoscience" for something which is known for centuries as "engineering".

I raised the question regarding all these misunderstandings and of course it was a scandal! They were shocked: How this poor guy is trying to teach us something and he even has no Ph.D. . . . .

OK, I confess: It was punch in my face, but I was not disappointed or offended. Simply I am used to it. The problem was that no one of these scientists try to think what I was speaking about and no one of them produced any contra-argument to my arguments.

As Engineering Training Manager in GAL ANS, I tried the opposite: To explain to the engineers and technicians that whatever they experience in their daily activities regarding maintenance of the ATM/CNS/MET system has a huge scientific background. And I cannot be so proud how much I was successful there.

I tried to write this book as a popular book about the knowledge needed to provide safety of the engineering Complex Systems. Clearly, it is a book dedicated to Safety Professionals where I am providing analysis and remedies (as my opinion) for the

---

[2] IMEKO stands for International Measurement Confederation.

uncertainties caused by the non-linearity of the Complex Systems which they work each day with. This is Sisyphus job and I am not sure that it will be successful . . .

I would like to warn the readers:

The book is titled *Learning and Relearning Equipment Complexity* because most of the Safety Professionals have earned the knowledge presented in this book during their university education. Attending their studies, they have met most of the things which are explained in this book, but simply, this knowledge was lost (forgotten) somewhere, in between their professional engagements. So, do not blame the method or methodology for Risk Assessment if something unexpected happen in your system (operation, process, activity, etc.). Methods or methodologies depend on your knowledge how you can use it, but it depends (maybe more) on the data you include in it. And the knowledge about data could make difference.

Also, this book is going into something qualitatively different and new from that what Safety Professionals have studied in their safety education and their safety trainings. This book is more mathematical than other safety books and it deals mostly with missing knowledge about engineering Complex Systems, as big part of providing Safety of the operations in the industry.

Although this book is an engineering book, I have not paid attention to some engineering disciplines. Let's say, I did not cover here the IT networking (servers, routers, switches, etc.) for transfer data as one of the engineering Complex Systems which is multinational and worldwide spread. The same things happen with the electricity grid where there is Pan-European (Pan-American, Pan-Asian, etc.) grid (a Complex System!) for power interchanges for our homes, cities, industries, etc. Tap water distribution in the cities is the similar case of Complex System. All these Complex Systems are not considered here, but I cannot say that most of that what is mentioned here could not apply to them also. This is a book for engineering Complex Systems which are used as machines (or other types of equipment) used to manufacture some product or to provide some service, mostly in Risky Industries.

This is a book about a "gray area" of safety of engineering Complex Systems and such a "gray area" is not easy to comprehend, even for the scientists. That is the reason that I will try to use some simple formulas and simple examples with intention to explain the Control Theory, the non-linear dynamics, and the Theory of Chaos to non-mathematical persons. Of course, I cannot present all Complexity of systems behavior here, but I have put a lot of efforts to achieve simplicity. Please forgive me if I was unsuccessful.

I hope that this book will occupy your attention and I hope this book will provoke more research in the behavior of Complex Systems (processes, operations, activities, etc.) from the point of view of the non-linearity in the future.

Sasho Andonov
Skopje, North Macedonia

Author Biography

Sasho Andonov is Graduated Engineer of Electronics and Telecommunications, and he has a Master's Degree in Metrology and Quality Management. He has 32 years of industry experience and most of it in aviation. He is dedicated to Quality and Safety Management, especially in Risky Industries. His research interest is in Quality, Safety, Metrology, Non-linear Processes, Complex Systems, and Calibration. He has teaching experience from time spent at the Military Technological College in Muscat, Oman; in GAL ANS in Abu Dhabi, UAE, where he taught aviation subjects (EASA Part 66 and ATSEP); and as Faculty Member of Higher College of Technology (Khalifa Bin Zayeed Air College) in Al Ain, UAE. He has contributed to many conferences with papers on aviation, safety, quality, and metrology. In addition, he has published four books with CRC Press covering areas of Quality Management, Safety Management, and Quality/Safety Auditing.

Acronyms and Abbreviations

7QCT	7 Quality Control Tools
ABS	Anti-lock Braking System
ADC	Analog to Digital Convertor
AGC	Automatic Gain Control
AI	Artificial Intelligence
ALARP	As Low As Reasonably Practical
AM	Amplitude Modulations
AOT	Actual Operating Time
ASIL	Automotive Safety Integrity Levels
ASSET	Assessment of Safety Significant Event Teams
ATCOs	Air Traffic COntrollers
ATM	Air Traffic Management
BIBO	Bounded Input—Bounded Output
BITE	Built In Test Equipment
BJT	Bipolar Junction Transistor
BPG	Baseline Practices Guide
CAD	Computer Aided Design
CAST	Causal Analysis (by using) System Theory
CNC	Computer Numerical Controlled (machines)
CNS	Communication, Navigation, and Surveillance
COTS	Commercial Off-The-Shelf
CPU	Central Processing Unit
CVR	Cockpit Voice Recorder
DAC	Digital to Analog Convertor
DAD	Data Acquisition Device
DAL	Development Assurance Levels (document DO-178C for software)
DAL	Design Assurance Level (document DO-254 for hardware)
DoD	Department of Defense
DoF	Degree of Freedom
EBD	Electronic Brakeforce Distribution
ESARR	European Safety Regulatory Requirement
ETA	Event Tree Analysis
EUROCAE	European Organization of Civil Aviation Engineers
FAA	Federal Aviation Administration
FAT	Factory Acceptance Test
FDR	Flight Data Recorder
FFT	Fast Fourier Transform
FM	Frequency Modulations
FMEA	Failure Mode and Effect Analysis
FMEDA	Failure Mode, Effect, and Diagnostics Analysis
FMS	Flight Management System (autopilot)
FTA	Fault Tree Analysis

GUI	Graphic User Interface
HMI	Human–Machine Interface
HP	Human Performance
ID	Interrelation Diagrams
IEC	International Electrotechnical Commission
ILS	Instrument Landing System
IM	Interrelation Matrix
IMEKO	International Measurement Confederation
INCOSE	International Council on Systems Engineering
IV&V	Independent Verification and Validation
JPD	Joint Probability Distributions
LCL	Lower Control Limit
LoC	Level of Criticality
LRU	Line Replaceable Units
LSL	Lower Specification Limit
MCAS	Maneuvering Characteristics Augmentation System
MIMO	Multi-Inputs–Multi-Outputs
ML	Machine Learning
MRAC	Model Reference Adaptive Controller
MRO	Maintenance, Repair, Overhaul
MSA	Measurement System Analysis
MTBF	Mean Time Between Faults
MTBO	Mean Time Between Outages
MTTR	Mean Time To Repair
NSATE	Nuclear Safety Analysis and Technical Evaluation
NSCCA	Nuclear Safety Cross-Check Analysis
OEM	Original Equipment Manufacturer
OHS	Occupational Health and Safety
OHSE	Occupational Health, Safety, and Environment
PA	Power Amplifier
PCB	Printed Circuit Boards
PIC	Pilot In Command
PID	Proportional Integral Derivative (controller)
PIO	Pilot-Induced Oscillations
PLC	Programmable Logic Controllers
PM	Phase Modulations
PPE	Personal Protective Equipment
PSD	Power Spectral Density
QA	Quality Assurance
QC	Quality Control
RAM	Random Access Memory
RCA	Root Cause Analysis
RF	Radio Frequency
RMS	Root Mean Square
ROM	Read Only Memory

RPN	Risk Priority Number
RTCA	Radio Technical Commission for Aeronautics
SAT	Site Acceptance Test
SCADA	Supervisory Control and Data Acquisition
SDLC	Software Development Life Cycle
SMS	Safety Management System
SoE	Sequence of Events
SPC	Statistical Process Control
SPF	Single Point Failure
SPICE	Software Process Improvement and Capability dEtermination
SQC	Statistical Quality Control
STC	Self-tuning Controller
STECA	System-Theoretic Early Concept Analysis
ToC	Theory of Chaos
UCL	Upper Control Limit
USL	Upper Specification Limit
UTA	Unsafe Control Actions
WEC	Western Electric Company

Introductory Explanations Regarding the Book

Maybe you will wonder why I am speaking about all these things here, but please understand that, later, all of these things will be used in explanations which will be given for particular aspects of Complex Systems.

Let's clarify the things in this place . . .

This book is divided into three parts: Part I is an introduction to Complex Systems from engineering point of view; Part II is about the dynamics in the Complex Systems with emphasis on non-linear dynamics; and Part III is about Safety of Complex Systems and how to achieve it.

I am using words here as "non-linear". Please understand that grammatically, in English language, the correct word is "nonlinear' (without hyphen), but in mathematics (and engineering!) mostly the word "non-linear" is used. In the Oxford Dictionary (which is pretty much good reference for the English language), you will find the word "non-linear" as the correct spelling and there is no other alternative. But if you try some of the US English dictionaries, the only word which you can find there is "nonlinear" (without hyphen). So, that is the reason that I am using "non-linear" (let's be more British . . .).

The Safety Management System, as a systematic way to provide safety, can be described as congregation of Equipment, Humans, and Procedures. The Procedures are those that connect the Equipment and Humans, and the Safety Procedures actually build the Safety Management System. There are also Operational Procedures which deal with operational (production) processes in industry.

When I speak in the book for Humans, Procedures, and Equipment as constitutive parts of any management system, I will use capital letters at the beginning. For all other cases, I will use small letters for these three words at the beginning.

I will use the word "Complex System" mostly as a word to describe the Equipment (industrial systems and machines) which will be aggregation of Hardware and Software. In addition, a word "system" can also be used for the management system (where the Equipment is constituent). However, most of the explanations will also apply to the processes, operations, activities, etc. So, this wide interpretation about the "system" in the book should prevail.

In general (and to simplify the future explanations), nevertheless, the Chaos is created by functioning of the physical laws and it started mostly as mathematical concept, there are many aspects of human behavior and in Nature which can be associated with the Chaos.

Also, I will speak about "behavior" of the system and it can also be associated with the "functioning" of the system. Anyway, every "functioning" of the system is characterized by particular "behavior" of that system.

I will use the words as "fault" and "failure". The word "fault" will be used for abnormal functioning of the parts (Subsystems) of engineering Complex System and the word "failure" will be used for abnormal functioning of operation, process, or activity of the Complex System. Having in mind that Complex System conducts some kind of operation, "fault" of the system could produce "failure" of operation, but "failure" of operation does not necessary mean "fault" of Complex System. The reason is simple: Failure of the operation can be caused by wrong command or data submitted to the Complex System by the humans and by bugs or missed data in the Software.

Let me support this by very good and expensive example . . .

In January 1999, NASA had launched its Mars Polar lander which arrived on Mars in December the same year. Unfortunately, after the landing, the lander did not respond to signals sent from the NASA Control Center and soon the engineers realized that the lander crashed.

The investigation was not easy, having in mind the distance and lack of data, but NASA experts thought that they have found the reason for the crash. They do believe that reason was a spurious signals created by the craft's legs when they were deployed before the landing. The landing engines, which needed to slow the descent of the lander before landing, actually stopped early. Automatic deployment of the craft's legs produced this spurious signal, signaling to the onboard computer that the lander touched the Mars surface and the engines were turned off. Sensing of the lander was based on a sensor which misinterpreted the noise from a Hall effect sensor, simulating that lander has landed, although it was 40 m above the surface. The point is that no component or Subsystem failed and everything was perfect except the assumption that legs deployment will not affect the landing sensor signal.

I will use the terms "equation" and "formula" in the book with the same meaning: It is a written statement used in mathematics that two amounts (two symbols or groups of symbols representing an amount) are equal.

I will use the terms "estimation" and "prediction" in the following meanings:

- The "estimation" is usually connected with setting (adjusting, seeking) the optimal value of some parameter (variable, factor, etc.) given the particular set of data is connected with this parameter (variable, factor, etc.); and
- The "prediction" is actually guessing of the future state of our system (operation, process, activity, etc.).

Very often, I will use term "system in use", this is actually the system which we have designed to provide some operation which could bring us benefit for something. This could be a "plain" system and with intention to make it better, we add some controls or additional devices which could help with use of the system (we produce Complex System).

Another thing to clarify is that by the terms "operation" and "process", I will understand the normal operations or normal processes and by the phrase "adverse event" (I will not use it very often), I will understand the incidents and the accidents.[3] Or to be more specific, the phrase "adverse events" in this book is a collective name

---

[3] Not all abnormal operations (or processes) will result in incidents or accidents. Some of them may not be even noticed (nothing will happen), but anyways this is not a normal operation (process).

for incidents and accidents. So, the different types of incidents and accidents belong to the outcomes of the same adverse event. Anyway, in this book, all incidents and accidents are other names for particular abnormal states of Complex Systems and as such when I speak about the Complex Systems, I will include also the processes, operations, or activities executed by these systems.

I will use the term "signal" with two meanings. Sometimes it will be used as electronically transmitted information or data and sometimes it will be used as input or output (row material, parts, tools, etc.) of the system. I do believe that the context of the sentence where the term is used will clarify its meaning.

I will use interchangeably the terms "Control System" and "Controller" with the same meaning. Usually, but not as a rule, if in the sentence there is already the word "system", I will use "Controller" as a second term in the sentence with meaning "Control System".

I will use the terms "speed" and "velocity" in the book. The difference in physics is that the speed is scalar quantity (just a number) and velocity is a vector (which has direction, magnitude, and point of attack).

Additionally, I would like to emphasize here that in my life I have not found any book on introduction to non-linear dynamics or Control Theory which is less than 400 pages. So, please note that just the essential basics are mentioned in this book regarding the analysis of non-linear systems and Control Theory. If you need more understanding about non-linear dynamics as mathematical (or better say Engineering) discipline, you may attend some of the courses offered on the Internet or else you can refer to plenty of books and articles available there.

Let me emphasize the fact that during the writing of this book for almost five years, I had read a large number[4] of books and articles for many of the topics included here. The result of all these readings is that I strongly understood that there are many different views on almost any topic which is mentioned in this book.

Please understand that from all these readings, I have chosen a side and this book is produced in synchronism with all these choices which I had to make. I fully understand that many of the readers will like or dislike my choices, but please note there is nothing about "like/dislike" in this book. But there is plenty which can trigger your advanced opinions and new ideas about the things which I am presenting in this book. And believe me, I did not make these choices taking care for my glory, but rather I have made them on the arguments which I thought fit reality and the present status of Safety in industry.

I may be wrong, but I do have enough courage to confess it!

Please note that I am using the term Functional Safety as something different from Occupational Health and Safety (OHS). This is not a "Functional Safety" which is regulated by IEC 61508 series of standards and some similar standards from International Electrotechnical Commission (IEC).[5] So, when I speak about "Functional Safety" from IEC 61508 standards, I will use quotation marks.

---

[4] This number is huge, so I gave up the idea to provide the references for all these books and articles. I focused myself to build my own attitude on the topics explained here. Internet is full with free books and articles and I encourage the reader to make his own investigation of all these items.

[5] I would also include here the ISO 26262 series of standards (Road Vehicles—Functional Safety).

There is a good reason for that and I would try to explain it here . . .

The IEC 61508 series of standards were brought in reality where the OHS(E) was moving into "Technological Era" regarding improvements in industry. This is a time when the science and technology developed very fast and considerably, so the humans started to try to improve OHS(E) by advanced technology. This is the era which was pretty much present in aviation where improvement of the technology built into the aircraft was seen as the only way to provide safety of the flights. In industry, this "Functional Safety" applies to the systems (production equipment) used for manufacturing the goods and offering the services. The IEC standards cover OHS(E) in industry and I think they are pretty much complex because they had totally undermined the human involvement into manufacturing processes. To be more precise, there is attitude that humans will be changed with systems and it will bring OHS(E) improvement.

I do believe that many of the readers will be shocked by this statement, but I cannot neglect the full connection of the IEC's "Functional Safety" with OHS(E). I do not say that this part of safety is not working, but I think the name is wrongly assigned: Instead to call it "Functional Safety", I would call it "Process Safety", simply because that is a type of safety which contributes very much to the safety of the industrial processes. So, whenever in the book I mention "safety" of the Complex System, it is about Functional Safety established to provide safe product or service, not as IEC 61508 "Functional Safety" (which provides safe processes in the factories).

Anyway, please do understand that this is a book about forgotten engineering knowledge of system's Complexity, especially about complex behavior of the systems which is introduced by their Complexity. Maybe when you read this book, you ask yourself:

- Have you thought about some of the things mentioned in the book previously?
- Can some of these things in the book be connected to safety events in your organization in the past and present?

Answers to these questions could provide good merit for the value of this book to you and to me . . .

To whom is this book intended?

This is good question, but the answer is simple: To the Safety Managers, Safety Theoreticians, and Safety Practitioners in Risky Industries,[6] or in other words, to the Safety Professionals all around the world!

In my humble opinion (which is based on my 30 years' experience), most of the Safety Managers are people with considerable expertise in industry and very poor knowledge in Safety. The Safety Management Systems (SMS), which they have created and maintained in their companies, are usually prone to fundamentals flaws, simply because these guys do not know that there is structured and systematic way to deal with safety issues (from the basic ones to the advanced ones). This problem

---

[6] Risky Industries are industries which could endanger humans, assets, and the environment in huge scale by their operations, activities, processes, services, and products. Such industries are aviation, railway, nuclear, chemical, pharmaceutical, medical, etc.

is increased by the employees in the regulatory bodies in Risky Industries, because actually they need to "regulate" the Risky Industries and they cannot do it. The simple reason is that all these company's Safety Managers (which I have mentioned in previous sentences) are the human resource for inspectors and auditors in Regulatory Bodies in Risky Industries. If they do not have proper safety oversight training, they will miss the point of being a Regulator. And believe me, there are more "bad boys" than "good boys".

In general, this is a complex and brave book . . .

It is "complex" because the chapters look uncorrelated to each other, which is wrong. The overall connection between the chapters could be found by reading the whole book. I found very important to put in the book everything which can help to understand the problems regarding safety of Complex Systems and their remedies in the industry.

It is a "brave" book because I am not very sure that I will be really successful, having in mind the Complexity of area which I decide to explain in a more engineering than scientific manner. In addition, the Complexity is subject of research in science and very much in philosophy, but I have focused myself to produce a book which will help engineers who are at Safety Manager's positions in Risky Industries and deal with engineering Complex Systems there.

Still, I am not sure, if the writing of this book was a wise step . . .

But I am sure that this is a holistic book and as such, it can be judged by one who will read it from the first to the last page. It will not be easy (I afraid . . .), but I think it will be worth.

I found all these explanations at the beginning of the book very important for the readers who have decided to read the rest of the book. I hope you will read this book with due diligence and you will appreciate plenty of my opinions after you finish it . . .

Part I

Engineering Complex Systems

1 Introduction to Complexity and Complex Systems

1.1 INTRODUCTION

Complexity . . .

This is a word which sounds as expression of beauty to me . . .

Its beauty and its extravagance are not connected only with the good "soundings" into my ears, but also with the meaning which (by my humble understanding) describes something special, something which poses a meaning of beauty, which cannot be correctly understood by everyone. I do believe that humans must have particular advanced level of education, knowledge, and mental power to be able to understand the particular meaning of this noun which is mostly used as adjective in combination with other nouns.

This is a word which is very much used to explain some advance in science, technology industry, sociology, psychology, etc. As such, it has been used very much in the past and in the present and it is associated with one unordinary paradox: Even today, the scientist, engineers, philosophers, and ordinary human beings could not agree about correct definition for this word and its use.

I would not try to solve this paradox with this book. In general, it is not my intention to deal with Complexity from the philosophical point of view. Having in mind that I prefer "engineering creativity" to "scientific rigor", I am dedicating this book to ordinary engineers and Safety Professionals with intention to help them to deal with engineering Complex System for the purpose of providing Safety to humans, to assets, and to the environment.[1]

This is the reason that I will try to introduce my pragmatic (engineering) view on engineering Complexity, not undermining all other efforts (social and philosophical). And all this is done with intention to help the Safety Professionals to deal with faults and failures of the engineering Complex Systems.

1.2 HOW DO WE (HUMANS) DEFINE COMPLEXITY?

If you go on Google (Internet) and try to find a definition for Complexity by just writing: "Complexity meaning", you can be encountered by 1 billion and 440 million results. At least it happened to me when I tried it . . .

---

[1] In general, the Complexity could be associated with different phenomena, aggregates, assets, behavior, organisms, structures, events, or problems from any areas in the Nature!

DOI: 10.1201/9781003404811-2

Many of these results are using different areas as subjects for definition and most of them are part of social science and philosophy. There are approximately 20% of all these results which are connected with engineering systems.

From this huge bunch of popular (not scientific) results, the closest one which I would like to use as definition for the purposes of this book is the one which defines Complexity as a state of a system characterized by having many parts, many interactions, and many interdependencies[2] between these parts and as such, being difficult to understand or to find an answer how it works.

The humans always tried to understand the functioning of the Nature and it is intrinsic to all these tries, to find a mechanism which will explain it. Sometimes, it was easy and logical, and sometimes it was not so easy and not at all logical. For those humans who tried to deal with these "not so easy and not at all logical" things, we use the name *Scientists* and all methods, tools, and efforts which they use are associated with the area called *Science.*

As one of the most important characteristics of the Science, I can mention the systematic approach in dealing with the search for "the truth about our reality". Obviously, if there is no simple explanation, then we need to find different approach and we use many "non-conventional" methods, tools, and efforts to explain truthfully the functioning of the reality and the events which happen there.

So, using the explanation, as mentioned previously, in accordance with these methods, tools, and efforts, we can divide our reality as Ordinary (easily understandable) and Complex (not easily understandable).

There is scientific discipline called *Complexity Science* which is dedicated to research regarding the different Complex Systems and their behaviors.

1.3   DEFINITION OF ENGINEERING COMPLEX SYSTEMS

The point is that the correct border between these two areas (Ordinary and Complex) could not be determined. Sometimes, something which looks very much Ordinary could be very much Complex and vice versa.

Having in mind that the definition of Complexity (first) is taking care for "many parts and their interactions", the humans gave a name to this "many-parts-assembly" and they call it System.

When I used the same "process" on Google to find the definition of System, I encountered approximately 5 billion and 160 million results. In general, these results can be associated with two areas: One which can be associated with something "physical" (touchable) and one which can be associated with something "imaginary" (non-touchable).

For the purpose of this book, I would define the "physical" system as a set of "tangible" things which are assembled to do some job together. I would define the "imaginary" system as set of "intangible" principles, procedures, regulations, or rules which are used to do some job. The elements of these two sets and any other combination of the elements from both sets, I will call Subsystems. The simple

---

[2] Somewhere in the literature dealing with complex system, you will find term "coupling" for interactions and interdependencies.

example of "physical" system is all mechanical or electrical apparatus (equipment) in our lives and example of "imaginary" systems is all management systems (safety, quality, financial, logistic, etc.). In the scope of this categorization, I would include the Software as "imaginary" Subsystem of "physical" systems.

But this is not enough . . .

Defining an engineering Complex System as a set of Subsystems, it could happen any of these Subsystems (or even all of them) to be also (by itself) Complex Systems. So, it is possible to have Complexity inside the Subsystems of the Complex System and depending on the system or its parts, this Complexity could be also considerably big.

So, having in mind the above-mentioned, for the purpose of this book, I would define Complex System as a system assembled by many parts (combination of "physical" and "imaginary") which are interdependent and interact between themselves on a way which cannot be easily understood, explained, validated, and verified.

How can we determine that the engineering system is "complex"?

In engineering, any system can be determined complex by the following factors:

- Its internal structure (organization), which means there are a lot of different parts (Subsystems) to build and to operate the system. Complexity actually could arise from complex structure of these parts in the system.
- The processes inside, which means that the system by its structure could be simple, but the processes (operations, activities) executed by the system are complex. So, there is need for additional efforts (knowledge, skills, time, resources, etc.) to control and maintain them.
- The changes (variations, dynamics) of the system in time and space during its operation, which is actually opening a door for non-linear systems to be called Complex Systems.
- The Nature of interactions of the system with humans and the environment.

Of course, the combinations of all these four subjects are not only possible, but also, very often, inevitable, so it would increase and "complicate" the Complexity of our system in total.

More critical Complexity is the one which can be found in last three points. Internal processes in the system and their variations in time and space, together with external interactions and interdependencies between them and with humans and the environment, are also subject of Complexity. All these things can be very much unknown, uncertain, and unpredictable, and as such they would be more critical!

Having in mind that the Complex Systems interact with humans and the environment makes them to belong to the class of dissipative systems. These are systems which exchange the energy (storing and dissipating) and this exchange depends on the operation of the system and conditions of the environment.

Somewhere in the literature you can find that the Complexity of any system can be expressed and measured by the amount of information which needs to be presented to explain the function of the Complex System. The information for the system is actually the knowledge used to produce the system because without the deep knowledge of engineering and scientific laws and principles, it could not happen. Here I must state that there is a lot of more knowledge needed to design a Complex System than to operate it. I hope everybody will agree to that because it is obvious that the knowledge needed to design and manufacture a car is considerably bigger than the knowledge to drive the car.

And this is the biggest problem in safety of the Complex Systems: Although the level of knowledge needed to maintain the Complex System is somewhere between that to design it and operate it, this amount of knowledge is close to designing the Complex System. But reality is not supporting this . . .

I had a chance to meet excellent maintenance engineers and very "poor" maintenance engineers, and in Risky Industries, to have "poor" engineers is dangerous. So, starting from this definition about Complex System, it is clear that the guys who design, operate, and maintain such systems must be excellent engineers. This statement also applies to the Safety Professionals!

There is something which must be mentioned here: Each of the engineering Complex Systems which is designed and used in our professional and private lives is mostly assembled with a purpose: The particular input needs to be transformed into particular output. So, they are made for purpose, not for beauty. This should not be forgotten!

There is another aspect of the engineering Complex Systems and this is connected with the analysis of these systems. There are plenty of simple systems which are subject of complex analysis to explain them. So, there are systems which are simple by construction and complex by their behavior, and as such they are not easy to understand. The non-linear dynamics know many such systems.

In general, the science has coined the name "Complex System" to describe particular phenomena, industrial structures and aggregates, biological organisms, social events, or problems in Nature that are not easy to understand or to explain.

The Complexity of the systems sometimes make them unpredictable, but this unpredictability is more often an expression of our poor knowledge about the system than a real problem with Complexity. Anyway, this is not always the case. As I will explain in this book, there are some natural aspects of this unpredictability which are caused by innate uncertainty.

The Complexity is connected very much with our wish to control the systems. Actually, our "wish to control" is very much a reason to make systems more complex. There are many processes, operations, and activities which we need to control and we use an engineering for such a control. The level of control is usually contributing very much to the additional Complexity of the systems. Usually, simple controls provide smaller Complexity and vice versa, smaller Complexity needs simple controls. Leveson in her book *Engineering a Safer World: Systems Thinking Applied to Safety* defines these control measures as constraints and I am happy to accept this definition.

With such a configuration and Complexity, we need a Team to take care of a Complex System. Especially this is required when we need to support or maintain a complex operation. The meaning of the Team is that there are members of the Team who have dedicated particular task (out of many) to maintain particular part (out of many) of the engineering Complex System.

The Philharmonic orchestra is a Team. There are plenty of musicians, and each of them is master to his instrument (and only to his instrument) and they, together, produce beautiful performance of the music. If some of the musicians are ill and cannot contribute to the orchestra, then they will be replaced by the same type of musicians. Simply, the Violinist cannot replace the ill Oboist.

In each Team, there is need for a person who will synchronize the Team members and oversee them. In Philharmonic orchestra, this person is the Conductor and in industry, this person is the Manager.

One of the most important questions is how we can quantify (measure) Complexity (if it can be quantified/measured). There is an approach which is based on asking simple questions:[3]

(a) How the Complex System is hard to describe?
(b) How the Complex System is hard to create/design?
(c) How the Complex System is hard to operate?
(d) How the Complex System is hard to maintain?
(e) What is the degree of organization of Subsystems inside the Complex System?

For the purpose of this book, the measure of Complexity is irrelevant. Dealing with safety of Complex System in engineering area does not need to quantify the level of Complexity, but a good understanding of hierarchy, interdependencies, and interactions inside the system is very important.

1.4 COMPLICATED SYSTEMS AND COMPLEX SYSTEMS

There are Complicated Systems and there are Complex Systems[4]...

Whatever the system is, complicated or complex, in the everyday life we assume that this is a system build up by many parts. In accordance with this, we can say that Complicated System is simply a sum of all its parts, but Complex System is more than that.

It means that there is significant difference between Complicated System and Complex Systems, nevertheless they both may have many parts. In reality, we can find both of them (usually) combined, producing "hybrid" systems. Roughly speaking, I can provide to you these explanations:

- The Complicated Systems are systems where building parts have particular autonomy. Removing one part of this system will not stop system from functioning. It will only produce loss of some characteristics or benefits not necessary for the primary reason why the system is built. For example, a car is Complicated System. There, losing the lights will not endanger the transport of humans and goods by the car, but it will decrease their safety and comfort, especially at the night.
- The Complex Systems are system where if one of the building parts is removed, the system will stop to provide its function.

---

[3] If you are interested to know more about conditions for a good measure of Complexity, you may consult Seth Lloyd. He is Professor of Mechanical Engineering at MIT, USA.
[4] Whatever is written about Complex System in this book applies to the engineering Complex Systems in industry. There are things regarding engineering Complex Systems which are quite different for social, human, or natural sciences and they will not be presented here. Please keep in mind that whatever is mentioned in this book regarding Complex System is for the purposes of this book only.

Having in mind the aforementioned definitions, we can assume that Complication can produce degradation of the functioning of the system (it will affect the "peripherals" of the system) and Complexity can cause end of functioning of the system (it will affect the "core" of the system). It means that we can make complicated systems simpler, but we cannot make Complex Systems simpler. In general, the things are not so simple, but I can say that the Complexity is usually embedded (innate) into the system and the Complication is usually a marginal consequence of designing the Complex System.

The human body is an excellent example of "hybrid" system: It is complicated and it is complex. Losing the hand or leg will not cause death, which means it is Complicated System, but losing heart, liver, or brain will produce death, which means it is Complex System. Going further with the human body, we can say that it is a "hybrid" system between "peripherals" (parts which build Complication) and "fundamentals" (parts which build Complexity).

The same duality of the systems exists also in industry. There are systems which if they lose one (or few parts) will degrade its operations and if they lose another (let's call it vital part), the functioning of the system will stop. To "kill" the Complicated System, you need to destroy many parts and to "kill" the Complex System, you may destroy only one single part. Having in mind that both things could happen, it is wise to treat each system in industry as a "hybrid" system.

One of the most important things regarding the Complex Systems is the fact that parts inside the system (working together) usually provide total behavior of the system which is qualitatively and quantitatively different than the behavior which would be provided from the parts if they are alone. This is called Emergent property of the Complex System.

The Complexity in these systems arise due to many Subsystems which actually impose constraints (limitations) in the functioning of the systems aiming to keep the operation of systems stable and controlled. However, from the safety viewpoint, as mentioned earlier, the Complex Systems may have a Single Point Failure (SPF) and this is a very dangerous situation.

Whatever be the difference between the Complicated and Complex Systems, we need to provide full functionality of any of them. Anyway, I will stick in this book to the engineering Complex Systems, simply because whatever is mentioned here about Complex Systems totally applies also to the Complicated Systems. But be careful: The opposite is not working.

However, it is clear that from the Safety viewpoint, there is no big difference in treating Complex Systems versus Complicated Systems. The point is that there is no clear distinction between them and most of the systems based on new technologies are mostly "hybrid" systems.

1.5   GENERAL SYSTEMS DISCUSSION

Any "hybrid" system[5] could be represented by Figure 1.1.

We use the systems to solve some problem, to provide some benefit, to produce some product, to provide some service, etc. As such, the systems can be characterized

---

[5] Actually, this figure can also be used for "imaginary" systems, but let's stay with "physical" system for the purpose of this book.

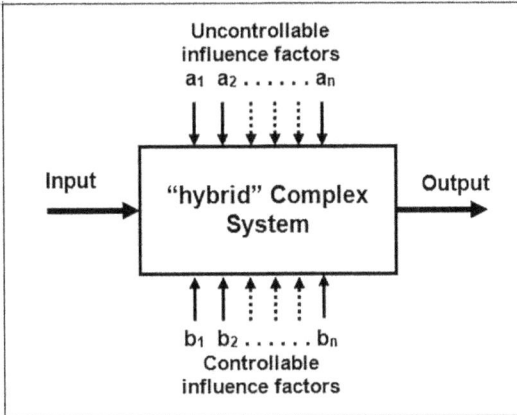

FIGURE 1.1 Complex System presentation.

by some input and some output. Systems use one or many processes to transform the input into output and this is presented in Figure 1.1.

Depending on the construction of the system (number of parts or Subsystems) and Complexity (interdependencies and interactions) of the processes inside and outside, we can decide if the system is complex or not. The main point is that you cannot find in the literature clear-cut criteria for a system to be complex or not.

Figure 1.1 shows a simple system with one input and one output, but the Complex System may have few inputs and few outputs. Such systems are known as MIMO (Multiple Inputs–Multiple Outputs) or multivariable systems.

Usually the systems, in addition, support some process,[6] which, not necessarily, should be complex. Supporting the process, the system must adapt some of the factors inside and outside the system. The factors are presented by some parameter which is a quantification of some adaptable characteristics of the system. The system (especially the complex ones) could have many factors (parameters) to be adjusted, but not necessarily all of them are used for the particular process. As it can be seen in Figure 1.1, these factors are influencing (inside and outside) the behavior of the system and they can be simply categorized as Controllable and Uncontrollable.

There could be produced another criterion for Complexity of the systems which can be based on the number of controllable parameters. If this number is considerably big, then the system can be pronounced as Complex. The problem with this criterion is that the trend in the industry is (without taking care of the total number of factors in the system) to use just few of them to control the process and the system. So, again, this criterion does not have objective form: The system can be complex, but during its use only few parameters are used to control it (which makes the system simple) or the system can be simple with just six factors (for example), but all of them are used for control.

---

[6] More generally, it can be production process, measurement process, building process, analyzing process, etc.

The simple example for this is the car. There is considerable complexity of designing and building a car, but the drivers are using only few factors (parameters) as pedals and steering wheel to drive it. The manufacturing of the car is a complex process, but driving the car is a simple process and you do not need university degree to do it.

So, whatever the system is, during the design process, we can decide how many factors could be controlled. The control can be provided manually or automatically, but in general both ways have particular flaws and particular benefits. Manually, the system is prone to all imperfection of humans and automatically, the already Complex System becomes more complex, at least by adding the single Control Subsystem.

In general, the total number of factors which can change behavior of the system could be considerably big, but most of them are not used for control or maintenance of the system. Most of these factors are adjusted during manufacturing process of the system and, as such, they are kept fixed inside some values determined by the tolerances[7] during the operation of the system.

1.6   HOW THE ORDINARY SYSTEMS BECAME MORE COMPLEX (COMPLICATED)?

In dealing with general aspects of Complex Systems, there are three approaches to the study of Complex Systems:

1 How the interactions and interdependencies produce complex patterns of behavior?
2 What is the available space of possible states?
3 How the Complex System can form through pattern formation and evolution?

These things are for general Complex Systems (in sociology, biology, environment, etc.), but having in mind that this book is about engineering Complex System, I would propose considerably different approach of generating Complex Systems.

Let's go through the phylogenesis[8] of ordinary car which is used today in our lives. Humans are very unique!

They pose geniality and creativity and eternal affection to make their lives easier.[9] Such an attribute can also be found with other animals, but not at such a level as with humans. At the dawn of human evolution, the humans carried their properties by themselves. After that, as the number and volume of the properties increased, they used the animals to help them (mostly horses, donkeys, camels, oxen, etc.). Innovation of the wheel was another step forward: Now it was possible to combine

---

[7] Tolerances are allowed value of variability of some parameter. It means that if the parameter changes its value in the scope of these tolerances, the process (operation, activity) will not be affected. This will be explained in Chapter 3.

[8] Phylogenesis is a description of overall development of a family of systems (species in the biology). There is also ontogenesis which is description of overall development of one system (one individual organism in biology). For the sake of this book, I will use phylogenesis and ontogenesis as terms connected with engineering systems.

[9] Honestly speaking, the purpose of any good engineering system is to ease the everyday burdens of humans either directly or indirectly.

horses (donkeys, camels, oxen, etc.) with carriages and, the number and volume of goods which could be transported increased more. Of course, they realized that these carriages can be used not only for the transport, but also for other things, for example, as war carriages and later war vehicles.

Anyway, the development did not stop there. There was need to find something more sustainable, effective, and efficient than animals and the industrial revolution started with steam engine. It was natural to continue, so the benzine and diesel engines were innovated. Later, there were electric motors, but due to lack of batteries, they could not be used at that time. Today, we use electrical cars and their development will continue.

Nevertheless, in the beginning, there were only the chassis, engine, seats, and wheels for the car. Soon, they realized that they need brakes to stop the car and steering wheel to control the direction of driving. Thereafter they replaced the wooden seats with most comfortable seats made by leather and added a roof. As the use of car spread all around the world, things continued to be improved: The lights for night driving were added; only the roof was not enough, so the closed cabin was added, etc.; later the air-conditioning, security bells, sensors, and actuators,[10] electronic controls, on-board computer, software, ABS and EBD, radios, GPS, etc. were also added. With all these "additives", the computerized Engine Management System was established which monitors all engine parameters. Later, it was upgraded to Car Management System to control everything in the car.

The point is that the cars eventually became more and more complex. So, starting with a simple car (carriage with the horses), it became a Complicated and a Complex System at the same time. A Complicated System because not all of these things contribute to the role of the car (transport of humans and goods), so malfunction or failure of many of them will not affect the driving capabilities of the car. A Complex System because many of these systems really contribute to better driving capabilities of the car and malfunction or fault to any of them will put a car in a garage.

When I was a child, my father and many of our neighbors did not bring a car to the mechanic for ordinary maintenance: In simple words, they were capable to handle all these preventive maintenance by themselves. But today things have changed: To do preventive maintenance of car, you need more specialized tools and instrumentation. All these things are too expensive for the individuals to buy themselves and those tools, those instrumentation, and the knowledge and skills to use them are not available to everyone today. So, the Complication and the Complexity included inside the cars affected the overall use and maintenance of the cars. Now, although the life with modern cars is better, there are other things which have become more complicated and complex . . .

Another case which contributed very much to the development of Complex System was Safety. In aviation, people realized that if there is a problem with some device inside the aircraft during the flight, nothing can be done. If this device is a part of systems supporting the flight, then aircraft will crash and chances for anyone

---

[10] Sensor is a device which detects or measures particular physical property of interest and usually transduce it into electrical quantity (voltage current, etc.). Actuator is a device in a system which receives commands from Controller in a machine to move and/or control the system in use.

to survive are very small. It is understandable: If the only aircraft engine stops work-ing, no one can help. At that time, it was a wise choice to build the aircraft with two engines, so that if one of them fails, the other one will provide the power for flight. Now, there are two engines in the commercial aircraft and everything else inside the aircraft (supporting these engines and other devices) should be doubled (redundant).

The same situations also happened with navigational aids on the ground and in the aircraft: If any navigation transmitter or receiver fails, the navigation data will not be available and, having in mind that there are no roads or signs in the sky, the aircraft will be lost!

Of course, the concept here is quite different than with the aircraft engines. With engines, both of them work from the beginning until end of the flight, and with navi-gational transmitters and receivers, the second one should start when the first one fails. This is an operational concept known as Redundancy!

To make redundancy applicable in aviation operations, there must be someone who would realize that one piece of the navigational system has failed and the second one should be switched on. In the past, it was done by the pilots, but today all these things are done by Subsystems which do particular monitoring by gathering data, processing these data, and decision-making when and what to do, if there is a prob-lem. Of course, all these things are computerized with particular Software inside; so, the "imaginary" Subsystem increases the Complexity. Today, not only systems in aviation are highly computerized, but our life is also full with such Complex Systems.

Redundancy is good for operation of the Complex System, but cannot be always achieved. However, in the systems where it is implemented and the fault happen on one Subsystem, the monitoring system will react and automatically, the trans-fer of operation to the redundant Subsystem will happen. There will not be cease of the operation, but with this change, our Complex System will enter Degraded Mode of operation. "Degraded Mode" because the system itself is faulty: Not all its Subsystems are working, because there is one which is faulty.

So, the due diligence shall be done to this situation, because the reason why the first system failed, could affect the redundant Subsystem also. If it fails, then disaster could happen. This means that there shall be an immediate reaction to check what the fault is in the first Subsystem and it can be expected that operation will continue by the redundant Subsystem.

Another reason that make the systems more complex is a need for adaptability. Evolution of life in Nature is based on constant adaptability to changes imposed by the changes in the environment. In Nature, the living things have the ability to adapt themselves through the internal process of adaptable self-organization, which is although not always successful for every creature. Inevitably, some of them will fail to adapt and they will die (extinction). But those which are successful, they will be stronger than previously and their future will be bright until another need for change is not imposed.

Humans realized that although the engineering systems are based on physical laws and as such they are not changeable or adaptable, it will be very nice to provide some adaptability to the engineering systems "copying" Nature and living subjects there. Of course, there was need to implement some monitoring device (similarly as per redundancy) and to have some criteria where computerized device would change the variables and parameters of the system with the intention to adapt the operation to the changed reality. Of course, this will increase the number of Subsystems, their interactions, and their interdependence: The Complex System is born!

We do not bother how they work and how they provide the benefits for our lives, but for the people who operate and maintain them, the life also becomes complex . . .

The few simple examples from above are important, but this case with humanity does not end here: It applies to all areas of our private and professional lives. So, I can simply say: The Complications and Complexities are embedded into human life, so all these Complicated and Complex Systems are product of the innate human nature.

Today, the engineering Complex Systems contain not only "technical systems" produced by technology and machines, but also the organization in the company, humans employed there, and the environment where the Complex Systems operate. The reason for this "holistic definition" is that all these things are included in operation and as such they affect the system performance. Many of the constraints which are making systems complex are actually imposed to decrease the influence, interdependence, and interactions with these three influence factors. So, if we do not include them in our analysis, something will be missed.

1.7   DIFFERENCE BETWEEN ENGINEERING COMPLEX SYSTEMS AND OTHER COMPLEX SYSTEMS

Maybe this is a place to explain what is the difference between *engineering* Complex Systems and *other* Complex Systems which can be met in Nature, societies, companies, economies, etc.

Maybe the biggest difference is that these other Complex Systems pose capability of self-organization. Self-organization can be defined as a spontaneous order and regularity in interaction and interdependencies of building elements during operations inside the Complex Systems. These building elements could be members of societies, employees in companies, stars in galaxies, weather, animals in herds. birds in flocks, etc. This self-organization is pushed by the forces of Nature and by the laws and regulations in societies, but it is missing in engineering Complex Systems made by humans. Imagine a "flock" of aircraft at the sky: They do the same job as birds (fly over the sky), but their efforts to self-organize themselves will be a disaster. So, they can achieve particular organization, but for such a purpose, they need a help from Air Traffic Control[11] (ATC).

In other Complex Systems, the self-organization provides sustainability even in the cases where something catastrophic happens, but in engineering Complex Systems, any accident will destroy the system and it will not be used anymore. That is the reason that we must pay attention to engineering Complex System, because if not monitored and maintained, they would be critically damaged.

However, the human involvement in operation of design, testing, installation, monitoring, control, and maintenance of engineering Complex Systems includes self-organization, so we cannot just throw it outside the great picture.

---

[11] The Air Traffic Control does not "control" the aircraft on the sky. They take care that particular separation between the flying aircraft and the ground is achieved.

Another difference is spontaneous Degeneracy in the Complex Systems in the Nature. These systems are capable incidentally to change the processes in their lives through the degenerative changes provoked by single sudden disturbance in their structure (mutation of the genome). Sometimes these changes would be beneficial and sometimes disastrous for some parts of the system, but the system itself will continue to exist. That is how evolution survived ...

Adaptation is another difference between engineering and other Complex Systems. In Nature, the biological system may adapt themselves to the changes. This adaptation needs time, but it is a matter of life and death. The economic, financial, and other sociological Complex Systems may also adjust themselves. But engineering systems cannot do that. Maybe the newest ideas in Information Technology, such as Machine Learning (ML) and Artificial Intelligence (AI), can change something in this area in the future, but for the time being, it is too early to provide any evidence.

The point is that the humans build their "systems" looking the picture of the Nature. The human laws are copying the natural laws, but due to the human changeable behavior, it is not always successful. The humans are prone to feelings which provide different behaviors in same situations and this could not happen to the laws of Nature: The Ohms law will provide the same currents in the circuit if the voltage is applied to the circuits with the same resistances and it will not depend on the feelings of the humans who operate these circuits.

The engineering designers use three approaches to make the engineering Complex Systems close to other Complex Systems:

1 *Coordination:* It is an effort made by designers to copy the coordination in Nature through achieving coordination of interactivities and interdependencies between the Subsystems and components inside the engineering Complex Systems. This coordination may be presented in the form of synchronization (exact timing of the processes in the Complex System) and through partitioning (grouping the Subsystems, components, and processes depending on their interactions and interdependencies). Partitioning could also be based on the hierarchy which will be explained in the next paragraph.

2 *Analogy:* The designers look at the Complex Systems in Nature (herds, flocks, swarms, etc.) and they try to provide the same analogy in engineering Complex System's operations.

3 *Selective plasticity:* This is actually an approach of controlling the Complex System operation in a similar way as Complex Systems are controlled in Nature: They learn how to survive, choose appropriate living conditions through migrations, adapt themselves to the harsh environment, etc.

In general, the designers always try to make the engineering Complex Systems close to other Complex Systems, simply because there is a lot of data and knowledge how these systems operate.

Here, maybe it is good to ask a question: What makes a Complex System good system? My answer would be as follows: This is a system that is made to fulfil particular task which would provide benefit and it has to be done by operating excellent in a particular problem-rich environmental situation. The good Complex System would be capable to adapt itself to a variety of situations and to be resilient to harsh conditions.

1.8 CHARACTERISTICS OF ENGINEERING COMPLEX SYSTEMS

Let's now move to explain some common characteristics of engineering Complex Systems in industry which contribute to their performance.

1.8.1 SET OF SUBSYSTEMS

The first characteristics is that the Complex System can be defined as a set of plenty other systems (called *Subsystems*) and the Complexity of the Complex System is not coming only from the Complexity of structure or structural connection between these Subsystems, but also from the Complexity of interactions and interdependencies between them.

1.8.2 INTERDEPENDENCIES AND INTERACTIONS

Not only there are interactions, but there is also interdependence between the operations of Subsystems. This interdependence is the second characteristic of the Complex System. All Subsystems inside follow particular rules (I would say "algorithms") and humans operating the Complex system use other rules (I would say "procedures") to provide operation of the Complex System. For example, there is a procedure which applies to changing the gears during car driving: The car engine must be separated from the wheels to change the gear when the car is moving. So, the procedure says that the driver must press the clutch (which would separate the engine and wheels) and then change the gear. Without pressing the clutch, the rear box, connected to the car engine could be damaged.

There are many such interdependencies which need to be regulated by procedures.

Interdependence is very much important during looking to solve the fault in Complex System. Maybe the obvious fault of one Subsystem is not the first one which happened, but it can be a result (consequence!) of the fault of some other Subsystem which is not so obvious (it is hidden inside the structure).

1.8.3 HIERARCHY

As third, I can mention that when the Designer design a Complex System, the Subsystems inside are not connected only "horizontally", but also "vertically" (Figure 1.2).

It means that there is particular "hierarchy" in the structure and in the importance of activity of any of those Subsystems which contributes to total operation of particular Complex System. The most important level of hierarchy (Level 1 in the case from Figure 1.2) will be the Subsystems which actually need to execute the intended operation and all other levels will be those who provide some other service in executing the operation. So, the "hierarchy" means that some of the Subsystems

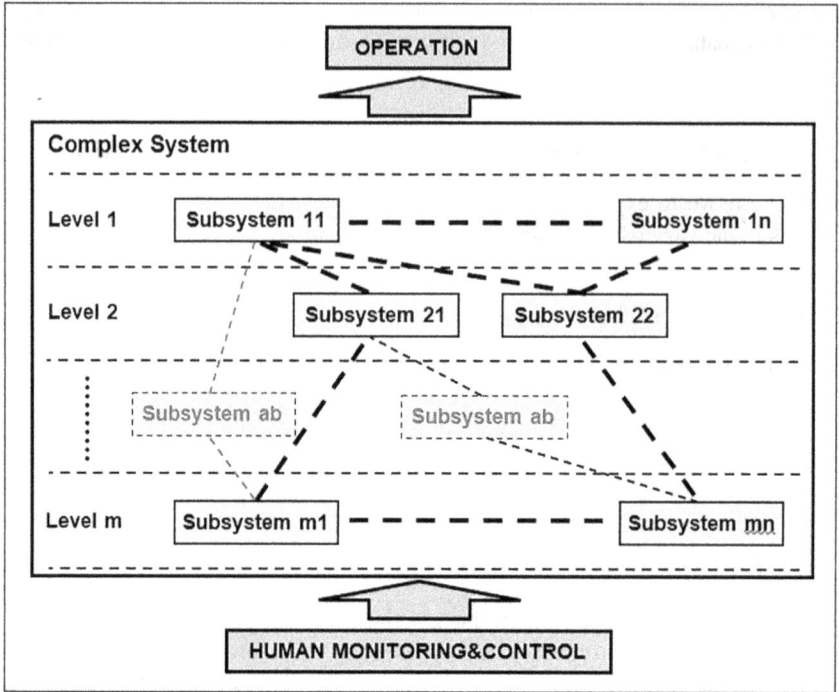

FIGURE 1.2 Complex system "hierarchy" levels.

are not contributing directly to the intended operation of the Complex System, but they are used for monitoring, control, adjustment, oversight, or maintenance of the system and/or operation.

However, the Subsystems in one hierarchy level usually communicate with each other and, in addition, some of them must communicate with Subsystems from lower or upper hierarchy levels. This communication can be expressed as interchange of data, information, commands, and settings used for successful conducting of the operation. Also, the communication may contain monitoring, measurement, and feedback information and commands used for control of the operation.

It can be said that this "hierarchy" is actually contributing to most of the particular systems to be called Complex Systems. The number of Subsystems and their "hierarchy" are not connected only with Complexity of intended operation, but also with the significance and importance of this particular operation in achieving particular goals in our lives or businesses.

From the point of Safety, a good-designed hierarchy is very important. If there is something wrong with the Subsystems from the lower levels, depending on the structure and design of the Complex Systems, the control provided by the higher levels will help automatically with solving the problems. Eventually, if the fault or the failure is not maintainable by the system, the higher levels will stop the operation and contain the damage. It means that in the case of fault, the system will *Fail Safe*.

The point is that if the fault of some of the higher levels endanger the operation, then it could be a systematic error and, as such the system cannot handle it. Here, there is reason to mention that at higher levels of hierarchy, we can also put the Software, human operators, company management, Government, and Regulatory Bodies. These are generally the subjects which are mostly obliged to do decision-making and oversight activities.

Maybe this is a place where a role of Regulatory Bodies in Risky Industries can be explained. It has been noticed that for most of the companies, the profit is more important than anything else and it results very often with decreasing the costs for good quality and safety of the products. Quality can be noticed immediately with the product or service, but the safety costs are not so clearly visible. They can be noticed only when something bad happen. So, the companies no longer have the decent ability to determine and control the risks and this is the reason that the Governments must step in and legally ask for greater companies' responsibility through laws, regulations, and various forms of oversights.

However, putting Regulatory Bodies in the chain could be beneficial for industries and also damaging. Whatever deficiency or error is done in regulation by Regulatory Body, it will affect every company in the State (World).

The hierarchy in the Complex Systems is very much important from one more aspect: Humans. Usually, human mind is capable to understand complexity by providing abstraction of the Subsystems and categorizing them in a particular hierarchy. The abstraction is actually embedded into a method known as Reductionism for trying to understand Complex Systems (explained in the next chapter). The hierarchy provides a material for top-down process for understanding interdependencies and interactions inside the Complex Systems.

This was the hierarchy of the Complex System, but I need to state here that there is also a hierarchy of the process supported by the Complex System. This process hierarchy is usually based on three levels:

1 *Task Level*, where there are tasks executed in the frame of process supported by the Complex System. This is the highest process level.
2 *Function Level*, where all functions for task executions and their appropriate controls and constraints are defined and placed. This is the middle process level.
3 *Resource Level*, where all resources needed for the process are submitted to the operation. This is the lowest process level.

It is important to mention again that these three levels are also interdependent and interact between themselves, and as such they may not be neglected during the safety analysis.

1.8.4 EMERGENT PROPERTIES

The fourth characteristic is that each Complex System has its own emergent properties[12] which differ from the emergent properties of the Subsystems. It can be said that particular combination of the emergent properties[13] of the Subsystems together

---

[12] Already mentioned and defined in Section 1.4.
[13] Somewhere, in the literature, these "emergent properties" can be found under the name "collective behaviour" of the Complex System and there is nothing wrong to use this term.

with the building structure of the Complex System are responsible for the emergent property of that Complex System. In other words, the Complexity (one of the emergent properties) of the Complex System can be built by the combination of emerging properties (if any) of the Subsystems. For example, the emergent property of aircraft engine is to provide a force which will move the aircraft, but the wings are those that provide a lift to the aircraft. So, to give one example regarding new emergent property, I will use the simplification of aircraft: The combination of the force from the engine and lift from the wings provide the flying (emergent property of the aircraft). The engine and the wings alone could not provide flying, but together it is possible.

Having in mind what is said in this paragraph, I may conclude that the Complex System, for sure, is not just a sum of its parts.

1.8.5   CONTROL SUBSYSTEMS

The fifth characteristic of any Complex System is that everyone has one or more Control Subsystems which actually monitor its performance, control, and maintain its operation. In addition, these Control Subsystems provide requested and required constraints in the operation with an intention to keep the normal operation as much as possible.

Especially important part of these Control Subsystems is the monitoring and control of the variables and critical parameters. Critical parameters are those that if changed could cause failure of normal operation. The monitoring shall be supported by different levels of notifications. There are usually Normal, Warning, and Alarm notifications which could provide visual and audio signaling.

- The Normal notification provides assurance that Complex System provides its normal job and it is usually marked by green color. There is no audio notification for Normal operation.
- The Warning is situation where some of the variables or critical parameters changed and they are close to overcome the tolerances. It is marked with yellow or orange color. Sometimes they could be accompanied by sound. This notification could be a sign for pre-fault condition and it needs to be investigated as soon as possible.
- The Alarm means that there is a fault in the system and the system has stopped its normal operation. There is urgent need for human (operator or maintenance) intervention. The notification is marked with red color and there is always a repetitive sound until not switched off. Alarm means that the failure of operation is caused by irregularity which could be highly damaging for the Complex System, humans, assets, and/or the environment.

All these Control Subsystems could be so much complex by itself and as such, they considerably contribute to the Complexity of the Complex Systems.

1.8.6   THE NECESSITY FOR A TEAM

The sixth characteristic of Complex System is that we need a Team to provide particular operation of Complex System. Having in mind the Complexity of the structure and structural connections between the Subsystems together with complex interactions ad interdependencies between the Subsystems, there is need of a few members of the Team, each of them with a different kind of expertise. This Team would implement, maintain, and make the operation of Complex System (each of them in different area) effective, efficient, and sustainable.

Taking again into consideration a car as a Complex System, someone can say: I can drive it alone and I do not need a Team. But the point is that when there is a problem with the car, you will bring it to a workshop and there, depending on the type of the problem (mechanical or electrical), different types of experts will dedicate themselves to solving the problem. Sometimes even one person cannot handle it, so, for example, the overhaul of the car engine will need few persons just to take out the engine from the chassis.

1.8.7 ADAPTABILITY

The seventh characteristic of Complex Systems is their adaptability. It is actually human try to copy other Complex Systems in Nature or societies. Today's systems are mostly "community" of Hardware and Software. For required function to provide the controls in operation of Complex Systems. the Software is very much important. The Hardware gathers information through feedback loops inside the Subsystems and the Software algorithms are used to process these data. Depending on the situation, when the variables or parameters under control deviate outside the tolerances, the Software, through particular applications, is responsible to "adapt" the parameters to be within the tolerances. If the "adaptation" is not possible, the Software would switch off the system to protect it or to stop the scrap on the output.

1.8.8 SYSTEM BOUNDARIES

Every engineering Complex System has its own boundaries and it is their eighth characteristics. The boundaries define the dimensions of the system (how big is the system outside and inside).

There are two types of boundaries: external and internal. Both boundaries can be defined and redefined depending on the requirements of analysis which needs to be done.

The external boundaries define the system from outside. These boundaries are places where the system can interact with the neighboring environment. My home is a Complex System where I live and its external boundaries are the walls of my apartment. There are doors and windows in the apartment and I am using them to interact with my environment.

The internal boundaries are parts inside the Subsystems. They are also known as *resolution* and they are defined as smallest parts from any system which will be taken into consideration during the analysis of the system.

The point here is, due to effective and efficient maintainability of the Complex Systems, the internal boundaries are usually Subsystems built in the form of Line

Replaceable Unit (LRU). So, it is enough for maintenance staff to define which Subsystem (LRU) is faulty and to change it, to restore normal operation during corrective maintenance.

The nature of the boundaries of the Complex Systems will define it a system open or closed. The open system is system which interacts with the environment (the boundaries are not strictly impermeable) and the closed systems are those which do not interact with the environment (the boundaries are impermeable).

1.9   VARIABLES, PARAMETERS, AND WORKING POINT OF THE SYSTEMS

There are variables and parameters in dynamical systems. The variables and parameters are actually the same thing which we call "factors".[14] Any change of factors can cause change in the state of the behavior of the system. When this behavior becomes abruptly sudden and qualitatively different, we say that the system changed its state. Each system could depend on many factors and, for particular system of interest, we use some of these factors to change its state and some of them we use to keep the state constant. Small changes of some of the factors could produce big changes of the state of the system and vice versa.

The factors within the system which we choose to keep constant, I will call parameters and the factors which we chose to change, I will call variables.

Usually, when the factors which could provide small change of the system change a lot, they are called parameters (they help to keep operation constant) and the factors which will provide big change of the system by small change of itself are called variables.

The reason is that during operations we strive to use less resources to change variables. Having in mind that variables are used to control the operation of the system, using less resources for their change will make the control of the operation of the system more efficient. So, I will keep the parameters stable, after they are used to create a state of a Complex System which is called *Working Point* of the system. Later, they are not used anymore to adjust the operation. But during the operation, I will use the variables to govern the system and to control this (already defined) Working Point of the system.

So, the Working Point of the system is chosen in the factory by adjusted values of the parameters and the variables are used during operation. The human operator, by changing them intentionally only at particular intervals, will guarantee the smooth operation of Complex System. What is important to be mentioned here is the fact that both parameters and variables must be chosen in such a way that they may not change more than determined tolerances.[15]

The simple example of the adjusting Working Point can be adjustment of the currents and voltages of bipolar junction transistor[16] (BJT). Depending on the chosen parameter (determined by the resistors), the BJT mode of operation will be

---

[14] Remind yourself of Figure 1.1!

[15] You can find more on tolerances later in the book!

[16] Please note that bipolar junction transistor is a non-linear device used in electronics.

"switch" or "amplifier". Another example of Working Point is choosing best adjustments of the parameters during first stage of Statistical Process Control[17] (SPC) in manufacturing. This is the stage when we adjust parameters to provide the smallest variations of the process parameters with intention to get the highest quality (smallest variation) of our product. If we would like to keep the adjusted normal operation, the adjusted parameters should not change during operation more than their tolerances, but the process could be controlled (if necessary) by the change of the variables. In general, the variables and the parameters are used to tune the Working Point of the system.

Do not be confused here with *parameters* and *variables* of the system . . .

In engineering actually, the parameters in each system (operation, process, activity, etc.) are also variables which once adjusted (to produce Working point) in the factory are not changed during the use of the system (operation, process, activity, etc.).

As example, look at the car. In the Complex System as the car is, there are thousand parameters to be adjusted in the factory to make a car function: Aerodynamics, fuel and air quantities filled into the engine, pressure of fluids in brakes, air pressure in tires, etc. But all of these are also variables which once adjusted, drivers do not adjust again during driving. So, we call them "parameters". That what I (as a driver) adjust during driving are variables, such as throttle, steering wheel, clutch, and brakes. So, these factors for driving a car are "variables" (I "vary" them to drive my car) and all other factors are "parameters" (adjusted in factory and I do not touch them during driving).

In the scope of next chapters and paragraphs regarding non-linear dynamical systems, the Working Points of the systems must be stable fixed points.[18] In such a situation, operation of the system will stay always in vicinity of the stable fixed point (or already adjusted Working Point).

However, due to some other (external or internal) reasons, parameters may change more than their tolerances. This change affects Working Point of the system and this change of the parameters beyond tolerances produces unexpected behavior of the system.

1.10   UNCERTAINTY AND COMPLEX SYSTEMS

Reading (mostly) philosophical papers about Complex System you may find the term "uncertainty" strongly connected with the definition of the Complex Systems. There is one philosophical aspect of Complexity which says that the Complex Systems are systems which operate in the area between random and ordered operations. Having in mind that randomness is uncertain and ordered operations are deterministic, one can understand that the Complex Systems operate in the area between randomness and determinism.

From the engineering context of Complex Systems, we cannot justify the failures of operations by randomness. Yes, randomness can be present in our systems, but to

---

[17] Statistical Process Control provides values of Upper Control Limit (UCL) and Lower Control Limit (LCL) of each parameter in the process and controls them during operation.

[18] See Chapter 7 for more details about fixed points.

understand that engineers are designing Complex Systems using the physical laws and the systems are pretty much deterministic.

And yes, I can again agree: The randomness can show up also there, no matter how deterministic our systems are!

Let's say that Quantum Mechanics is full of randomness and Chaos. Although it is strongly deterministic, it is pretty much full with uncertainty. However, sometimes you can find a description of Chaos as "random-in-appearance" phenomenon.

In our engineering Complex System, their design is based on physical laws which are very much certain, but the interactions and interdependence of the Subsystems inside, in the scope of designed systems, could be uncertain (be a problem to understand). This is valid especially for their combinations in the scope of their operations.

Having in mind that this is a book about knowledge, I think it is honest to state here that, in any case where randomness is dominant, the knowledge about interdependencies and interactions of the system is futile regarding determining its future state. This is especially valid in the cases where the probability distribution regarding the states of the system is uniform[19] (which could be taken as definition of randomness). This is like tossing a dice: We have extremely good knowledge about the dice, but any of the possible six numbers has the same probability to show up. Simply, we cannot determine in advance which one will show up.

All of the operations of the engineering Complex Systems are inherently and unavoidably uncertain, although the systems itself abide by the physical laws. The uncertainty arises through the problem of not having full understanding of the interactions and interdependencies between Subsystems inside the Complex System and, more specifically, from the uncertainty of human behavior and environmental factors (weather in aviation or space radiation for spacecrafts, for example).

In the presence of such uncertainty, triggered by human behavior or harsh environments, the correct mathematical and physical relationships between the system parameters and variables are hard to understand and control. The nature of uncertainty can sometimes be different, but the processes created by the system could be intrinsically uncertain, very often due to lack of proper information, especially in highly dynamical environment. As mentioned earlier, it is the presence of these uncertainties that is responsible for the implementation of constraints for the control of the system and for the control of its operation and that make these systems complex.

I will use the process of measurements in industry to make a point about the presence of uncertainty. As many of you know, there is need for measurement in all industrial processes. The measurement is basic of automation and control and the required and implemented level of automation and control is strongly connected with the Complexity of the engineering systems.

The first things which students in engineering study regarding the measurement is that the measurement results are full of uncertainty. During laboratory experiments in measurements, every student could notice that although each of them uses the same instrument, the same method, and the same measurand,[20] the results

---

[19] It means that any state of the system has equal probability to show up.

[20] Measurand is characteristics of the unit (thing, characteristic, etc.) which is subject of measurement. For example, speed of the car is a measurand and the results of the measurements are presented on the tachometer at the panel in front of the driver.

are different. The reason is simple: Metrology, as a science of measurements, is clear that the measurement is an operation (activity, process) which cannot produce accurate result. There are plenty of reasons and all of them are connected with some type of variability of the assets used during measurements: instrument, measurand, environment, measurement method, operator, calibration, time, temperature, pressure, etc. All these variabilities will contribute to the particular uncertainty of the result.

If we cannot measure something with particular accuracy, it means that we cannot determine its real (true) state. So, whatever observation of the states in Nature we are doing in the presence of uncertainty, we cannot be sure what is it, although this "true" state must exist. In the theory of science, it is known as uncertainty caused by "observation noise".

It is not known that the "true" present state of the system will also affect the future of its next states. But here the uncertainty gets bigger due to dynamics of the change. It is reasonable to count on noise in dynamics also. Every movement measurement will be uncertain due to "dynamic noise", although the movement can be observed, but cannot be determined (measured) accurately.

That is the reason why the results of the measurements are always statistically calculated and, during presentation of these results, they are expressed as statistical values where (roughly speaking) the mean (μ, the arithmetic mean value) is a measure for "accuracy"[21] of the quantity measured and the extended standard deviation (usually 3σ) is measurement for "precision". Having this in mind, the "precision" (standard deviation σ) could be defined as a measure for uncertainty.

This uncertainty of the measurements will be small under the following assumptions:

(a) The measurement system is well-maintained and regularly calibrated.
(b) The laboratory is maintained as controlled environment (no changes in the temperature, humidity, pressure, etc.) during the measurements.
(c) The measurement method is validated and verified to be good for this type of measurement.
(d) The laboratory staff is trained and competent of using the measurement system and the measurement method.

Using statistics means that we do a series of measurements and we process this series statistically. It means that we express our knowledge about the states of the system as probability. The bigger the series and the more the data used for statistics, the better the forecast based on probability. So, whatever the system is (complex or ordinary), we use probabilities to express our knowledge about future states of our system.

---

[21] The words "accuracy" and "precision" are in quotation marks because the most important document produced by BIPM (Bureau International des Poids et Measures), known as GUM (JCGM 100:2008, Evaluation of Measurement Data—Guide to the Expression of Uncertainty in Measurement), does not use the words "accuracy" and "precision". There is only "uncertainty" and there is clear reason for that: If something is uncertain, it can neither be accurate nor precise. This document is actually the "constitution" of the World Metrology, but unfortunately, not known by most of the engineers . . .

Speaking about measurements (of the locations, parameters, etc.), the expression of probability inside the measurement result R of any state will be presented by the following equation:

$$R = \mu \pm 3\sigma$$

where μ (arithmetic mean value) and σ (standard deviation) would be calculated by using these equations:

$$\mu = \frac{\sum_{i=1}^{n} M_i}{n}; \sigma = \sqrt{\frac{\sum_{i=1}^{n} (\mu - M_i)^2}{n-1}}$$

where M_i are individual repetitive measurements (of the same series) of that state (location, parameters, etc.).

The probability of expression is hidden in the integer numbers of multiplying the standard deviation σ from result R or inside the expression $\pm 3\sigma$.

Actually, there are three areas of probability around the value of the mean (μ), which can be expressed by the multiplication of σ with 1, 2, or 3:

1 The probability that the single measurement (out of the set of infinite measurements) in the area marked by $\mu \pm \sigma$ will be 68.2%.
2 The probability that the single measurement (out of the set of infinite measurements) in the area marked by $\mu \pm 2\sigma$ will be 95.5%.
3 The probability that the single measurement (out of the set of infinite measurements) in the area marked by $\mu \pm 3\sigma$ (known as extended standard deviation) will be 99.73%.

In our case, the possible source of uncertainty in operation and functioning of Complex System could be not only our ignorance about these physical laws and about their interaction and interdependence inside the system, but also the uncertainty about choosing the correct Working Point of the system during installation and after every overhaul or corrective maintenance. This uncertainty could also affect (very much) the sensor's part of installed Subsystems for monitoring, control, and automation and the actuator part for executing the control commands. In addition, the ubiquitous uncertainty will also be present during operations, preventive maintenance, and fixing the faults (corrective maintenance).

These are uncertainties in operation of the system, but during the design phase, there are also uncertainties which are triggered by the structured and unstructured uncertainties. The structured uncertainties are those characteristics which are included in modeling and simulations, but we oversimplify them or we do not clearly understand their influence on the operation of the system. Unstructured uncertainties are those which come with the inaccuracies or underestimation of the system characteristics and their hierarchy.

The uncertainties in both the modeling and simulation during the design must be addressed and two methods for dealing with them (especially in non-linear systems) are *robust control* and *adaptive control*.

1.11   SYSTEM ENGINEERING

In today's reality, all these Complexity of the new systems resulted from a try to establish theoretical background for analysis and synthesis of these Complex Systems. In (approximately) the last half of the 1950s and the first half of the 1960s, in few scientific books, a new discipline dealing with that was mentioned as System Engineering.

Harry H. Goode, Robert E. Machol, and Arthur D. Hall could be mentioned as "fathers" of System Engineering. In 1957, Harry H. Goode and Robert E. Machol published the book *Systems Engineering: An Introduction to the Design of Large-Scale Systems* and in 1962, Arthur D. Hall published *A Methodology of System Engineering.* These books were the first two published texts having the phrase "System Engineering" in the title can be seen.

However, I do believe that Hall could have advantage, because in 1962, he published a second paper on Systems Engineering where there were included three new areas of interest:

1 The concept of value in the decision-making process
2 The basics of holistic integrated general methodology for Systems Engineering
3 The fundamental concepts of Systems Engineering

The second area (about formal general methodology) for the analysis and synthesis of Complex Systems has significantly bigger impact on future developments than the other two areas. The point is that the creation of Complex System must be done through systematic approach and it needs to be conducted as a project. Hall has mentioned five phases of the projects based on System Engineering:

1 *Systems Studies:* In general, it deals with the theoretical planning about the task which system in design can be done and how it can be done.
2 *Exploratory Planning:* The planning regarding the technologies and materials which can be used, as well as the companies which can help in building the Complex System.
3 *Development Planning:* More advanced, more detailed, and more focused planning about how the design of the system would be achieved and how the system (following the design) could be built.
4 *Studies during Development:* Of course, there is no need to be blind in following the accepted design. If there are some issues which cannot help with realizing the plans for design or building and testing the Complex System, the redesign could be implemented. It will help with gathering a new knowledge (which could help with other Complex Systems).
5 *Current Engineering:* How the system would be operated and maintained? This could also produce new knowledge about technologies and theories used to design, build, operate, and maintain the next generation of Complex Systems.

These five project[22] phases will be later improved by other scientists and engineers. Today, we can use the following five phases:

1 *Project Planning:* This is the beginning of the project where it is explained how to conduct and complete a project during the project life cycle. There is need to define the stages, time frame, and the resources. It could contain the following activities: setting the measurable objectives regarding the stepstones, identifying deliverables for each stepstone, scheduling the activities, planning every task, etc. Project planning have to produce clear picture about human resources needed, communication methods between them or with the customers, and risk management regarding possible problems.

2 *Analysis and Sustainability of the System Requirements:* Output of this phase have to give an answer what you can do to satisfy the requirements by the system and what are the features, functions, and tasks that need to be completed for a project of building a required Complex System to be successful.

3 *Assessment of Alternative Paths to Build the System:* This would be actually determining activities for elimination or mitigation of the risks identified during the risk assessment process. It is wise to have alternative for any such problem.

4 *New Methods for Design and Building the Complex System:* Of course, you will use already known design and production methods to design and produce required Complex System. But maybe there is a new technology which need to be elaborated and used for the first time, so there is need to think in advance for change of some methods and how to implement the new ones.

5 *Validate and Verify that the Finished Complex System Meets the Requirements:* This could be done by experiments and testing to prove and ensure (first by yourself and thereafter by the Regulator and the customer) that the Complex System will satisfy all requirements.

All these efforts could help only if the multidiscipline design team accepts a structured and systematic approach for problem-solving during the design, operation, and maintenance with intention to save time, costs, and resources and, at the same time, to provide the best results.

There is another discipline named System Theory which gives a theoretical basis of practical activities with regard to the System Engineering. This discipline will be explained in one of the next chapters. The point is that the System Engineering is more technical and applies only to manufacturing industry and the System Theory is more holistic and, additionally, it provides background for Complex Systems in natural, scientific, and social areas.

System Engineering also establishes a new way of thinking in industry which is known as System Thinking. System Thinking is an approach to considering

---

[22] Each project is based on three parameters which need to be monitored and controlled during the project: resources, time, and costs. Having in mind that the costs are not a part of engineering, I have neglected them here.

operations (processes, activities, etc.) of Complex System by paying attention to the behavior of the whole Complex System instead of the individual operations (processes, activities, etc.) of the individual Subsystems used to build the system. There are many interactions between Subsystems and components of a Complex System and their effect on the operation to be conducted must be subject of analysis to provide clear understanding how the system functions. It means that although the Subsystems and components are taken into consideration during design and operations, they are treated as "black boxes" and only their external interfaces are considered. In addition, the environmental and organizational issues (external factors) of the place where the Complex System will operate must be taken into consideration to provide smooth operation and required performance for the system during its life cycle.[23]

Since the 1960s, as the discipline evolved, the definition of System Engineering has changed. My favorite is the one published on the website[24] of INCOSE (International Council on Systems Engineering). There you can find the following text[25]:

> Systems engineers are at the heart of creating successful new systems. They are responsible for the system concept, architecture, and design. They analyze and manage complexity and risk. They decide how to measure whether the deployed system actually works as intended. They are responsible for a myriad of other facets of system creation. *Systems engineering is the discipline that makes their success possible—their tools, techniques, methods, knowledge, standards, principles, and concepts.* The launch of successful systems can invariably be traced to innovative and effective systems engineering.

Of course, the System Engineering evolved with time to add the "customers" and their requirements to the Complex Systems developments, but this is more economic or business move than something which is connected with the development of the engineering or technology.

However, System Engineering is not an engineering discipline (although it contains the word "engineering"!), but a tractable way of management of Complex Systems during their life cycle[26] which supports multidisciplinary team of experts dedicated with task to provide benefits with such a Complex System. In such a way of management, the economic aspect must also not be neglected. So, System Engineering takes care to provide the right balance between the cost, time schedule, resources, quality, and operational and safety risks against the operation and performance of the Complex System. It means that necessary compromise must be done in particular phases of the management process.

---

[23] System life cycle contains all phases of a system design and operation that can be broken down to different phases.

[24] www.incose.org/systems-engineering

[25] Definition reprinted with permission of the International Council on Systems Engineering (INCOSE), San Diego, CA: International Council on Systems Engineering.

[26] Life cycle of a Complex System can be presented as different phases, from the beginning to the end of its functioning. The phases are design, validation and verification, installation, manufacturing, testing, support, maintenance and repair, upgrade, and decommissioning.

The engineering Complex Systems need a holistic, multidisciplinary, and systemic approach to reach the understanding of the technology problem involved in building the systems and there is need for contextual framework presented as intended operation of the systems. Only this holistic and multidisciplinary approach could provide satisfactory results. For such a task, the systems engineers must have knowledge, experience, skills, and attitude to provide systems based upon scientific, technological, and engineering principles and on validated and verified methodologies or methods.

The connection between the Complex System and System Engineering is innate. As stated earlier, the approach of System Engineering is actually used to build Complex Systems. It is worth mentioning here that the System Engineering is very much used in Software development than in Hardware development. Unfortunately, there are engineers who still abide by the old-fashioned ways of designing the systems. The Software engineers are younger, more keen, and more adaptable to accept the news in technology and management, so it was easy to accept it.

Having in mind that this book is about the safety of the engineering Complex Systems, System Engineering is very important because the safety starts with the design. The average size of the book in System Engineering is approximately 200 pages, so, if you need more on information on System Engineering, please use Internet. There are many useful resources in the form of books papers, videos, etc. I would personally like to recommend the NASA book *Engineering Elegant Systems: Theory of System Engineering*, written by M.D. Watson, B.L. Mesmer, and P.A. Farrington (available on the Internet).

2 Understanding Complex Systems

2.1 INTRODUCTION

There must be a difference in analyzing the Complex System, having in mind what was said in the previous chapter. Obviously, the development of science and technology provided us with systems which are different from those we used in the past and our present social development (adjustment) is not so fast to get used to these new technological systems.

One of the most important paradigm's shifts which must be done to understand the normal operation of any Complex System is to update our focus from individual faults of components or Subsystems toward the failures in interactions and interdependence between the components or Subsystems. It does not mean that the faults should be neglected. Not at all! It means that we may not neglect the interdependencies and interactions of the components and Subsystems inside the Complex System which could also be a cause for failure of operation. I hope that most of the Safety Professionals are aware that there is a possibility for each component and Subsystem to function perfectly, but still there can be failure of operation. Some of such examples will be mentioned later in the book.

This chapter explains the methods used for understanding the Complex System which are not designed by us. If the systems are not designed by us, then we miss the knowledge about their structure, their operations, and how they function. In addition, in many companies which design and manufacture Complex Systems, they provide a testing of their ideas and operations of their systems using the methods explained in this chapter. When these companies sell their product to any company, they provide some of these methods for training dedicated to operational and maintenance staff who will use these systems.

In general, there are two approaches in understanding Complex Systems.

The first one is "bottom-up" approach based on understanding the function of each Subsystem used to build the Complex System and later try to understand the interdependencies and interactions between them when they operate together. The next section is about that.

The second approach is "top-down" approach. It is a method which starts from the operation of the system needed to provide a product or service and go down to understand what devices, commands, controls, and data submitted to the Complex System are needed to build the product (provide the service).

Whatever the approach used, the interdependencies and interactions of the Complex System with the operators (humans) and the environment where the system operates must not be forgotten.

DOI: 10.1201/9781003404811-3

Let's see some of the methods used to comprehend the Complex Systems today . . .

2.2   REDUCTIONISM (CLASSICAL METHOD)

There is a method in science which is used to investigate and to do research on Complex Systems. It is called *reductionism*[1] and, in scientific literature, it can also be found under the name "analytic reduction". It is the first method used by humans to understand complexity and it is also known between the engineers as "classical method".

This method uses simplification of the system to establish its functioning. Each part (component or Subsystem) inside the structure of the Complex System is investigated alone and when a particular knowledge about these parts is gathered, the interactions and interdependence of the structure of the parts are investigated. It means that the "physical" system is reduced to its Subsystems and the knowledge gathered regarding functioning of these Subsystems is later used to understand their interactions, their interdependencies, and the holistic functioning of the whole system.

This method is actually a "heritage" of the Newtonian Mechanics and it was very much popular during industrial revolution in the 19th century.

In general, this method works good for Complicated Systems, but not always work as required for Complex Systems. Sometimes interaction and interdependence of some Subsystems with other Subsystems is neglected and this can considerably change the behavior of the system. In addition, within the Complex System, the Software is treated as an "imaginary" part (Subsystem) used to "govern" the functioning and interactions of the Hardware inside the Complex System and that is the reason that the Software cannot be investigated (understood) without other "physical" parts (Hardware) of the system.

Another problem is with the Feedback Subsystems . . .

The automation and control over Complex Systems are mostly achieved through Feedback connection of some control signal from sensing devices and as such it will bring considerable non-linearity in the system.[2] Again, similarly as Software, the Feedback functioning is based on interaction and interdependence with other Subsystems and as such, it cannot be tested without other Subsystems affected by this Feedback.

The point with the Software and Feedback Control Systems is that both in combination with other Subsystems provide emerging self-organizing properties and these properties cannot be registered and fully investigated if we use reductionism. As I have explained previously, these emergent self-organizing properties are one of the main characteristics of Complex Systems.

The reductionism has been used in physics, civil engineering, and structural mechanics and to be honest, the science has made great progress by taking things apart in these areas. However, it did not give good results in computers,

---

[1] There is big use of reductionism in philosophy, but there are few disciplines which depend on the subject where reductionism applies.

[2] More about Feedback and non-linearity are provided in later chapters.

telecommunication, automation, and especially, where functioning of the systems depends on Human–Machine interactions and where we do have considerable uncertainty regarding the interactions and interdependences.

In all these cases, the benefits of the reductionist investigations are of no use even today.

2.3 MODELING

We cannot always gather knowledge of Complex System by investigating them on the site of operation. Sometimes, it is simply impossible and sometimes it is simply not feasible. We cannot go to Andromeda galaxy to explore it, so we use different types of telescopes.[3] With today's development of technology, we can explore the planets and other bodies of our Solar System by spacecrafts made for that purpose, but it is not always feasible.

So, to help us understand the Complex Systems, we use modeling and simulations. In science, modeling provides a chance to study Complex Systems which is not possible with other (direct) methods of study. In the industry, modeling is used during preparation of a new product and it has considerable economic effect on decreasing time, money, and the number of the resources needed for prototypes and testing.

It is clear that the first step in designing a Complex System is to write down the requirements. The second step is to provide theoretical model of the Complex System where all operational and regulatory requirements are satisfied. The third step is to use this model to check its behavior in the range of interest through bunch of computer simulations. Finally, the fourth step is to build the prototype and submit it for extensive testing in laboratory and in the real environment.

For modeling, it is essential to determine the so-called *generalized coordinates* which will be used to express the model. This is usually Cartesian coordinated system (e.g., x, y, z) or Polar coordinate system (e.g., r, φ, and θ). We need coordinate system to adequately describe the position and orientation of all Subsystems in the Complex System of interest. The minimum number of coordinates required to describe any model of the Complex System is known as Degree of Freedom (DoF). DoF can be calculated as the number of chosen coordinates minus the number of constraints imposed on the system model.

The modeling is a process of building (usually very simplified) model of the system (to the best of our knowledge or wishes) which we would like to study. We study the characteristics of this system using this model for simulating real system behavior. The model is a tool used by simulation to provide data on the behavior of complex things. Processing data gathered through simulation of the model could provide us with knowledge that is missing for the real system and its behavior in the real world. This is something which is very much feasible with today's computers and information technology.

Regarding the modeling of Complex Systems, I would recommend using the so-called Causal models, which are mathematical models dealing with causal

---

[3] Optical (ultraviolet, infrared, etc.) and radio (gamma and X-ray radiation, radio and microwave frequencies, etc.) telescopes are systems which are used in astronomy.

relationships within an individual system. They are very much used in computer science and they are able to take care of external influence factors. These models help with inferences about causal interactions and interdependencies inside and outside the Complex Systems using statistical data. They are based on probabilities (as everything which uses statistical data) and they have also been applied to topics of interest not only in engineering, but also in philosophy, logical decision theory, and the analysis of actual causation. The root cause analysis (RCA) is one place where it can be used.

These models are capable of providing predictions about the behavior of a system taking care of the truth value (using Logic: True = 1 and False = 0) or they can be based on the probability to predict the effects of human or machine interventions.

Although modeling is a very good method, there is need to be cautious when working with any of the models. The biggest problem is that a model usually represents our understanding based on our knowledge about the system. It means that two scientists will not always produce the same (let's say: unique) model of the system under study. Model can be not so good, simply because we use it to investigate something which is usually not known (a new technology, new system, etc.) and as such, the model could be full with uncertainties. These uncertainties come due to the following:

- Intentional simplification of the idea presented by the model
- Things which are unknown to us, because this is a new thing and we do not have enough knowledge
- Neglecting structural modes and non-linearity in the real system.

There is additional issue with the modeling and it has to do with the unknown aspects of the system. Having in mind that the model is based on our understanding and our knowledge about the system, there are few questions which need to be considered:

- What to do with the unknown things about the system (known unknowns)?
- How do we know that there is something which we do not know, but it will affect the system (unknown unknowns)?
- What if these unknowns have profound implications on the system behavior and they are not covered by the model?
- How will we know to explain the behavior of the system with a model which is not holistic?
- Etc.

The biggest problem with modeling occurs if you confuse the model with the real system. This is a prescription for disaster!

Another problem is to use modeling for investigating behavior of Complex System without understanding that the model is a simplification of the Complex System. In such cases, it is urgent to be careful about how you will do simplification and how will it affect the functioning of the model (compared with the functioning of the system). Here, increased Complexity of the model could help to study the Complex System.

In general, there are two types of modeling which can be used in the analysis of Complex System:

1 *Input/Output Models:* These are models which look for the mathematical description of the relationship between input signal and output signal. Usually, the ratio is presented in a diagram where the input changes are presented on x-axis and the output changes on y-axis. For this type of modeling, the inner structure of the system is not important. These models are known as Classical Models.

2 *State Space Models:* These models take into consideration the structure and processes inside the system as well as the behavior modeling of the variables and the parameters inside the system. The problem here is that there are different ways to choose these variables and parameters for the same system which could produce different models for the same system. These models are basis for Modern Control Theory.

However, there is a need for all decision-makers to understand imperfection of modeling, and as such they must not hide behind modeling for its bad decisions.

Luckily, with today's computers and their ability to compute millions of operations in small duration of time, modeling can be easily achieved. Use of computers in modeling is the reason that today we speak mostly about computational modeling.

In general, I do believe that modeling should be cautiously used for exploration of unknown systems, but if it is used (as it is used today), you shall take care of possible misunderstandings explained earlier. I agree that the models can help us to visualize, predict, optimize, regulate, and control Complex Systems and in this area they can be very useful. In addition, I do believe that modeling and simulations should be used to study some aspects of behavior of well-known systems. Or, as George Box[4] has said:

All models are wrong, but plenty of them are useful (if you know how to use them).

The lack of certainty with the models can be solved by trying to conduct experiments in accordance with the model of the system.

For example, it was common in aviation industry to understand the aerodynamics of a new aircraft, to use a scaled model of aircraft in a wind (aerodynamic) tunnel. These days, however, the model is created within a computer program which simulate the airflow at a fine level of detail, allowing immediate calculations and registration of all malfunctions and deficiencies. But modeling and simulations will not stop the use of wind tunnel for aerodynamic checks of aircraft (in real size) or the experimental flights of the prototype of aircraft (when it is built). Have always in mind that the models are good, but the experiments are better!

However, you may not forget that the final testing of the designed Complex System shall be done in the environment where this system will be used and not in the laboratory. That is the reason that aircraft, after testing in the wind tunnel (laboratory),

---

[4] Georg Edward Pelham Box (1919–2013), a British statistician, worked on Quality Control, Bayesian interference, and Design of Experiments.

are tested in the air through test flights. In simple words, you cannot simulate the air temperature at 10,000 meters in the atmosphere in the aerodynamic tunnel.

Another thing not to forget is that by using an experiment, usually only satisfaction of the requirements posted to the system are tested. The experimentation shall also confirm the reliability of the system and the availability of its normal operation under different external and internal influence factors.

The testing through experiments was a method which was pretty much used in science in the past, but it is very often "simply impossible and sometimes it is simply not feasible". To be honest, this deficiency of experimenting move science toward the modeling.

Modeling is something which we use in Safety to identify the hazards and for Risk Assessment, so due diligence shall be provided during these activities. Using FTA (Fault Tree Analysis), ETA (Event Tree Analysis), or FMEA (Failure Mode and Effect Analysis) for Risk Assessment is actually based on a model of our system which we produce.

2.4   SIMULATIONS

As I said above, the main point is to build a model of something and use this model for simulations. There is a very good reason to do simulations and it is the fact that theoretical analysis supported by computer simulation is the cheapest way to analyze the non-linear Complex Systems before, during, and after design. The bad thing is that if there is no good mathematical theory to be applied to the model and to the simulation, the simulation is useless.

The simulation can be defined as imitation of something (person, system, process, behavior, etc.) and as such, we use the previously produced model to do simulation. The simulation will provide data which will be the subject of considerable analysis. How you build a model and how you conduct simulations are very important, but analysis of the results is also no less important. The analysis of the results from simulation will provide insight (maybe even crucial evidence) about behavior of the system in reality.

But be careful with the analysis, and especially with the results. Sometimes they can be inconclusive and in such a case, the model or the simulation must be changed. Sometimes even the results will say that modeling and simulation were wrong steps in studying the capabilities of our designs of system. In such processes, the experience, intuition, and instincts are very much important.

Unfortunately, you cannot always rely on feasibility of the simulations, simply because you cannot always obtain guarantees regarding the stability, operations, and performance of Complex Systems. Especially, this is valid for the systems that have considerable amount of non-linearity inside. In such cases, the safety critical cases may be missed.

The use of modeling and simulations in our search for better systems put the focus on our search for better understanding. Our attempts to use this understanding will be to provide rational decisions in the world where there are different simulations under imperfect models.[5]

---

[5] This can be found as Burns Effect in Leonard Smith's book *Chaos: A Very Short Introduction* (published by Oxford University Press in 2007).

As said in the previous paragraph, the extensive testing on the real system could provide better insights into the operation and possible problems in its performance and safety. The simulation is just an analytical tool to obtain formal mathematical proof that the system is OK. What we need is an experimental proof that the Complex System will operate in reality as intended.

The design of Complex Systems is always based on analysis techniques. Since design methods are usually crucial for modeling and simulations, it is almost impossible to create good system without first studying the mathematical model. The simulation, as a tool, also allow us to assess the draft-designs, as soon as they have been made. If the performance of the draft-design model during simulation is inadequate, the simulation may also help with the directions of modifying the draft-design models.

One of the most used simulation methods is Monte Carlo method. It is a stochastic method which was developed during Manhattan Project (building an atomic bomb in the United States in the 1940s). The intent was to investigate the distribution of neutrons during Uranium atomic fission. It can be executed on computer by using random inputs for particular algorithms. The random nature of the method is very important having in mind that randomness in approach provides better information about the behavior of Complex System. Method is used in all cases where analytical or numerical solutions are too expensive or are too difficult to implement. Having in mind that it is a stochastic method, the computer must have pseudorandom generator for producing the random inputs. It also helps to investigate behavior of the system under model with changeable variable or parameter belonging to different probabilistic distributions. The good thing with this method is that we can also investigate the future assumed behavior of the system, because we can assign different algorithms for execution.

It is not a statistical method, so before implementing it, there is need to check the statistical properties of possible inputs to determine probability distribution. After that, the pseudorandom generator generates as much as possible random inputs which belong to the determined probability distribution. Computer will calculate all possible outputs and the results should be again statistically analyzed. The method gives approximate results, but the error in calculation decreases by $1/\sqrt{n}$, where n is the number of generated random numbers (inputs). In other words, generating less than 1 million random numbers (inputs) makes no sense.

This is probably a good place to mention another use of simulations . . .

I am not sure how the things are going in other Risky Industries, but in aviation, the simulations are extensively used for training the operators to use Complex Systems. The simple example are pilots of the aircraft and Air Traffic Controllers (ATCOs) for providing ATC services.

Both these categories of employees, during their education which starts with theory of flights and Air Traffic Management (ATM), eventually finish it by considerable training on appropriate simulators. The first "flights' of the pilots are done on simulator and the first ATM operations for ATCOs are also done on simulators. However, even after training on simulators, they need on-job training at a particular time, which is a requirement for practical part of examination.

In addition, the pilots and ATCOs are subject of recurrent training at least once per year. This is not an ordinary training, where the pilots are trained how to fly the aircraft or the ATCOs are trained how to control the separation of aircraft on the sky. This is a training for emergency procedures, which do not happen very often with these Complex Systems, so they can be forgotten. These simulator recurrent trainings for pilots and ATCOs are also subject of examination.

2.5   REVERSE ENGINEERING

There are two definitions of Engineering. The first one is a practical "brother" of science where the science knowledge is used for designing, manufacturing, and maintaining products used for different purposes. The second one is as conducting a process by maneuvering or managing the materials and tools used for that process.

Sticking to the first definition (important and applicable to this book), the question is: If we can use this, conditionally named, "bottom-up" approach to design and manufacture "new things" based on scientific principles, then why we cannot do the opposite: Trying to understand the operations of these "new things" which are based on these underlying scientific principles by implementing "upside-down" approach? In the first case, the name of the discipline is Forward Engineering, and in the second case, the name of the discipline is Reverse Engineering.[6]

Reverse engineering can be defined as a process of trying to understand the operational principles of a system (object, asset, etc.) through analysis of its building structure and its way of functioning.

The point is that it applies to Hardware as well as to Software and to be honest, the applications to Software "decoding" are more popular, more complex, and more beneficial.

For the Hardware, the Reverse Engineering finds its application in industry in the area of industrial competitiveness. There companies buy products from their competitors with the intention to understand their functioning and to produce similar ones which cannot be affected with the legal patent rights. Also, the patent agencies use it to check the patent rights, especially where there is a legal issue.

In addition, it is used to produce spare parts for the systems which are not in production anymore and there are no spare parts. During the design, the physical model built in plastic (clay, wood, or anything else) can be used for computer-aided design (CAD) model and later, the simulations are done to this model.

For the Software (there patent rights are more "flexible"), cracking the code of successful applications is very beneficial. Formal definitions in Software area for Reverse Engineering can be found in ISO/IEC/IEEE 24675/2017 standard (Systems and software engineering—Vocabulary):

1   Determining what existing software will do and how it is constructed (to make intelligent changes)
2   A software engineering approach that derives a system's design or requirements from its code.

---

[6] In some literature, you can also find the names of Backwards Engineering or Back Engineering.

In our case, the Reverse Engineering could be used to better understand the Complex Systems or to test their deficiencies after the particular use of them.

Reverse Engineering is based on three activities:

1 *Data Gathering:* It is understandable that there is need for thorough investigation of the Subsystems to the given Complex System and their components. It must be done through deconstruction and decomposition of the Hardware and Software. The data gathered would be a starting point and quality of data is strongly connected with the quality of fulfilling the aim of the Reverse Engineering.

2 *Knowledge Gathering:* It is also understandable that there is need for analysis of data gathered and reaching particular conclusions about functioning of the system and about the interdependencies and interactions (hierarchical and relational) of its constituting parts.

3 *Knowledge Using:* Gathered knowledge can be used for improvement of the system, fixing the deficiencies or improving maintenance and operational activities.

Most important thing in Reverse Engineering is deconstruction (parsing) and decomposition of the Complex System, because overall data gathering and analysis depends on it. There are many tools, methods, and methodologies, depending on the case, which can be used.

In general, Reverse Engineering can find its application in gathering knowledge from everything which humans have made.

2.6   TRIAL-AND-ERROR METHOD

Long time ago, trial-and-error method was very much used in the industry. The lack of enough mathematical knowledge and methods for calculation provide the construction of all industrial systems to be based on trial-and-error method: The engineers (at that time there were no specialized category of engineers) made a plan with the system and built it. If the system was OK, they were using it. If the system was not OK, they tried to fix the errors or to build another one taking care of the lessons learnt from the previous errors.

Today, all design processes are based on modeling, simulations, and experiments (testing) of the design. But the trial-and-error method is still used in designing Complex Systems with high non-linearity through building the materialized models (touchable) and extensive testing of them in different conditions. Also, it is used very much in Software development, because you cannot destroy the software as a result of any trial or any error. The knowledge, experience, intuitions, and instincts are very important in these activities.

In addition, the trial-and-error method is very much used for management systems and, for big organizations, especially those in Risky Industries, these management systems can be very much complex. This method is used for making procedures.

2.7 INTERRELATIONS DIAGRAMS

The Interrelations Diagrams[7] (ID) are one of the seven new tools[8] produced by coordinated work between the Society of Quality Control Techniques Development and Japanese Union of Scientists and Engineers in 1976.[9] Although it was designed to help understand the interdependent causal-and-effect relationships of a complex problems in companies, it can also be used to understand the causal-and-effect interactions and interdependencies of Subsystems activities in the Complex Systems, including Software and Human (operator) interactions with them.

It is a semi-quantitative method and it was dedicated to help in the management and planning of bigger and complex projects. However, it is an excellent tool for understanding the operation of Complex Systems during design and/or Safety Assessment before making it operational. It really helps to identify, analyze, and classify the interactions and interdependencies in Complex Systems.

In management and planning, there are cards where the causes and effects are presented and the interrelations are presented by arrows. The arrow can enter (IN) or leave (OUT) the card. In engineering, the cards could be activities of the Subsystems and there could be three types of arrows showing the following:

- Data transfer from sensors into Control Systems or information for operation of each Subsystem based on Software Algorithms (Firmware)
- Commands (Control) toward Subsystems from CPU based on the Software Algorithms (Firmware) or from Control Systems
- Materials and product movements for production purposes.

As it can be noticed, this is similar to three types of internal buses inside the computer: Data Bus, Control Bus, and Addressing Bus. The difference in ID is that we are more interested about interactions and interdependencies between the Subsystems and the cause and effect between them than in simple digital communication between them.

The simple example for ID tool is shown in Figure 2.1.

Figure 2.1 shows a system for maintaining the water at a particular level in the tank. There is a water flow through the Water Pipe which is controlled by a Valve receiving commands from the Flow Controller. There is a CPU which receives data from all Subsystems and, based on the level of water in the Tank and on the flow of water through Water Pipe, maintains the same level in the Tank. The interactions and interdependencies are presented by different arrows explaining the "flows" of Data, Commands, and Materials. Usually, the arrows are drawn from the source (a cause) toward the end (an effect), or from means toward the objective; or, simply, they present just flaw of materials or products during its production process.

As you can notice, I have added numbers associated with each arrow which are actually the Level of Criticality (LoC) for the Complex System's operation:

---

[7] You can find it in the quality literature also under the names Relations Diagram and Digraph (Diagrams and Graphs).

[8] In the literature, you can find these new seven tools under acronym 7QCT (7 Quality Control Tools).

[9] To be absolutely correct, these seven QC tools were provided as legacy of the previous work of Shiguru Mizuno (who wrote for these tools as part of Total Quality Control.

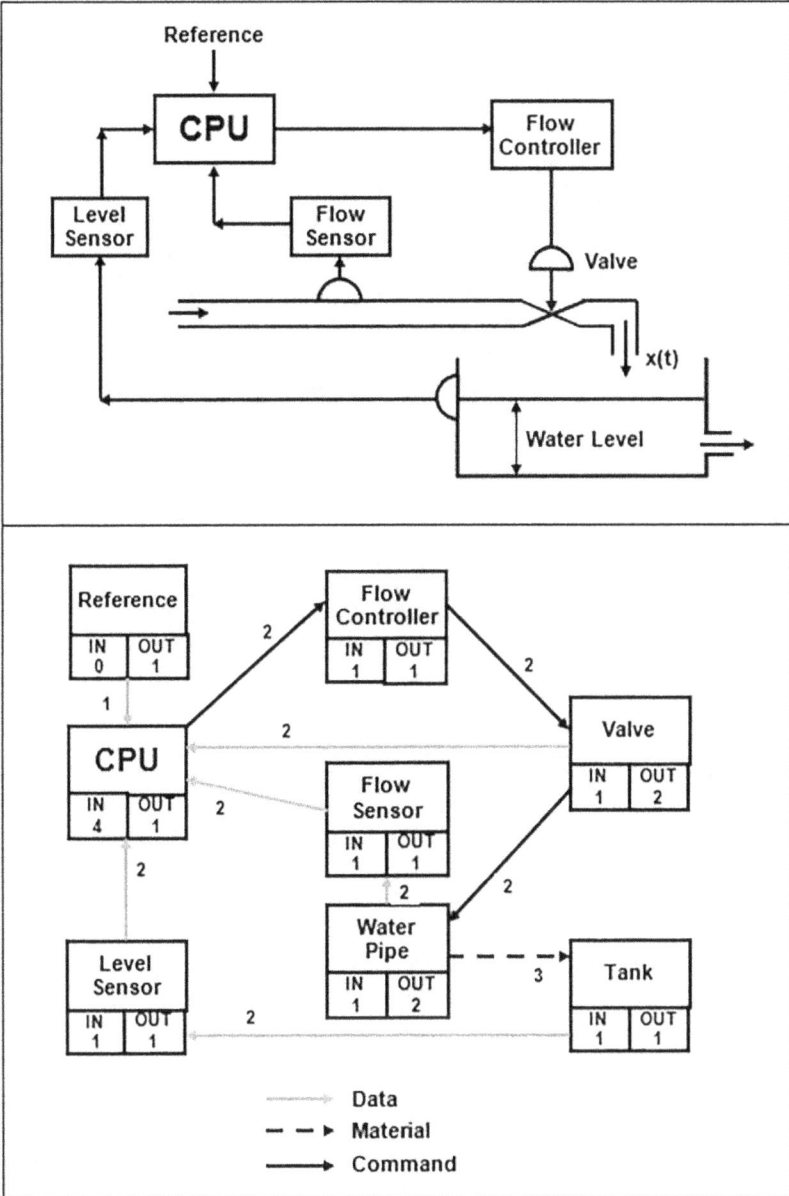

FIGURE 2.1   ID for controlled-by-the-level water flow into reservoir.

- LoC = 1 is a level which will have small influence on the operation. In our case, it is Reference and if this arrow is "broken", the operator must find some other reference which will be added manually.
- LoC = 2 is a level which will provide degraded operation and there is need for maintenance engagement.

- LOC = 3 is a critical influence on the operation and if this arrow is "broken", the operation will stop.

Please note that I did this LoC just as example (not a rule) and you can do any kind of levels similar to these.

The IN and OUT rows in each card explain the quantity of interactions and interdependencies between these cards (Subsystems), which implicitly point to the main devices in the systems. Usually, the cards with more "OUT" arrows than "IN" arrows are causes, while the cards with more "IN" arrows than "OUT" arrows are effects. But be careful determining what is what: Sometimes, the different combinations are very much present!

The cards with names of the devices and arrows could be designed differently and it is for the Safety Manager to realize what is important in the operation of the Complex System and what is needed for proper understanding of that operation. Associated with the arrows, there can be short sentences or phrases to help understand what is that arrow about.

In addition, the ID may be presented as quantitative or qualitative.

In the qualitative format, the cards are just connected to each other and the understanding of interactions and interdependencies is achieved by intuitive understanding. In the quantitative format, there are numbers associated with each arrow explaining the strength of interactions and interdependencies between the Subsystems, so those with higher number present more important or critical relation.

There are few recommendations about order of the steps when you try to create an ID:

1 Collect (as much as you can) information from a variety of sources about the system and the science and engineering used to build it.
2 Try to use short phrases or sentences as explanation of the cards representing the Subsystems.
3 Draw draft-diagram and give it to the multifunctional Team which is used for Hazard Identifications and Risk Assessment. Expect the inputs from any of the Team members.
4 Make changes in the draft-diagram after discussions with the Team regarding the proposed cards and arrows.
5 Pay attention to all identified cards and arrows marked as critical by the Team.
6 Pay attention to all internal and external influence factors that may affect the Complex System and its operation.
7 Have in mind that this ID should be "live" document: If something changes in the operation or in the system, the ID must be changed accordingly.

It is good to ask "why" questions to clarify the relationships and to understand the process, so that Team members can critically evaluate, revise, examine, or discard influence factors. For example, in Figure 2.1, the question, "Why the Valve is open?, will point to the Command issued by the Flow Controller. Going further, the question. "Why the Flow Controller commanded Valve to open", will point toward the CPU which provided a command to the Flow Controller how much the Valve should be opened (based on the inputs received from other cards).

The variant of ID is an Interrelation Matrix (IM) presented in Table 2.1.

TABLE 2.1

Interrelation Matrix for Case Presented in Figure 2.1

In / Out	Reference	CPU	Level Sensor	Flow Sensor	Flow Controller	Water Pipe	Valve	Tank
Reference		I1						
CPU					C2			
Level sensor		I2						
Flow sensor		I2						
Flow controller							C2	
Water pipe				I2				M3
Valve		I2				C2		
Tank			I2					

C stands for command, I for information, and M stands for material.

In the first column of IM (presented in Table 2.1) are the outputs from the cards and in the first row of IM are inputs from the cards. In other fields are the type of interactions and interdependencies (I, M, or C) associated with the Level of Criticality.

As it can be noticed, there is no symmetry within the matrix, so it can explain why the ID and IM provide tools for analyzing non-linear interactions and interdependencies inside the Complex Systems. That is the reason that approach with ID is considered a "lateral thinking" instead of "linear thinking".

If we would like to classify ID, it can be said that it belongs to TQM (Total Quality Management) approach in Quality, but I propose here to use it in Safety. I do believe that it is an excellent tool for understanding the interactions and interdependencies within the Complex Systems. As such, I personally prefer ID than IM, but you can decide what is more appropriate in your case.

2.8 MARKOV CHAINS

Markov chains, together with the Markov processes, are part of the so-called, Markov analysis developed by Andrei A. Markov.[10] Only difference between the Markov chains and Markov processes is that the first one applies to the systems which change their states discretely and the second one to the systems which change their states continuously.

You can see in this section that this is something similar to Interrelations Diagrams from the previous section, but Markov chain method is fully quantitative and fully mathematical by using probabilities.

A simple definition of Markov chains would be as follows: A method to investigate sequences of transition of random events where any future event is caused by particular probability of previous event. It means that Markov chain method is a

---

[10] Andrey Andreyevich Markov (Professor at Saint Petersburg State University, 1856–1922), a Russian mathematician, is best known for his research in the area of so-called Markov chains.

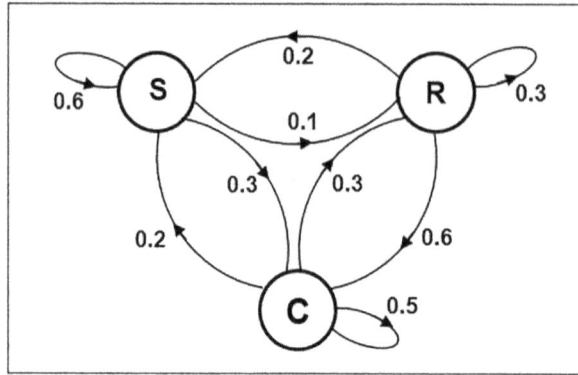

FIGURE 2.2 Graphical presentation of probabilities for the weather tomorrow knowing the weather today.

memoryless method, because it considers only the (last) previous event, and it does not consider events before this last event or any other combination.

I will emphasize here the Markov chains from the aspect of safety of the Complex Systems. Here, the probability of faulty states can be determined based on the probabilities of problems assigned to particular previous states in the process conducted by the Complex System. As such, the Markov chain could help us understand the mechanism of faults in our Complex System and, the so-called Fault Propagation. If we can state the faults happen randomly in the Complex Systems, then, based on the historical data gathered, the Markov chain can help by providing probability that our system which works normally today will be faulty tomorrow.[11]

There is no problem to implement the Markov chains on the states in the normal operation of the Complex System. The normal state of operation will be determined by its Working Point adjusted by parameters and controlled (changed) by variables. So, you may use them in calculations to better understand the process of normal operation, but although there are a lot of uncertainties determined by the tolerances, we can assume that normal operation is a pretty deterministic process. However, the probabilities for random fault happenings give more meaning to the analysis.

It can also be used for decision-making process, based on the probabilities given at the end of the analysis.

I will explain Markov chain by a simple example of what will be the weather tomorrow knowing the weather today. It is assumed that the weather today can be sunny (S), cloudy (C), and rainy (R). The same possibilities can happen tomorrow and let's see how it will look graphically (Figure 2.2).

The numbers given in Figure 2.2 are the probabilities that a particular type of weather today (S, C, or R) could transform into a particular type of weather tomorrow (S, C, or R). These probabilities are a result of gathering weather data changes in the past, processed statistically. You can notice that all arrows getting out from a particular "type" add to 1.

The graphic presentation is also presented in Table 2.2.

---

[11] You will see later that the Markov chains are something different from Reliability . . .

TABLE 2.2

Probabilities for the Weather Tomorrow (TMR), Knowing the Weather Today (TDY)

TMR \ TDY	S	C	R
S	0.6	0.3	0.1
C	0.2	0.5	0.3
R	0.1	0.6	0.3

Again, all numbers (particular probabilities) in all horizontal rows gave a sum equal to 1.

Or it can be presented as equation (as matrix):

$$W_{TMR} = W_M \cdot W_{TDY} \rightarrow \begin{vmatrix} S_{TMR} \\ C_{TMR} \\ R_{TMR} \end{vmatrix} = \begin{vmatrix} 0.6 & 0.3 & 0.1 \\ 0.2 & 0.5 & 0.3 \\ 0.1 & 0.6 & 0.3 \end{vmatrix} \cdot \begin{vmatrix} S_{TDY} \\ C_{TDY} \\ R_{TDY} \end{vmatrix}$$

So, the prediction that a particular "type" of weather will be tomorrow, based on the numbers in the matrix and Table 2.2, can be described by the following system of equations:

$$S_{TMR} = 0.6\, S_{TDY} + 0.3\, C_{TDY} + 0.1\, R_{TDY}$$

$$C_{TMR} = 0.2\, S_{TDY} + 0.5\, C_{TDY} + 0.3\, R_{TDY}$$

$$R_{TMR} = 0.1\, S_{TDY} + 0.6\, C_{TDY} + 0.3\, R_{TDY}$$

Considering all written above, it can be assumed that if today is sunny (S_{TDY}), then tomorrow's weather will be sunny with the probability of 0.6 (60%), cloudy with the probability of 0.3 (30%), and rainy with the probability of 0.1 (10%). Similarly, you can calculate the probabilities that tomorrow would be cloudy (C) and rainy (R) day.

I will repeat again: There is a need for historical data availability and the use of statistics to calculate the particular probabilities used above.

I hope this example will help you understand Markov chains. For more details, you may refer to articles, books, and videos available on the Internet.

2.9 SYSTEM THEORY

There is a lack of one general definition of System Theory, but few different definitions can be found. All of them depend on the area of interest and paid attention to some aspects of systems, which were important for the persons who provided these definitions. Again, I would provide my definition, for the purposes of this book and the one which I feel confident with.

The engineering Systems Theory[12] can be defined[13] (by me for the purpose of this book) as a unified group of specific and multidisciplinary approaches, which are considered part by part and together (as a whole), as aid in understanding the Complex Systems. Based on that, the System Theory is the way to treat Complex Systems in a different way than the previous methods. Contrary to the reductionism, the method is multidisciplinary and holistic: The system is not "reduced" to its Subsystems, but it is investigated as whole: All the Subsystems and their interactions, interdependencies, and hierarchies are studied together.

Let's explain it with one example from thermodynamics of gases.

Imagine there is one container with some type of the gas inside. If I like to know the average energy of the one molecule of the gas and if I use Reductionism, I will consider every molecule in the gas (as a Subsystem) and its kinetic energy. Gas molecules move all the time and their kinetic energy is a half product of their mass and their speed squared. The molecules in the container will have same mass (it is the same gas), but due to their collisions with the walls of the container and with other gas molecules, their speed and direction will vary. It means that different molecules may have considerably different kinetic energies (due to different velocities). So, I will try to find out any of the energies of the individual gas molecules, but I have to use statistics and I will calculate the average velocity of all these molecules together.

This is a scientific approach, but it is impossible to do it. The molecules are very small, they move all the time, and their number is so huge that I simply cannot measure their energy. Simply, I cannot use the Reductionism for this calculation.

Using the System Theory, I will solve the problem very easily. I will measure the temperature of the gas and it will show me the total kinetic energy. I will measure the pressure on the walls of the container and it will give me the approximate number of molecules (bigger pressure will come from more molecules). So, dividing total energy by the approximate number of molecules, I will calculate the average kinetic energy of one molecule. In this process, keep in mind that the temperature and pressure of the gas are characteristics of the gas: The individual gas molecules do not have such characteristics.

So, the well-known General System Theory emerged first in the natural sciences or, to be more accurate, in biology, by the ideas and research of Ludwig von Bertalanffy.[14] Later it spread all around in other natural sciences, in mathematics, engineering, sociology, environment, philosophy, etc.

The engineering inherits the definition and principles from mathematics and physics. Actually. George Jiri Klir[15] gave the first mathematical principles which

---

[12] To be honest, I am using here expression "engineering System Theory", but such thing actually does not exist formally. To be more honest (again), the System Theory in engineering is very poorly developed and there is a lack of research and principles in this area. So, in this section, I am just trying to support Leveson with my ideas in her efforts to use System Theory for safety purposes. However, System Theory is part of System Engineering and as such it was mentioned in Section 1.11.

[13] This is the definition which I would use for the purpose of this book. As I have already said, in reality, Systems Theory is a phrase which does not have clear and unique definition. It is still a phrase that has been used in different areas (natural, technical, and social sciences) regarding the systems. This is the reason that often the System Theory definition is a subject of misunderstanding when used outside engineering.

[14] Ludwig von Bertalanffy (1901–1972) was the originator of general Systems Theory.

[15] George Jiří Klir (1932–2016) was a Czech-American computer scientist and professor of systems sciences at Binghamton University, New York.

can be used in engineering. By these principles, there are "physical" and "imaginary" Subsystems inside and outside the Complex Systems which are compiled from equipment, environment, humans, processes, behavior, states, transitions, other elements, connections, hierarchy, management, Software, etc. That is the reason that the Complex Systems need to be assessed by a team of multidisciplinary experts for different areas.

Another scientist who contributed very much in engineering System Theory is Jay Forester.[16] He was the first who used computer modeling and simulations to analyze social systems and tried to predict the future states in his models. He is actually known as the founder of System Dynamics. This is the approach which looks at the system's failures dynamically, trying to develop formal models of the processes inside and outside (based on first-order differential equations) and trying to discuss them with a team of multidisciplinary experts.

As mentioned earlier, the engineering System Theory in Complex Systems can be easily explained by the System Theory in Thermodynamics. There, the big picture and assumed interactions of the molecules between themselves and with the walls of the container are used to get knowledge of the gas. So, we can say that the System Theory is a theoretical perspective where the system is analyzed as a whole.

The system could be engineered by different and discrete (specialized) "physical" and "imaginary" Subsystems which do have different roles in the operation of the Complex System, but it does not matter. In the engineering System Theory, the Software, the environment, and humans (interacting with the system) are part of the Complex System.

There is an interesting story how the System Theory developed in engineering. With increasing Complexity of the equipment, scientist started to investigate the possible ways how to implement scientific approach in Complex Systems with intention to gather proper understandings of their operations, behavior, and functioning. They "borrowed" the approach from natural and social sciences and following those steps, they produced three categories[17] of the Complexity believing that they will offer better understandings.

The first category is Organized Simplicity. These are systems where Reductionism can apply. The second category is Unorganized Complexity. Here, the Reductionism cannot provide good results because these systems are complex, but at the same time, regular and random. It means they can be analyzed using statistics and probability. The third category is Organized Complexity. These are systems where the Software and humans are very much involved and, as such they need System Theory.

It is worth mentioning here (again) that the System Theory applies also to Complex Systems in engineering, but it is far more important as it is used in both natural sciences (biology, physics, chemistry, etc.) and social sciences. In general, it can be used as a scientific theory that can help understand and describe the biological, physical, and social systems.

However, Leveson is relying very much on the System Theory in safety area, so it is worth mentioning in this book. In my humble opinion, System Engineering is more

---

[16] Jay Wright Forrester (1918–2016) was an American computer engineer and systems scientist.
[17] This categorization was adapted by Gerald M. Weinberg and you can read more about it in his book *An Introduction to General Systems Thinking* (published by John Wiley & Sons, Inc., in 1975).

applicable to Safety, but although I think the System Theory is somewhat abstract for engineering area, regarding the Complex Systems, it can be cautiously used.

2.10   THE MOST RECENT METHODS FOR UNDERSTANDING COMPLEX SYSTEMS

Science and technology advance rapidly! As many other fields, Complex Systems cannot ignore the rise of Artificial Intelligence (AI). Actually, I would like to take a moment and make the distinction between several buzz words surrounding AI. AI, ML, and DL are terms we hear often, but what do they actually mean and how do they differ from each other?

As mentioned previously, AI stands for Artificial Intelligence and is used as an umbrella term for all algorithms, approaches, and ideas that deal with Artificial Intelligence.

The European Commission adopted the definition of AI, based on the study conducted by it, as follows:

> Systems that display intelligent behaviour by analysing their environment and taking action—with some degree of autonomy—to achieve specific goals.

As such, AI is a super set of ML and DL. The field of AI is trying to find a way to make computers think, just like humans.

2.10.1   MACHINE LEARNING

ML stands for Machine Learning. From a scientific point of view, it is a combination of Calculus, Linear Algebra, and Statistics, but I will try to explain the general idea of it without going into the technical details. ML encapsulates all algorithms that try to learn patterns from data and use them in the future on unseen data. The main idea behind ML is to automatically adjust the parameters of a function given the input/output pairs of data. Let's explain this with an example. Imagine we have the following function:

$$f(x) = y = a \cdot x$$

Here a is a parameter and x is a variable. Now imagine that we do not know the value of a but we do have a lot of input/output pairs that we can observe in some way. For example, imagine we have the following pairs:

- $y = 4, x = 2$
- $y = 6, x = 3$
- $y = 10, x = 5$

You probably guessed it by now but in order for the above function to take those values of x and produce those values of y, a has to be equal to 2. And this is exactly the idea behind ML! We are trying to *learn* the function (or its parameters to be more precise) by finding patterns in the input/output (x/y) pairs that we have. And since this process is conducted by *machines* (our computers of course), it is conveniently named Machine Learning.

Now you might ask, how do we have some function whose parameters we don't know? Well, in most cases, you don't explicitly have the function, but you define one yourself in order to approximate or model some complex system. Let's show this with an example. Imagine we have a nuclear reactor. Nuclear reactors are complex systems which are maintained by a lot of scientists. Let's take one small part of maintaining a Nuclear Reactor, the temperature control. A human scientist would take a look at multiple sensor readings and based on their values would take steps to either reduce or increase the temperature of the reactor. This is an example of a part of a complex system that can be approximated or modeled by a function. Let's assume that we can model the system as follows:

$$T_\Delta = a_1 \cdot s_2 + a_1 \cdot s_2 + \cdots + a_n \cdot s_n$$

s_n are all the sensor readings that we obtain from the nuclear reactor, a_n are some parameters that we don't know at the moment, and T_Δ is the change in temperature we would like to achieve given the sensor readings.

Now imagine we have historical data of how human scientists adjusted the temperature given the different sensor readings in the past. This means that we have a lot of input $(s_1, s_2, \ldots, s_n)$ and output (T_Δ) pairs. As mentioned previously, the idea of Machine Learning is to use these pairs in order to find values for the parameters a_n such that we obtain a function that given those sensor readings, we are able to produce the corresponding changes in temperature. Now using ML algorithms, we can learn patterns in the input/output pairs and learn values for the parameters a_n such that we obtain a function that approximates the system we are trying to model as good as possible.

What I wanted to show here is that anything that can be modeled or approximated by a function can be modeled or approximated by Machine Learning. All we need is some domain knowledge that will help us define a function, inputs and outputs, and a ML model that will go through all the input/output pairs we have and learn the parameters of the function we have defined. Hopefully, this will help you understand how useful can this be in the context of Complex Systems.

Before we move to what actually DL is, I am guessing that some of you are a bit more curious and are interested in how exactly can a machine learn the parameters itself? What really is Machine Learning under the hood?

Well, how exactly different ML algorithms achieve this can be a book on its own and is therefore outside the scope of this book. However, if you are interested in some good ML books, I would recommend *Pattern Recognition and Machine Learning* by Cristopher Bishop.

2.10.2 Deep Learning

DL stands for Deep Learning. Just as ML is a subset of AI, DL is a subset of ML. More precisely, when we say Deep Learning, we refer to all ML algorithms that are based on Neural Networks. Neural Networks are a special family of ML models that are designed to mimic the human brain. Explaining how exactly Neural Networks look and work is once again outside the scope of this book; however, a good book on the topic is *Deep Learning* by Ian Goodfellow.

You might wonder how are Neural Networks and Deep Learning connected? Where does the name come from?

Well, Neural Networks (by design) have something that is called "layers". Research has shown that Neural Networks with a lot of layers are exceptionally good at learning functions, given the large amounts of inputs/outputs pair. Having that in mind, using computer science terminology, a Neural Network with a lot of layers is simply called a Deep Neural Network, hence the name Deep Learning.

In recent years, Deep Learning has dominated the field of Machine Learning. Almost all fundamental research and groundbreaking advances happen in the field of Deep Learning, simply because it has shown us that it is, by far, the best model for learning functions that we have at the moment.

However, Deep Learning comes with some requirements . . .

By definition, Deep Learning needs large amounts of data. We need input/output pairs on the order of millions or even billions. Now this might seem intimidating, but let's put this into the Complex Systems perspective. Most Complex Systems, like our Nuclear Reactor explained previously, have a lot of sensors that monitor the function of the system. These sensors produce readings, in arbitrary small steps of time, often in the scale of seconds or even milliseconds. This is how we obtain our inputs. Now for every set of inputs, the System (or the people managing the system) perform some actions. These actions are exactly the outputs we are looking for. As explained previously, we can look at the historical data and what these people did based on the inputs in order to create them. This shows how easy it is to reach millions and even billions of input/output pairs when we try to model Complex Systems.

To conclude, Deep Learning is simply a family of Machine Learning models based on Neural Networks that are extremely good at learning functions and can be used to model certain parts of, or even entire, Complex Systems.

2.10.3 SOME OTHER ASPECTS OF ARTIFICIAL INTELLIGENCE

Considering the Nuclear Reactor example and some of the explanations from the previous sections, you can probably guess how powerful AI is and how many potential applications of it exist. However, as the saying goes: With great power comes great responsibility!

Let's consider some additional AI applications that model Complex Systems:

- Autopilot
- Self-driving cars
- Diagnosing medical images

And a second category of applications:

- Netflix movie recommendations
- Advertisements recommendations

Now consider the following question: Do both of these categories of AI applications are associated with the same risks? Is the risk of misdiagnosing a medical image the same as the risk of suggesting a bad movie? It is easy to see that certain AI applications, based on the Complex Systems they are modeling, are associated with much larger risks than others . . .

Our society is governed by rules and laws which are there to prevent bad things from happening, as well as to help identify the responsible person when something goes wrong. But who do we blame if an AI system misdiagnoses a medical image? What about a self-driving car causing an accident? You might think that you have a quick solution, but the more you think about the ethical implications of AI and the risks that come with some Complex Systems, the more you'll realize how difficult the problem of assigning a blame to a machine is. This is exactly why you will only see an AI system modeling an entire Complex System, if and only if, the risks of doing a mistake are moderate.

Current AI systems for problems that bring great risks with them are developed in a way that they do not replace the human, but instead they try to help the human. This means that an AI system might analyze the inputs of some Complex System and produce outputs which are there to provide more guidance and more information to the human, but at the end the human will be the one making the final decision.

(a) In the case of diagnosing medical images, an AI system might inform the doctor that there is an area on the image that looks like it contains a tumor, but a doctor will take a closer look and run additional tests to confirm or deny the suspicion.

(b) Self-driving cars explicitly mention, in their terms of agreement, that you are not allowed to let them drive unsupervised and you should be ready to adjust and correct whatever they are doing at any given time.

(c) If we go back to our Nuclear Reactor example, remember how the AI system was analyzing the sensor reading and was outputting a temperature delta. It was simply outputting a recommendation. Using that recommendation, a human can then step in and perform certain actions to reduce or increase the temperature of the reactor.

(d) Etc.

The reason I'm telling you all of this is because I want you to understand that not everything that can be done should be done and you should always consider the legal and ethical implications. Like I mentioned in the beginning: With great power comes great responsibility!

So, make sure you use AI responsibly.

2.11 PITFALLS WITH TRIES TO UNDERSTAND COMPLEX SYSTEMS

In our efforts to understand the engineering Complex Systems, there are few pitfalls. Their "nature" is based on interdependencies and interactions between their Subsystems and components, humans and the environment, as well as on the synergy between Hardware and Software during their operations.

Two of the pitfalls which I would like to mention here are "narrative fallacy" and "platonicity".

Narrative fallacy can be defined as presentation of human's need to find and fit a similar story (pattern, case, example, etc.) to the new events (systems, facts, etc.) which we encounter for the first time. According to Nassim Nicholas Taleb and his bestseller *The Black Swans: The Impact of Highly Improbable*, it is one of the reasons that we are shocked by some events. Black Swans are synonym for events which are highly improbable to happen (unknown unknowns), and when they happen, they come with catastrophic consequences and in hindsight, they look very logical to happen.

Narrative fallacy is something which is very much used by professors and students to initially comprehend new teaching material during studies. The point is that in our minds, the professors (for the purpose of teaching) and the students (for the purpose of studying) create a simplified model of the Complex System (lecture, theorem, etc.) which could help understand the real system (lecture, theorem, etc.). Just imagine that such a narrative fallacy is simplification of the bird's flying by the aircraft: although both have wings, the operation of flying is quite different for both of them. Another example could be presentation of the human heart as hydraulic pump and the similarity of their operation, although they are significantly different by their construction.

This example also points to the danger of misunderstanding the emerging properties of the Complex System, so you need to pay attention not to do it.

The platonicity is very much similar to narrative fallacy. It can be defined as human efforts to explain the unknown things (which we have suddenly met) by known things (which we already have met). In general, it means that our study is limited with our present knowledge and our previous experience, so we struggle to use it to find better (and maybe, more appropriate) explanation of the things around us.

A simple example of platonicity is the linearization of the non-linear functions which I have mentioned in Part II of this book. There, the non-linear differential equations are very hard to solve, so to find solutions, the mathematicians use well-known methods of linear mathematics. However, this linearization of non-linear functions provides solutions which are limited only in the vicinity of the area where the linearization is used.

Another example of platonicity is modeling in science and industry. The modeling a is very powerful tool, but it is also based on our understanding of the known things about systems, which is pure platonicity. Modeling can provide better understanding, but we cannot run away from the fact that usually it neglects Complexity, which is essential for the understanding of the Complex Systems.

I do believe that many of the readers must have encountered many of the things provided in this book for the first time and please be open-minded in your effort to understand how it can help you to provide safety of the Complex Systems.

3 Complex System's Operations

3.1 INTRODUCTION

As we have seen in the previous chapters, all the systems in their functioning are accompanied with particular uncertainty. This uncertainty is quintessence in science, industry, and even in our personal lives. We struggle for certainty because we believe it can bring stability and whatever efforts we put to provide the certainty, not always it can be achieved.

In the industry, people are well aware about the uncertainties and imperfections of their equipment and processes, so they implement tolerances which can be defined as allowed deviations of the characteristic of the manufactured product.

I will spend some time to explain what goes on with the uncertainties in Complex System's operations and how we should take care about them.

3.2 TOLERANCES IN INDUSTRY

The industry is driven by standards!

Standardization is a good thing for the economy because having a standard and abiding by it will give your company chance to sell your products (or offer your services) all around the world (where this standard is accepted). Nevertheless, accepting the standards is voluntary,[1] the companies are willing to accept the standards and abide by them. There are also some companies that write their own standards which are more stringent than the available international standards.

Roughly, there are two types of standards. One type is dealing with tolerances of the products (services) and another type is dealing with management. The first one is dealing with Equipment and the second one is dealing with Humans. Having in mind that this book is mostly dedicated to the Equipment, only these tolerances will be considered in this section.

The standardization is connected with Quality. Usually, if your product (service) is in accordance with the requirements of some standard, it can be assumed that this is a product with good quality.

To prove the good quality of the product, the industry implements Quality Control (QC). There are lot of misunderstandings about QC, but in simple words this is a process of measuring the particular characteristics of the products (weight, speed, voltage, etc.) and comparing the measurement results with particular standard. The standards for particular characteristics are usually expressed as range of values

---

[1] Most of the ISO standards are voluntary, but there is exception in some industries and countries which have decided some of the standards to be a regulatory requirement.

DOI: 10.1201/9781003404811-4

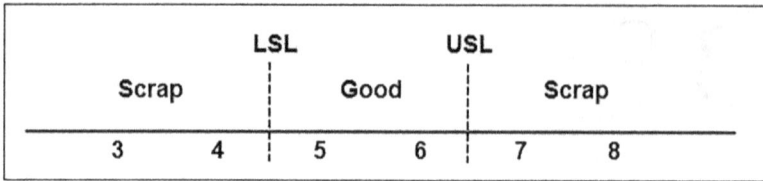

FIGURE 3.1 Concept of tolerance in industry.

where particular values of measurements of the product shall fit. These ranges of values are known as "tolerances".

The tolerances can also be defined as limits of some values and there are Upper Specification Limit (USL) and Lower Specification Limit LSL.[2] Whatever the result of measurement of the characteristics of the product is between these limits, the product is accepted as good. Whatever the result of measurements is outside these limits, the product is a scrap.

This tolerance is a two-side tolerance, because there are two limits: LSL and USL. However, there are also one-side tolerances which can be limited from the above (the value of the parameter shall not be bigger than . . .) as well as from the below (the value of the parameter shall not be smaller than . . .).

So, I can state that the tolerance is the acceptable (by standard) deviation from the specification of the product; in other words, it is the range of variation of some characteristics permitted in the process for producing a specified product. The tolerances are introduced into system during design process and they are actually constraints which need to be monitored and controlled to provide normal operation of the system.

Figure 3.1 presents the concept of tolerances and establishing the quality of the product. If the characteristics of the products are bigger than the LSL (4.5 in the figure) and smaller than the USL (6.5 in the figure), the product is good. If not, it is scrap and as such subject of repair (if possible) or else thrown away (scrap).

If the specifications of the products are not given by the standards, usually companies choose by themselves tolerances for their products. Of course, the smaller tolerances produce better quality of the products, but at the same time, they need better production equipment, more skilled and educated workforce, and better measurement equipment (to prove to themselves, to the Regulator, and to the customers that the stated quality is maintained).

The choice of tolerance mostly depends on the process and measurement equipment. To clarify how this should be made, let's go back to Section 1.10.

In Section 1.10, I have explained that measurement result is expressed by stating the mean of series of measurements (usually not more than 10) and calculated the extended standard deviation (usually 3σ). Mathematically, it will be presented as follows:

$$R = \mu \pm 3\sigma$$

---

[2] Somewhere in the literature you can find the names of Upper Control Limit (UCL) and Lower Control Limit (LCL) and this is not the same. In general, UCL and LCL are used in Statistical Process Control and USL and LSL are used for products (services). For LCL and UCL, see Section 3.4.

where R is the accepted measurement result, μ is the mean (arithmetic mean or average value) of particular series of measurements of particular characteristic under consideration, and σ is the standard deviation of series of measurements, as a measure for precision (uncertainty).

Usually the process of measurement of the characteristics of the product in manufacturing industry is conducted as a series of measurements of the same characteristic of different products, rather than many measurements of that characteristic of the same product.

The concept of "many measurements for the same characteristic of different products" means that the laboratory (inside the company) does sampling of, say, ten products out of the batch of products manufactured in one day (shift, process, hour, etc.). This sampling will provide approximate information how good is the batch of products. If the measurements say that "the same characteristic of many measured products" from the same batch are within the tolerance limits, it is assumed that it is a good batch. If some of "the same characteristic of many measured products" are outside tolerances, then they should be treated as scrap. There is a good reason to do just one measurement for any product. Using a series of measurements of the characteristic of only one product to gather information about its quality is not feasible in industry, or in other words, it is too expensive!

The word "feasible" here implies that there is actually something which will also affect the quality of the product. To provide more "accurate" information about the characteristic of the single product which is under consideration, we need time, measurement system, and trained and skilled humans which will do the measurements. If we do a series of measurements for the characteristics of the same product, then we need more time and more humans and the humans need to be paid. Increasing the cost in production (more time and more staff with salaries) is not good for the profit, hence more measurements are not cost-effective (although they are, very much, quality effective!).

The best companies in the watch industry follow a different rule: They do not do sampling, but they measure each characteristic of each part included into watch production process, before they are used for single watch production. In addition, when the watch is assembled, they do thorough tests of all characteristics of the produced watch. That is the reason that these companies offer impeccable quality for their products and the customers must pay "impeccable" price for these products. Anyway, the customers do not complain about that: Having a watch from any such watch manufacturer is not about measuring a time, it is about prestige!

All other companies in other industries use previously mentioned procedure. For example, if today they produce (approximately) 1,000 product, they choose randomly 10 of them and measure the characteristics of only these 10. Assumption behind this procedure is that all the products are produced by the same process, so their characteristics should be same. This is actually statistics and probability implemented in the production process, so random choosing is needed to provide unbiased information.

Those three sigmas (3σ) in the equation above speak that 99.73% of the products will be within the tolerances. For those 1,000 produced products, it means that there is a probability that 27 of them will be outside the tolerances (they will be scrap). The bad thing is that company could not know which 27 of all batch of 1,000 would be a scrap. For such a purpose, they need to measure every product from these 1,000. The companies are aware about this probability to have a scrap, so they try to handle this by offering warrantees for their products for 1, 2, or 5 years. They have calculated that this is feasible for them.

So, they produce each day 1,000 of products and they test 10 of them each day. It is reasonable (and plausible) to assume that values of μ and σ calculated each day will not be the same. It could be a problem, but for the manufacturing purposes, only the variations of μ are important. Value of σ is always very small, so its variations are not calculated.

Industrial statistics has already shown that average variations of μ (expressed as σ_μ) can be calculated by the following formula:

$$\sigma_\mu = \frac{\sigma}{\sqrt{n}}$$

where n is the number of measured (controlled) products each day from each batch. In the example from above, $n = 10$.

Having this in mind, the establishing of tolerances (for internal purposes where there are no standards) can be expressed by the following formulas:

$$LSL = \mu - (3\sigma + \sigma_\mu) \quad \text{and} \quad USL = \mu + (3\sigma + \sigma_\mu)$$

The graphic presentation is given in Figure 3.2.

Of course, the company will strive to make the tolerances smaller, as much as it is possible, because it will provide product with better quality. So, looking at Figure 3.2, they can assume that it can be done by trying to decrease the variability of the production process or, in other words, they need to provide $(3\sigma + \sigma_\mu)$ to be smaller, as much as it is possible.

To achieve that, they need to put their effort to stabilize the process used to produce the product. If the process uses Complex System for producing the product, it means that the operation of the Complex System must be stable. In general, if

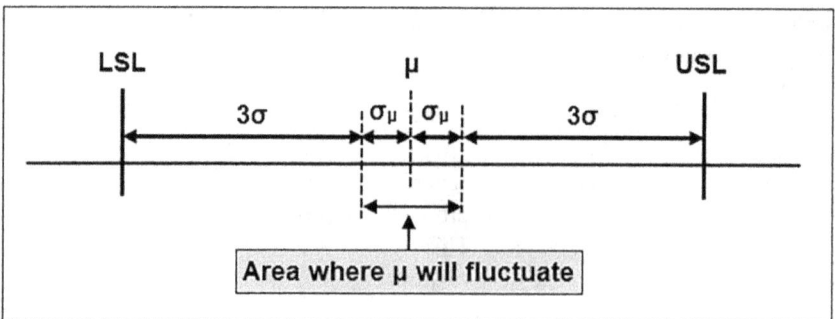

FIGURE 3.2   Choosing LSL and USL.

properly designed, adjusted, maintained, and monitored, the Complex System would be stable in its operation. However, the reason to provide Complex System's stability is very much important.

3.3   MEASUREMENT SYSTEM ANALYSIS

But before I move toward stability, let's mention one very important thing which must not be neglected: The used measurement method in combination with the used measurement system for Quality Control of the products.

The measurement system used to determine the value of the characteristic of the product (measurand) must be appropriate for the intended measurement process. So, there is so-called measurement system analysis (MSA), which should also be done as a requirement of the standard. For example, in automotive industry, there is IATF (International Automotive Task Force) standard that determines the requirements which must be satisfied to provide good products. It is standard known as IATF 16949 and there you can find requirement for executing MSA for all instruments included into product development and product manufacturing for automotive industry.

A Measurement System Analysis (MSA) is a practical and theoretical assessment of the measurement process and everything which is used for it (instrumentation, method, tools, calculations, analysis, training and education of employees, etc.). Usually, it is done through a set of experimental measurements of the same or different products and by different employees. The reason why MSA must be conducted is to realize whether the measurement system and the associated measurement method are good enough to provide good quality.

As mentioned earlier, the Quality Control is a process where we measure the required characteristic of the product with an intention to confirm that it is within the limits of applicable tolerances. It is clear that if your company is producing equipment for aviation or space industry, the tolerances are very small and the chosen measurement system must have particular resolution to provide requested "accuracy" and "precision".

Depending on the tolerances, we would choose measurement system and measurement method. The problem is that MSA must be conducted in the company premises when the measurement system is already purchased and the measurement method is determined. So, if it is not suitable, nothing can be done, except to change the method or tolerances and/or to purchase a new measurement system.

The measurement system itself can also be very complex. But with today's computerization, it is possible to make the overall measurement process pretty much easy. The measurements are also needed for the sensors used in control and automation Subsystems and, as such the measurement circuits (sensors, processing the data, actuators, etc.) inside the Complex Systems are reason for increasing the Complexity (as it has been stated earlier).

I would not provide more details how to do MSA, but if you are interested or if you are in automotive industry, maybe you will need to do it by yourself. However, there are plenty free resources on the Internet and you can find what you need there. Let me warn you, it is not so simple, but it can be done by yourself!

3.4 STATISTICAL PROCESS CONTROL

Long time ago, the "Quality Gurus"[3] realized that the process (provided by the system) of manufacturing the product is most important to provide a good product. In simple words, a good process will produce a good product.

Anyway, it is not so simple . . .

In general, the process (system) is defined as "Black Box" where there is input and output (Figure 3.3).

The "Black Box" actually transforms inputs into outputs. For example, the input can be a particular ore, the "Black Box" would be smeltery, and the output will be associated metal. Of course, there are also parameters which affect the process. Some of these parameters can be changed (controllable parameters = variables) and they are used during the process to control it. Some of them can be adjusted before the starting of the process and later, they will not be changed (adjustable parameters = parameters).

Process can go wrong when it loses its stability. It can mostly[4] happen due to sudden and unexpected change of parameters or variables which are under the influence of environmental or some other factors that are intrinsic to the process, to the materials used, or to the environment.

Going back to the beginning of this section ("a good process will produce a good product"), instead of controlling the product, it is better to control processes. So, for each of the variables and the parameters in the process, we need to establish tolerances. In this case, the names of the tolerances will be Lower Control Limit (LCL) and Upper Control Limit (UCL). If the variables and the parameters are outside the LCL and UCL, the product will be out of tolerances (out of LSL and USL).

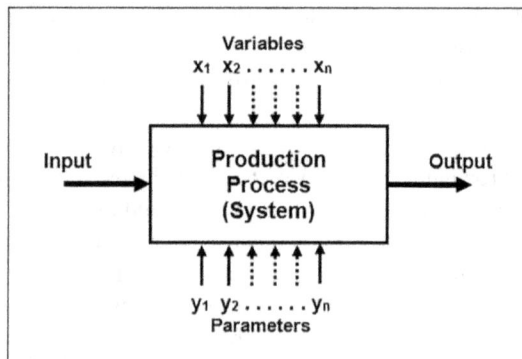

FIGURE 3.3 Description of the process.[5]

---

[3] "Quality Guru" is a name dedicated for the pioneers of Quality: Shewhart, Deming, Juran, Crosby, Taguchi, etc.

[4] Of course, that process will lose its stability due to inappropriate changes of controllable parameters, but it could also happen due to error inside or mistake of the operator.

[5] This type of description of the process was first time introduced by Genuchi Taguchi.

The variables and parameters can be temperature, density, concentration, pressure, atmosphere, and anything else which are highly dependent on many factors. Adjusting them is actually adjusting the Working Point of the process and it is done by testing the final product: We test different specifications for our product for different values of variables and parameters.

To find the limits of values of variables and parameters, the values of the process are changed and at the same time, the product characteristics are tested. The particular values of variables and parameters of the process, for which the tested characteristics of the product are within the limits (between LSL and USL), should determine the Working Point of the process. There will be few such values for each variable and each parameter, so we choose the average values for them. It will provide buffer zone for possible variations of that variable or parameter.

These values of variables and parameters are making the Working Point for Optimal Process for the given product tolerances. When the Working Point for Optimal Process is adjusted by each of the parameters, then there is need to implement Statistical Process Control (SPC) for the variables maintaining the Working Point in stable position. Of course, it is assumed that the parameters are kept within their tolerances.[6]

During the tests for adjusting the Working Point of the process, the measurements of variables are statistically processed and for each variable a mean value (μ) and standard deviation (σ) are established. These values are later used by SPC to control the process.

The SPC appeared in the beginning of the previous century. We can say that "the father of SPC" was Walter Shewhart who started first to use it at Bell Laboratories in 1924. The method was accepted later with other "Quality Gurus". Somewhere in the literature, it is can be found under the name Statistical Quality Control (SQC), which was the name used in the beginning.

The SPC is a method[7] of control of the manufacturing process which allows Operator to keep variations of the variables within the limits of tolerances for a good product. The SPC is an excellent tool to provide good product and it is very much used in Quality to provide high-quality products and in Safety to provide safe product (or service).

Once the good process is established, all variables and parameters shall be kept stable by human or by automation circuits. Variables are also part of the process, but they are changed intentionally during the process by the Operator(s). As such, having in mind that Operator(s) monitor them "live" during the process, it is not necessary that it is controlled by the SPC.

The SPC takes (periodically) the values of each variable and parameter at a particular time during the process (Figure 3.4). The results are statistically analyzed and the output of the analysis provides data how to proceed.

Figure 3.4 presents optimal value (dashed line in the middle) limited by ($3\sigma \pm \sigma_\mu$) lines (LCL and UCL) for one particular parameter and variable in the process. If we choose the

---

[6] Do not forget that variables are used to control the process and the parameters are kept steady during the process.

[7] I prefer to call SPC methodology, having in mind that it is sublimation of many methods for control, oversight, and automation, but mostly in literature it is named as method.

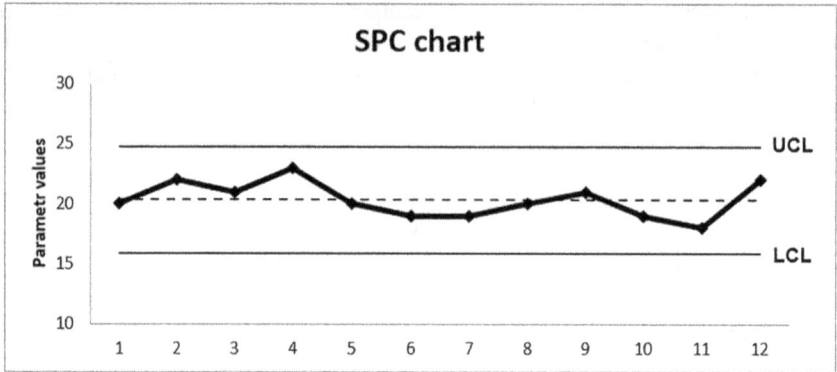

FIGURE 3.4 Variation of the adjusted parameter during the process execution.

tolerances of every variable and parameter of the process to be as explained in the previous section and we keep all values of them between their LCL and UCL, we can expect that the 99.73% of the products will be within the tolerances (will be good products).

But things are not so simple. Due to random variation of the parameters and variables, there are rules which will tell the Operator that the process is out of control.

The first book which dealt with SPC was *Statistical Quality Control Handbook*, published by Western Electric Company (WEC) in 1956, and the four rules for SPC mentioned there are valid even today. Later, in October 1984, Lloyd S. Nelson published his eight rules in *Journal of Quality Technology*. The four WEC rules are between Nelson's rules (but more detailed) and in addition four new rules were introduced (based on the experience with implementation of SPC).

I strongly recommend Nelson's rules for application of SPC in any area in the industry!

But there is more . . .

The diagram in Figure 3.4 is actually a process "law" which needs to be maintained automatically. So, there is need for the SPC to be conducted by few Subsystems which are usually capable to extract necessary data about variables and parameters (their values) from the ongoing process, execute processing of the extracted data, make decision whether the variables and the parameters need change or not, and execute the change. It means that automatization of SPC is another factor which contributes to the Complexity of the system needed to execute, monitor, control, and maintain the production process. The use of SPC is actually creating a Complex System!

Let's be honest: The SPC is long-known tool for achieving a good quality for the products, and as such it is very much used in the quality area. But do not forget: The good quality of the systems is a precondition for good safety!

3.5 RELIABILITY

By definition, the Reliability is a probability that the fault of any engineering component, built into any system, will not happen for a particular period of time. In some of the literature, for the cases in industries where this time (t) is determined, you can find it also under the name "probability of survival" of the component for time t. So, I can say that Reliability is probability for successful operation of the component for a particular period of time.

On the Internet you can find a lot of definitions of Reliability, but my favorite is the one mentioned in *Military Handbook 338B* (MIL HDBK 338B[8]). In this document, the Reliability is defined as "probability that an item can perform its intended function for a specified interval under stated conditions". Let me explain the meaning of "specified interval" and "stated conditions".

"Specified interval" is related to the time interval (t) which is considered for the component not to fail. It is the time period for which I would like to calculate probability that equipment will not fail.

"Stated condition" means the system is in operation in a particular environment and under particular conditions and its operation is stated as a "normal operation". So, calculating Reliability, I just need to find the probability that this "normal operation" will continue for a particular period of time (the "specified interval").

The very important thing is to have data to calculate this probability. The amount of data available for a particular system will show whether you will deal with frequencies or with probabilities. For particular risky industries, there are data available from Regulatory Bodies. The Risky industries are strongly regulated by laws, regulations, directives, rules, and standards, so the Regulatory Bodies have obligation to oversee the company's performances and to gather safety-relevant data at periodic intervals (at least once per year).

There are plenty of documents which deal with Reliability analysis and most of them can be found on the Internet. The most famous are NUREG-0492, the famous *Fault Tree Handbook*,[9] issued by US Nuclear Regulatory Commission and, as already mentioned, MIL HDBK 338B, *Electronic Reliability Design Handbook*,[10] issued by the US Department of Defense. Two other very useful books for electronic equipment are MIL HDBK 217F (*Reliability Prediction of Electronic Equipment*) and Telcordia[11] SR 332 (*Reliability Prediction Procedure for Electronic Equipment*).

Using such handbooks, during the design, the companies calculate the Reliability for their systems and it is submitted to the purchasers of the products in the form of MTBF values. The point is that this value is just a theoretical prediction under "stated conditions"; so, the companies which purchase these systems should prove these values during operation of the products installed on the sites. In many Risky Industries, these companies are obliged to keep a record of all faults of the systems and to use these data to calculate and confirm that the MTBF of that system, installed on their site, is approximately the same as the one calculated by the manufacturer. In many industries, the data should be gathered in a reasonable period of 4 years[12] during the

---

[8] Issued on 1 October 1998 by the Department of Defense, USA.

[9] Used mostly for teaching purposes!

[10] Guidance material only!

[11] The company called Bellcore changed its name to Telcordia (somewhere you can find it as Telecordia) and made revisions to MIL HDBK 217 for telecommunication purposes and published it as SR 332.

[12] In aviation, this period is two years for navigational equipment. In general, the longer the period, the better the results for calculation of Reliability in reality!

system's operation. If these two values (stated by the manufacturers and calculated by the purchaser) are not similar, then something is wrong with the calculations, with the system, or with the data.

The Reliability is used mostly in industry and there it is connected with equipment installed in factories. However, it is also fully applicable to products sold to customers. In this book, I will speak for both (Equipment and products); nevertheless, in this chapter I will use mainly the words Equipment or systems).

3.5.1   THE BASICS OF RELIABILITY

The Reliability ($R(t)$) can be calculated by the following formula:

$$R(t) = e^{-\frac{t}{MTBF}} = e^{-\lambda t}$$

where t is time and $MTBF$[13] is the Mean Time Between Failure. $MTBF$ is calculated as follows:

$$MTBF = \frac{AOT}{n}$$

where AOT is the Actual Operating Time (time during which I am using equipment) and n is the number of faults during the time t in Reliability formula from above.

As it can be seen from the above formula, the $MTBF$ is connected with the Failure Rate (λ) which is expressed by the following formula:

$$\lambda = \frac{1}{MTBF} = \frac{n}{AOT}$$

The calculated value for Reliability can be multiplied by 100 to obtain the probability expressed in percentage (for example, $R(t) = 0.52$ ($\times$ **100**) = **52%**). In general, the bigger the **MTBF**, the better the Reliability. Here I must say that you should understand that the $MTBF$ (as probability) is calculated as statistical value and it brings with it all the good and bad things of statistics.

To better understand the Reliability, I will use a simple example. Let's assume that time for which we would like to calculate Reliability is $t = MTBF$, so formula from the above will give the following result:

$$R(t) = e^{-\frac{t}{MTBF}} = e^{-\frac{MTBF}{MTBF}} = e^{-1} = 0.37\,(\times 100\%) = 37\%$$

This result can have two meanings. The first meaning is that if I consider all components to have been produced in one batch, then after the period equal to $MTBF$, only 37% of the components from the batch will be operational. If I consider only one component, the second meaning would be that after period of time equal to $MTBF$, probability that the component will not fail is only 37%.

---

[13] In the literature, the **MTBF** is used for repairable items and the **MTTF** (Mean Time to Failure) is used for non-repairable items. So, for repairable items you can use **MTBF** and for non-repairable items you can use **MTTF** in the Reliability formula.

TABLE 3.1
Reliability that Same Equipment with MTBF = 30,000 hours Will Have after 1, 2, and 3 Years

Time (t)	1 Year (8,760 hours)	2 Years (17,520 hours)	3 Years (26,280 hours)
Reliability	0.75 (75%)	0.56 (56%)	0.41 (41%)

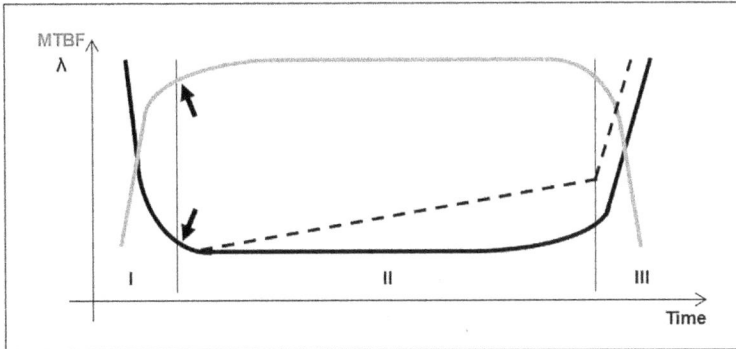

FIGURE 3.5   Reliability expressed by Bathtube curve for MTBF (gray curve) and λ (black curve[14]) of the equipment during its life.

To put in the context, let's provide few other calculations: for one, two, and three years of equipment which has **MTBF** = 30,000 hours. Results are in Table 3.1.

It can be noticed from the table that, with time, the Reliability decreases and it is normal for the exponential function with negative exponent which is used to calculate Reliability.

In Figure 3.5, the life cycle of one system is described by using **MTBF** and λ.

As you can often see, during the design testing of the system (region **I** known as Childhood), there will be plenty of faults which are connected with the flaws in the design of the system. Later, after purchasing and installation of the system in the factory, you can see that Reliability and **MTBF** are entering the region **II** (Life) where both values (**MTBF** and λ) are mostly constant. As it can be seen in Figure 3.5, there is one short period of time where the Reliability is still low (shown by the black arrows) and this is the period immediately after installation of the system in the company premises. This smaller value of **MTBF** (higher value of λ) corresponds to the need of the system to adapt to the environment of the company premises and to stabilize itself. This Reliability is not so perfect, due to the need for the operators and maintenance people to adapt to and get familiar with the system and its operation. So, during this "adaptation time", the faults are very much possible.

---

[14] A full black curve for λ is for electronic systems and dashed curve is for mechanical systems. Mechanical systems have increasing line due to wearing which is not so evident for electronic systems.

When **MTBF** tends to decrease (the Reliability is going down), as presented in region **III** (Retirement), there is a need to think about changing the system with new one due to the increased need and cost for maintenance. The region **III** is time when the number of faults increases and maintenance becomes too expensive. In addition, there could be also unavailability of spare parts, which can put the system for longer time out of operation.

Anyway, the Reliability needs to be monitored all the time, because it is a valuable source for information how the engineering systems behave, how the probability of failure changes, and when it is a time to buy a new system.

There are few other things which need to be mentioned here. The Reliability will be improved if **AOT**[15] is big and it can happen only if time for the maintenance of the system is short! The maintenance time is the time which maintenance people will spend to fix the faulty system and it is expressed as **MTTR** (Mean Time to Repair). Again, this is a statistical value and it is assumed not to be bigger than 30 minutes for industrial systems to be put again in the operation.

3.5.2 RELIABILITY OF COMPLEX SYSTEMS

The Complex Systems, which I am speaking here, have a plenty of Subsystems and to calculate the probability of overall Complex System, there is need to take into consideration the Reliability of each Subsystem. It is assumed that the Reliability of each Subsystem is already calculated by the calculating combination of the Reliability of each component built inside each Subsystem. The method presented here apply to the calculation of the Reliability to Complex System using Reliabilities of the Subsystems and to the calculation of the Reliability of each Subsystem by using the Reliability of each component.

There are three ways of connecting the Subsystems and components: Series, parallel, and combined (series and parallel).

For series connection of the Subsystems and components (Figure 3.6), there is situation when the output of one Subsystems and components is connected to the input of next one and so on. This could be a connection when I would like to amplify some electric signal, so I will put few amplifiers[16] in series to achieve the proper amplification without distortion.

For a series connection to keep the overall system operational, I need all Subsystems and components to be OK. It means that for series system the following formula applies:

$$R_S = R_A \cdot R_B \cdots R_N \iff e^{-\frac{t}{MTBF_S}} = e^{-\frac{t}{MTBF_A}} \cdot e^{-\frac{t}{MTBF_B}} \cdots e^{-\frac{t}{MTBF_N}}$$

---

[15] **AOT** stands for Actual Operating Time. **AOT** can be explained as the time of using an equipment. Let's say I have a car for 6 years, but I drive it only 2 hours per day (in average), so **AOT** of my car will be 2 hours × 6 years × 365 days = 4,380 hours.

[16] Actually, even the single amplifiers for electric signals are used by series connection of few transistors or operational amplifiers.

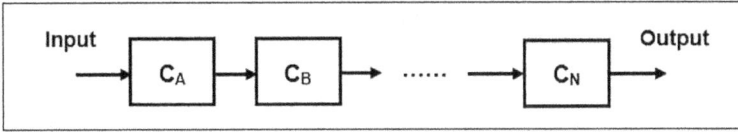

FIGURE 3.6 Series connection of subsystems and components.

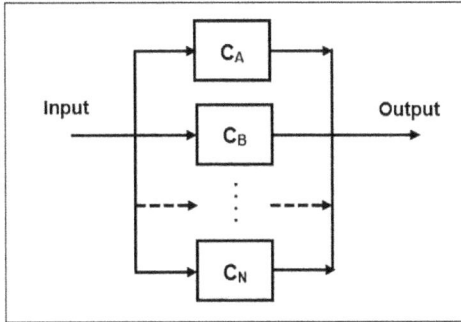

FIGURE 3.7 Parallel connection of subsystems.

Regarding **MTBF**, I will have

$$\frac{1}{MTBF_S} = \frac{1}{MTBF_A} + \frac{1}{MTBF_B} + \cdots + \frac{1}{MTBF_N}$$

It can be noticed from the formula above that $MTBF_S$ of the total Complex System will be lower than any other individual **MTBF** of the Subsystems in the formula, so series connection of Subsystems into Complex System will decrease the Reliability. But it is not all about Reliability: A series connection failure of any Subsystem will cause the overall Complex System to fail. It is called Single Point Failure (SPF) and it is critical not to be allowed in Risky Industries.

In Figure 3.7, *N* parallel Subsystems are presented (the same applies to the components also).

The situation with parallel connection is a little bit complicated . . .

To get the formulas for Complex Systems with parallel Subsystems, you need to use a double negation of Reliability. It makes the calculations much easier. The negation of Reliability is expressed with this equation:

$$P(n) = 1 - P(f)$$

where *P(n)* is the probability for normal operation and *P(f)* is the probability for faulty operation.

It is clear that if I have *N* non-identical components with different Reliabilities ($R_1 \neq R_2 \neq \cdots \neq R_N$), the equation will be as follows:

$$R_S = 1 - (1 - R_1) \cdot (1 - R_1) \cdots (1 - R_N)$$

If I have N identical Subsystems with equal Reliabilities $(R_1 = R_2 = \cdots = R)$, then the formula will be as follows:

$$R_S = 1 - (1 - R_1)^N$$

The equation above means that the Reliability of normal operations is presented through the full probability of the set (equal to 1) minus the faulty operations.

As it can be noticed, the Reliability of Complex System built by parallel Subsystems will increase.

This is actually a Complex System which is used as example for achieving redundancy of the systems, especially in aviation. There, redundancy (providing better Reliability) is provided by two transmitters connected in parallel (both to same microphone and same antenna), so if one fails (the operating one), the other one (the redundant one) will automatically continue with transmitting. Usually, there is another pair of Subsystems (duplicated Monitors) for automatic sensing and transfer of the transmitters to the microphone and to the antenna. In aviation, each system for CNS (Communication, Navigation, and Surveillance) in the aircraft and on the ground is duplicated. This is common also for other Risky Industries: To have redundancy in their systems and/or operations.

The equation for two Subsystems (A and B) in redundant configuration will be expressed as formula:

$$R_S = R_A + R_B - R_A \cdot R_B$$

Using Subsystems in parallel connection is a way how we can improve the Reliability of the Complex Systems. In addition, with parallel connection, failure of any Subsystems will not endanger the system and it will continue to operate, but with decreased functionality. This decreased functionality means that if there is another fault in the redundant system, the Complex System will be switched off and will not provide any more a normal operation.

A simple combined system made of few Subsystems connected in series and in parallel is presented in Figure 3.8 (the same applies to the combination of the components).

The Complex Systems, built as combination of series and parallel Subsystems, can be calculated using the formulas for Reliability of series and parallel systems.

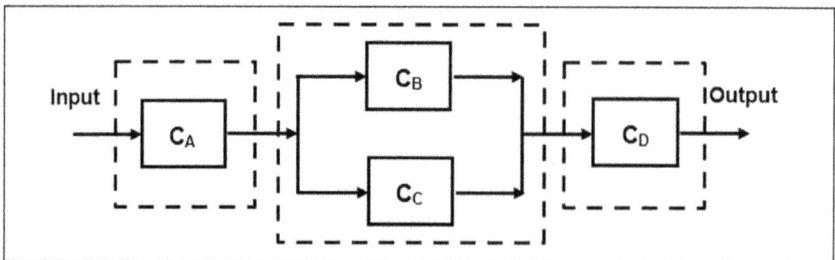

FIGURE 3.8   Combined (series–parallel) connection of subsystems.

Continuous operation will strongly depend on the fault of the Subsystems. If one of the Subsystems in the series fail, then (depending on the configuration and use of the Subsystems) the whole Complex System will fail. It is good if you can group the particular combination into series and parallel configurations and calculate all of them separately. Later you just calculate the overall Reliability using formulas for series or parallel combinations.

You can notice that all Complex System is "broken" into three components (dashed squares) and Reliability of each component can be calculated individually and multiplied later. So formula for calculating Reliability of this combined system from Figure 3.8 will be as follows:

$$R_S = R_A \cdot (R_B + R_C - R_B \cdot R_C) \cdot R_D$$

3.5.3 RELIABILITY OF COMPLEX SYSTEMS AND RELIABILITY OF THE SERVICE

The Reliability is connected by engineering systems, but let's speak about it a little bit . . .

If I have critical Complex System for functioning of the critical operation, then I need high Reliability. This situation is common in Risky Industries. As said earlier, in aviation, almost all of the CNS equipment in the aircraft and on the ground is doubled (redundant) with intention to increase Reliability of the Complex Systems used there. It means that if one of the transmitters fails, there is another (redundant transmitter) which will be switched on by automatic Monitor(s) and it will continue to transmit the required signal. These transmitters are in parallel and Reliability with two transmitters will be bigger than having only one transmitter. So, having in mind the previous section, I can calculate Reliability for this "doubling" of transmitters, which, together with doubled Monitors, will be my Complex System.

Let's assume that I have ILS (Instrument Landing System) which is sending navigational signal to the aircraft, which is used by the pilot to safely land on the runway. Both parts of the ILS (Glide path and Localizer) are having two transmitters connected by two monitors and appropriate automatics. Looking for each of them, I can assume that if one transmitter has Reliability expressed as MTBF of 16,000 hours, then his Reliability for 1 year time will be as follows:

$$R(t) = e^{-\frac{t}{MTBF}} = e^{-\frac{8670}{16000}} = e^{-0.5419} = 0.58 = 58\%$$

Putting another transmitter in parallel, assuming that both are of same type and same Reliability (0.58% or 58%), the Reliability of redundant combination of two systems will be as follows:

$$R_S = R_A + R_A - R_A \cdot R_A = 2 \cdot R_A - R_A^2 = 2 \cdot 0.58 - 0.58^2$$

$$= 1.16 - 0.3364 = 0.8236 \approx 0.82 = 82\%$$

To be honest and more accurate, I must say that in the formula above, I shall calculate also the Reliability of the doubled (redundant) Monitors. These Monitors will monitor and control the two transmitters and when one of them fail, it will change to other one. But, for the sake of the reality, I will get just a little bit smaller Reliability that is calculated in the formula and point of the context of explanation will be the same.

So, you can notice that Reliability was 0.58 (58%) of one transmitter and now is 0.82 (82%) for two transmitters in redundant (parallel) combination. It is almost 41% improvement. Expressing new Reliability through **MTBF**, I will have

$$R(t) = e^{-\frac{t}{MTBF}} \Rightarrow MTBF = -\frac{t}{\ln R(t)} = -\frac{8670}{\ln(0.82)} = 43,688.38 \approx 43,688 \text{ hours}$$

Looking on the result you can notice that new MTBF is now 43,688 hours which is 2.73 times better than **MTBF** of single transmitter.

But, by putting one more transmitter, I did not improve the Reliability of the transmitter! Both transmitters still have the same Reliability (0.58 = 58%) as systems. That what I improved is actually Reliability of the Complex System build by these two transmitters. Now these two transmitters are actually offering more reliable navigation service through the signal radiated by them.

The increase of the Reliability of the Complex System built by two transmitters (Subsystems) in redundant combination is evident, but this new **MTBF** is actually the calculated time between two outages of the system and a lot of manufacturing companies state that this is **MTBO** (Mean Time between Outages). Of course, it differs from **MTBF**. The outage is actually a failure of the operation provided by the Complex System. It means that failure is loss of the signal (signal is not available for the aircraft) caused by failure of both transmitters.

So, in general, the **MTBF** applies to single Subsystem and component and the **MTBO** applies to operation (service) conducted or supported by the Complex System.

3.5.4   RELIABILITY OF HUMANS AND ORGANIZATIONS

Reliability is connected with engineering systems, but in the previous section, I just explained that operations (services) may also have Reliability and it is expressed as **MTBO**.

In the operation, besides the Complex System, other Uncontrollable influence factors are also included such as environment, humans, and organizations. The main point, which is good to discuss here is, can I speak and can I calculate the Reliability of these influence factors which are inevitable part of the operations? Another thing is the Software: What about its Reliability? Can we determine it?

This is the question which is more connected to the latest development of Safety Management. Long time ago, safety science established that humans are weak links in all safety-related events. In aviation, approximately, 70–80% of all causes for incidents and accidents are human errors. This percentage in road traffic is for sure higher. But, investigating human factors responsible for these human errors brings new "player in the game": Organization of the work or, in other words, the company.

Here, emphasis is put on the fact that the bad management in the company, which produces bad working (quality or safety) culture, is the reason for most of the human errors. So, yes! I can speak for Reliability of the humans and especially of the organizations, but I cannot calculate it in the same manner which I am using for Reliability's calculation for equipment. It means that I need to find another way how to calculate it and this new way may come from the probability calculations based on the history of previous events.

Anyway, these calculations are never called Reliability of humans. Improving human Reliability can be achieved by the so-called *Poka-Yoke* methods, which are actually error proofing methods. Poka-Yoke is a Japanese term that means "error-proofing". These are methods for designing devices (systems) where inadvertent error by humans cannot happen. It is actually, any mechanism, designed and embedded in the system (operation, activity or process) that does not allow the operator of the system to make any mistakes and to produce defects by wrong way of using the system. But the point is that all of these "error-proofing" mechanisms need to be implemented during design phase. That is the task of the designers, who must have good knowledge about the influence of the HMI (Human-Machine Interface) on human behavior. Taking care for this, the designers could produce systems which can actually accommodate and enable humans to take care for their job.

With Reliability of organization, I can speak in the manner of safety culture built inside the organizations, but still there is lack of measurement methods also. The increasing of Reliability in organization is dependent only on capabilities, understanding, and personality of top managers. For the time being, there is no systematic solution, but Leveson is sure that it can be included into safety assessment of the Complex Systems. I will speak about that later in the book.

3.5.5 Using Reliability for Probability Calculations

Actually, Reliability is part of the specification of the components and the systems in industry and there are a lot of books written about it. The Reliability is well-known in industry and it is one of the main quality specifications. High Reliability means that the system will maintain the operation for longer period of time. In the Risky Industries, it is part of the regulatory requirements which I cannot say that apply for other industries. In the Risky Industries, the Reliability must be calculated during the design of the system, because every fault of the Complex System during operations will have considerable safety consequences.

As mentioned earlier, there are handbooks where all data for *MTBF* calculations are mentioned, so the engineers can calculate the Reliability for every system. These calculations are time-consuming, so I will not go in this book to explain them. As I said, it is not easy, but anyway it is regulatory obligation for the manufacturer to state the Reliability expressed in *MTBF*. When the equipment is installed, there is procedure which is used by the user of the system and it needs to prove that manufacturer's calculations about Reliability are true.

But here I would like to emphasize something else, which must produce more awareness for using Reliability in safety calculations . . .

Reliability is probability that equipment will function normally for a particular period of time, but in Safety, I am interested in probability that equipment will not function normally for a particular period of time. It means that in safety, the

Reliability is applicable by its negation. So, the probability (P_{faulty}) that the system WILL NOT be operative (will be faulty) will be given by the following formula[17]:

$$P_{faulty} = 1 - R(t) = 1 - R(t) = 1 - e^{-\lambda t} = 1 - e^{-\frac{t}{MTBF}}$$

For overall system with known Reliability, the formula above can be used to find the probability that system will not function normally (will be faulty). If I do not know the total Reliability of the Complex System, then I can use the same formulas for calculating Reliability of the Complex Systems (series, parallel, and combined).

---

[17] The important approximation which I can do is for $\lambda t < 0.001$ than $P(faulty) = \lambda t$, but this is applicable only for simple manual calculations.

4 Stability of Complex Systems

4.1 INTRODUCTION

Whatever be the system we use, we would like to keep its (adjusted) state (the Working Point) stable during its operation. This applies also to the Complex Systems. But fulfilling that for the Complex Systems[1] is not easy. As Complexity rises, we need more knowledge, more skills, more experience, and more resources to achieve that.

This chapter is dedicated to the linear systems and it can be used for non-linear systems only if proper linearization is applied. Even in such a case, this will be a forced simplification, because non-linear behavior is very much diverse and rich.

In science and in the life, there are three possibilities which systems may have, based on their stability. All of them are presented in Figure 4.1.

For the time being, ignore the dashed gray arrows in Figure 4.1.

As it can be seen, there is a rectangular piece of some material (not important which one) and there is a nail (screw, fixation point, axis, etc.) which allows the piece to rotate. Based on the place where the nail is put, there are three situations to be considered: A, B, and C.

In situation A, the rectangular piece, if moved from its normal position (intuitively you can agree), will come back again to its normal position. If you rotate the piece left or right around point O, due to the gravity, it will come back again. It could experience a small dumped oscillations around its normal position, but eventually it will settle in the same position. This situation is known as Stable.

It happens that this piece is in position where it has a minimum potential energy, so it is common to say that when mechanical systems find themselves in positions with minimum potential energy, they are stable. Actually, physics (as science) says that all bodies in Nature strive to achieve the level of minimum potential energy.

From engineering point of view, if we choose point A to be a Working Point of our system, with any change (internal or external) small or big influencing our system, it is highly appreciated if it comes back again to its working position. That is something which we like very much, and engineering such a system during its design and production is known as Resilience Engineering (RE).

In situation B, whatever change of position we implement on the piece (move it left or right around point O), it will stay in its new position. It means that for every new position which we choose for the piece, it will not come back to its previous position. So, the piece will stay stable in its new position until it is not affected with another

---

[1] What will be considered and presented in this chapter will apply to any engineering system, so I will use (mostly) in the text only term "system" instead the term "Complex System".

DOI: 10.1201/9781003404811-5

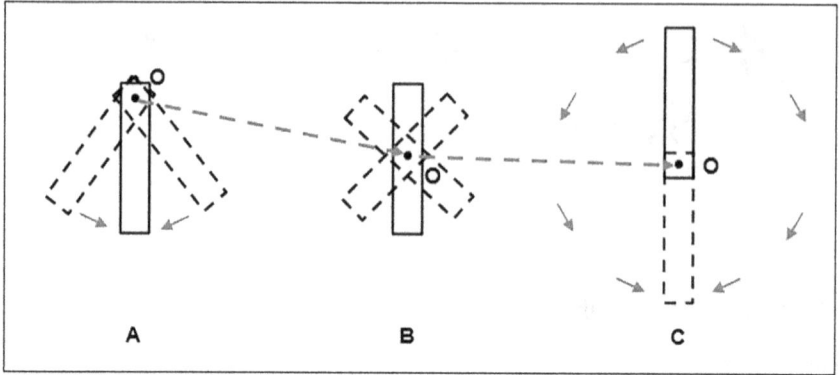

FIGURE 4.1   Three possibilities regarding stability of systems.

change. In this situation, the stability of its position depends on the input. This situation is known as Indifferent.[2]

From engineering point of view, it means that if anything (internally or externally triggered change) happens to our system, it will move out from its Working Point and it will stay in its new position. This is not good, because any of these new positions of our system (not necessary) could be an incident or an accident. However, in general, it does not mean that it will be beneficial or terrible for our operation.

In situation C, the piece is "upside down". Any change (millimetric small!) will make the piece to come to its "normal" position as mentioned for situation A. It is called Unstable.

With regard to the explanations about energy of our system, it is obvious that the piece in situation C is in a position with maximum potential energy, and as such any change (even the smallest) will cause movement and bring the piece down, into the stable position (where its potential energy is minimal).

There is another approach in stability science, which can be used to explain Figure 4.1. Stability can be defined by capability of the system to come back to its normal (stable) state after some disturbance has changed it.

Imagine an arrow "fired" from a bow. Due to its construction (weight in the "head" and feathers in the "tail"), the arrow is a stable system (situation A). Any disturbance during the flight of the arrow (due to cross wind, for example) will be compensated by the air pressure to the feathers on the tail.

The weight in the "head" puts the center of gravity toward the "head" and the feathers on the "tail" put the center of pressure toward the tail. For stable flight, the center of weight shall be in front of the center of pressure for at least two times of the diameter of the body of the arrow. If this distance is bigger (as in case of the arrow), it will produce bigger stability of the flight of the arrow and bigger resilience to the wind disturbances during the flight.

---

[2] Somewhere in the scientific literature (Stability Theory), very often you can find also the names "neutral" or "marginally stable". "Marginally stable" is very much used in theory!

If we move the center of pressure on the "head" and the center of weight on the "tail" and we try to "fire" the arrow, it will be situation C: The arrow will strive to adjust the center of weight in front of the center of pressure and the arrow (in these efforts) will flip around.

Taking the arrow where center of pressure is in same location as center of weight (feathers and weight are at the middle of the arrow), we will have situation B: The arrow will not fly under any circumstances, but it will "rumble-tumble" in front of the archer.

In more common words, our engineering Complex System (same as arrow) must "restore" its normal operation (which needs to be stable, by definition) always when some disturbance put it into abnormal operation. This is possible if we apply Resilience principles and I will speak about it later. However, if the Complex System cannot "restore" itself into normal operation after some disturbance affected its operation, this would be unstable system.

From engineering point of view, if any change (internal and external) occurs in our system in situation B, it will move from its Working Point and it will be in some other "normal" position, determined by its design. This is obviously a mistake in the design or in adjusting (settings) the Working Point of the system. It is not good, because this system obviously looks like it is "sensitive to its initial conditions". What does it mean, I will explain in Part II of this book!

Now let's go back to Figure 4.1 and look at the dashed gray arrows. It seems that we have the same system (a rectangular piece) and we change its stability by moving the nail (screw, fixation point, axis, etc.) from A to B to C (Figure 4.2). This change is presented by the dashed gray arrow.

So, it could happen in reality: We design the system, but based on our design or on influence of other internal or external factors, the system may experience all three situations of stability. Choosing the parameters, we need to investigate the stability of our system, and this chapter will explain how we assess the stability of engineering systems.

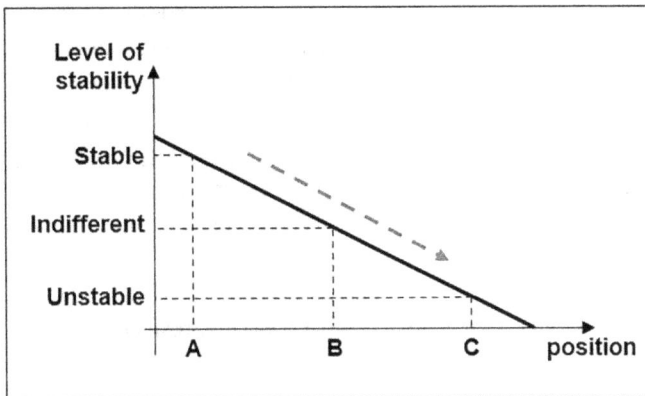

FIGURE 4.2 Level of stability based on the position of the nail.

4.2   STABILITY AND VARIABILITY

The stability of the Complex Systems can be divided into two areas of interest. The first area is stability of the Subsystems, or going further stability of ordinary elements which are used to build a Subsystem (which is used to build a Complex System). For example, every engineer of electronics is aware that for general-purpose resistors, their initial resistance variability may be of 5% and the variability in its resistance under full-rated power may reach 20%. Change of resistance means change of current inside. It means due to some internal or external factors (usually particular value of the current flowing through them), the resistor will change its resistance. In addition, those resistors have a high temperature coefficient (big change) of resistance and they also bring high noise levels in the device. For power resistors (used for power supplies, Control Systems, and voltage dividers), the operational stability of 5% variability is common.

As it can be noticed, the stability of the circuit-building element is usually expressed by variability. These two terms have connected meaning: Stable circuit (Subsystem, System, Complex System) is the one where variability of any of its characteristic is within the tolerances. If the element change its characteristics and it is not anymore within the tolerances, it can be damaged. In addition, it can produce fault of the Subsystem or failure of the operation of Complex System. In any case, the product will be scrap.

The variability is common everywhere in our lives and stability is something which must be achieved by the design. For the sake of truth, variability is quintessence of the Nature, so by designing a new system, we try to minimize the variability of the internal or external variables and parameters. The ideal solution (ultimate goal of design) is to achieve zero variability of the characteristic of interest in the Complex System. Unfortunately, in practice, especially for engineering systems, it cannot be achieved.

The second area of interest is the stability (variability) of interactions and interdependence of the Subsystems when they are embedded into the Complex System. Not only the building elements must provide some level of stability, but also the operations and communications inside the Complex System must be with minimum variability.

From engineering point of view, there are two very important questions which need to be investigated and answered regarding the designed system. First, "Is the system stable on unstable?"; second, "How big is the instability (if it is present)?" As it can be noticed, the second question is about quantification of the variability of the parameter(s) which is(are) important for stability.

4.3   GENERAL EXPLANATIONS ABOUT ANALYSIS OF STABILITY OF ENGINEERING SYSTEMS

4.3.1   Transfer Function

There are many methods in engineering by which the designed linear system can be assessed analytically (mathematically). Most of the assessment is done on the characteristic equation of the system, or better say on its Laplace transformation.

Let us explain this in more detail using Figure 4.3.

Figure 4.3 presents a simplified diagram of general system. I said "simplified" because compared to Figure 1.1, the Controllable and Uncontrollable influence

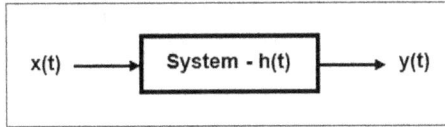

FIGURE 4.3 Simplified diagram of system with input and output.

factors are not presented here. The input and output could be single or multiple, but all these things do not matter.

The point is that the system is presented (mathematically) with its "transfer function" $h(t)$. This function is actually responsible for the change of input into output or, if we try to mathematically express it, the output will be a convolution between the input and the transfer function:

$$y(t) = x(t) * h(t) = \int_{-\infty}^{\infty} x(\tau) \cdot h(\tau - t)\, d\tau$$

So, knowing the transfer function $h(t)$ of the system and knowing the mathematical expression of the input, the output could be analytically calculated at any time of the operation of the system. Of course, the result of output will change as the values of input changes with time, but these changes will also be presented by the preceding equation.

The transfer function is a characteristic of the system expressing the relationship between the input and output signals. It can be assumed that the transfer function is a mathematical model of the system. The most important characteristic of transfer function is perhaps its bandwidth: The wider the bandwidth, the better the accuracy of the operation (transformation of input signal into output signal).

Having in mind that it is a dynamic function (dependent on time), it can be used to express completely the dynamics of a system with one input and one output. If the system had more inputs and outputs, there is a need for many different transfer functions to express each of the possible relationships between each input and each output. In that case, the above mathematical equation will be transformed into a system of equations presented by matrices.

The problem is that the operation of convolution is analytically very difficult to be solved; because there is an integral and considering that the system is complex ($h(t)$ will also be complex), the overall solution of the convolution is not analytically easy to obtain.

4.3.2 LAPLACE TRANSFORM

Having in mind that convolution cannot be solved easily, for mathematical operations with the convolution integral, the Laplace transform is used. The Laplace transform actually is used to model the system. It is derived from the linear and time-invariant[3]

---

[3] The phrase "time-invariant" explains the system which does not change its structure or its operation in time. It means that if (in any time) there is signal $x(t)$ at the input, it will result in the corresponding output $y(t)$ at any time. I will not use it anymore in this book, assuming that all the systems mentioned here are time-invariant (but not necessarily linear).

differential equations which are used to describe the systems using signals as input. It is said that the Laplace transform changes the domain of the consideration of the system from time domain ($h(t)$) to the s-domain ($H(s)$) or, in other words, to the complex domain.

The Laplace transform is described by the following two equations:[4]

$$L\left[h(t)\right] = H(s) = \int_{-\infty}^{\infty} h(t) \cdot e^{-st} dt$$

$$L^{-1}\left[H(s)\right] = \frac{1}{2\pi i} \int_{-\sigma-j\infty}^{\sigma+j\infty} H(s) \cdot e^{st} ds$$

The upper equation is for the Laplace transform ($H(s)$) of any signal: In this case, the signal is expressed by $h(t)$ and the lower equation is for inverse Laplace transform, which is actually a mathematical operation to obtain the function $h(t)$ from the already known $H(s)$.

Using the upper equation for Laplace transform and implementing it to the equation for convolution from above, it will yield

$$Y(s) = X(s) \cdot H(s)$$

where $Y(s)$, $H(s)$, and $X(s)$ are Laplace transforms of $y(t)$ (output of the system), $h(t)$ (transfer function of the system), and $x(t)$ (input to the system), respectively.

It is important to mention here that $s = \sigma \pm j\omega$ (it is complex). Usually, σ[5] can be explained as a damping factor and it is important for the stability of the system. It usually results (as I will show later) in exponential function in time domain. The ω can be explained as an oscillation factor (responsible for dumped oscillatory movement inside the systems) and it is actually the frequency domain of the Laplace transform. If s is presented as complex, it will always be a complex conjugate ($s = \sigma \pm j\omega$) and this happens with second-order systems. If σ is equal to 0 ($s = \pm j\omega$), we are speaking for harmonic oscillations which are with stable amplitude and with frequency $\omega = 2\pi f$.

As it can be noticed, convolution transforms itself into multiplication of the Laplace transformed elements. So, instead of calculating convolution integral, the Laplace transform transforms our operation from integration to multiplication, or from calculus to simple algebra.

The simplification goes further. The calculation of Laplace transform of the system is extremely hard to achieve, so engineers found a shortcut how to find it. Using the expression of Laplace transform of the convolution for our case, we can write:

$$H(s) = \frac{Y(s)}{X(s)}$$

---

[4] To be precise, this is general formula for Laplace transform used in mathematics and it is known as bilateral Laplace transform. In Control Engineering, the same formula is used, but the limits of the integral are changed from 0 to T (instead of from $-T$ to T as per the mathematical formula). This one is known as unilateral Laplace transform.

[5] Please note that the letter σ was also used for standard deviations. The reader must understand that, unfortunately, the same letter has different meanings in Metrology (standard deviation) and in Control Theory (Laplace transform, damping factor).

The Laplace transform of the transfer function is a rational function and to be physically realizable, it must have the denominator degree bigger (or at least same) than nominator degree. Also, it is independent from the input of the system.

The engineering method is simple: We produce the prototype of the system and we put a particular signal on the input ($X(s)$). At the same time, we measure the shape of the output ($Y(s)$). Their ratio will give us the equation of the inverse Laplace transform $H(s)$ of the transfer function $h(t)$ of the system.

Speaking from the point of engineering, if the input $x(t)$ is the Dirac impulse function $\delta(t)$, then the output $y(t)$ in that case would be actually the transfer function $h(t)$. It is known as Impulse Response of the system.

Or speaking from the point of Laplace transform, where $H(s) = $ L $[\delta(t)] = 1$, we will have

$$H(s) = \frac{Y(s)}{X(s)} = \frac{Y(s)}{L[\delta(t)]} = \frac{Y(s)}{1} = Y(s)$$

Here, I can add another definition of transfer function: It is a Laplace transform of the output of the system into consideration, if on the input we bring an impulse function $\delta(t)$.

For many elementary functions, the Laplace transforms are already calculated, so it helps very much the overall operation about calculations. Having in mind that it is linear transformation, it could help calculations when we have sum of different signals at the input.

4.4 GENERAL CONSIDERATION ABOUT STABILITY

Speaking about general criteria for the stability, we can make further clarifications.

In general, $H(s)$ could be presented as a ratio of polynomials. For example, it can be

$$H(s) = \frac{b_1 \cdot s^3 + b_2 \cdot s^2 + b_3 \cdot s + b_4}{a_1 \cdot s^4 + a_2 \cdot s^3 + a_3 \cdot s^2 + a_4 \cdot s + a_5}$$

In the equation above, a_n and b_m are polynomial coefficients.

For stability of the system, importance stays with the polynomial at the denominator of the fraction above. The roots of this polynomial can be found by solving the following equation:

$$a_1 \cdot s^4 + a_2 \cdot s^3 + a_3 \cdot s^2 + a_4 \cdot s + a_5 = 0$$

It is understandable why we are looking for these roots: If any of them is present as parameter value in our system, then the denominator of the transfer function $H(s)$ will be zero and as such it does not make a sense for any fraction in mathematics. Or better say, if any of these zero-solutions are present (the parameter takes a value of 0 or close to 0), then the value of $H(s)$ will be infinite. The infinite transfer function means the infinite output: The System is unstable (the BIBO principle[6] is not fulfilled)!

---

[6] The BIBO principle will be explained soon.

This polynomial is known as *characteristic polynomial* and these roots (solutions) are known as *poles*.

To better understand the poles, they can be mathematically presented by factorization of the characteristic polynomial from above:

$$a_1 \cdot s^4 + a_2 \cdot s^3 + a_3 \cdot s^2 + a_4 \cdot s + a_5 = (s + s_1) \cdot (s + s_2) \cdot (s + s_3) \cdot (s + s_4)$$

As it can be noticed, there are for four poles (s_1, s_2, s_3, and s_4) and some of them could be same. If all of the b_m coefficients are real, then the poles could be only real ($s = \sigma$) or complex conjugate ($s = \sigma \pm j\omega$).

There is another polynomial in the equation of $H(s)$ and it is the polynomial at the nominator. Its roots can be found as a solution of the equation:

$$b_1 \cdot s^3 + b_2 \cdot s^2 + b_3 \cdot s + b_4 = 0$$

These roots (solutions) are known as *zeros*, because if they are present, the output of the system will be zero. Output of the system to be zero means that the system is not working. It could be lack of power, lack of input (very rare), and some fault. Anyway, this is also important from the context regarding stability of the system, but I will explain their effect later. It must be repeated again that the order of the polynomial in the nominator must be smaller than the order of the polynomial in the denominator. In engineering words, there must be more poles than zeros to realize the physical system.

When we find the poles, it is wise to present them on the s-plane. The s-plane is a complex system of coordinates where x-axis presents the real part of $s(\sigma)$ and y-axis presents the imaginary part of $s(\omega)$. Figure 4.4 shows the s-plane with general criteria for stability of the system depending on the place of the poles.

As it can be noticed from Figure 4.4, there are four criteria for stability:

(a) If *all poles* are real and all of them are on the negative part of x-axis (gray area), then the system is stable.

(b) If there are poles which have only an imaginary component (complex conjugate poles, $s = \sigma \pm j\omega$), it means there are oscillations in the system. Regarding the complex conjugate poles (they always came in pairs symmetrical with reference of σ-axis), it is worth mentioning here that they could also be in other parts of s-plane, but the system behavior will depend on the fact that it is $\sigma > 0$ or $\sigma < 0$. If $\sigma > 0$, the system will produce unbounded oscillations and if $\sigma < 0$, then the oscillations will be damped.

(c) If there are *poles which are on the left side and poles on the right side* of the s-plane (even one!), the system is unstable.

(d) If there are *poles which are anywhere*, but there are some *pair of poles on* **Im(s)-axis**[7] (ω-axis), then the system is marginally stable (indifferent).

---

[7] If the system is not intentionally designed to be an oscillator, then all poles will be on the real axis (σ). It means, there will be no complex conjugate poles ($s = \sigma \pm j\omega$).

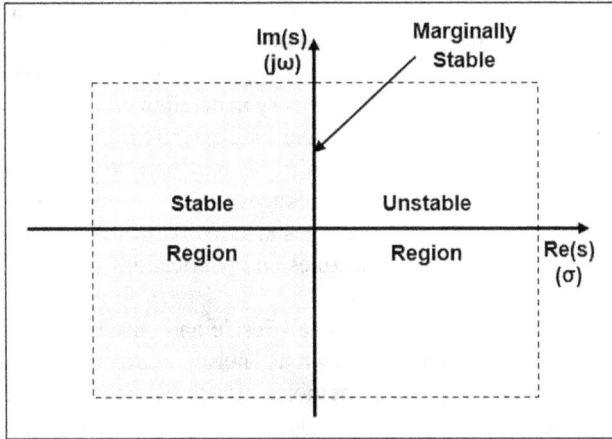

FIGURE 4.4 *s*-plane of the system with criteria for stability.

The presence of poles in this area means that the output of the system does not increase to infinity nor it goes to zero as the time passes, but it can go and stay to some particular state. The oscillations also belong to this area of stability. There are particular devices (oscillators) which are intentionally designed with only imaginary poles ($\sigma = 0$; $s = \pm j\omega$) to provide stable oscillations.

These criteria are actually a reason why $\mathbf{Re}(s) = \sigma$ is known as Damping factor and $\mathbf{Im}(s) = \omega$ is known as damped natural frequency (oscillatory behavior).

An important thing in engineering systems is whether the output signal remains limited or increases in infinity as a result of the application of an input signal. The system can be defined as stable if the output signal is limited (bounded). So, the engineering system must have mechanism inside (a Subsystem) which is responsible to limit the output for any input. Of course, this is not easy to achieve, so designers usually provide limitations of the maximum input which is allowed to enter the system. I have mentioned already that these limitations are known as tolerances for the input signal.

In engineering, it is known as BIBO (Bounded Input–Bounded Output) stability and it is the main characteristic of any designed Complex System. During operation, such a system, if any bounded signal is present on the input, will result in a bounded output. The system in such a case is known as absolutely stable or unconditionally stable. The BIBO stability can be theoretically calculated by examining the Laplace transformation of the equation describing the ratio between the output and input signals of the system under consideration. The point for worry is the fact that *every system* begins to behave in a non-linear way, if there is an amplitude limitation (bounding) of any signal inside the system.

This is the mathematically so-called *Stability Theory* which is based on Stability Analysis to determine which system is stable and which one is not. The Laplace transformation is very much used in designing and analysis of Control Systems. Anyway, I will not speak about Stability Theory in detail in this book. There is plenty of literature on the Internet, so curious reader may extend his knowledge in this area.

All these calculations are often known to be very hard to execute manually, so there are many commercial Software packages which are available for this task today. You can find these Software packages also in commercial ads on the Internet or as advertisements in technical magazines and publications such as *IEEE Control Systems Magazine* and *IEEE Spectrum*.

There are also plenty of undesired signals inside and outside the systems (results of unwanted interference) which are known as "noise". Sources for noise can be different, but the common thing is that the noise affects the operation and stability of the Complex System, So, one of the tolerances which needs to be defined during the design of the system is the allowed level of noise inside the system. Knowing that, there are many methods to deal with it, depending on the area of engineering (electronics, mechanics, chemical, etc.). These methods are known as Noise Reduction methods.

To give a little bit wider explanation about the noise, I can state here that even the uncertainty in measurements and aperiodicity of the strange attractors (will be explained later) can be explained as presence of noise in our systems and instruments. However, the noise (internally or externally generated) could be one of the disturbances which will affect the operation of our Complex System.

4.5   TRANSFER FUNCTION IN TIME AS INVERSE LAPLACE TRANSFORM

It is very reasonable to explain the stability through the four criteria mentioned in the previous section. If you know how the inverse Laplace transform will look, it is clear that the system will be constrained in some limits. For production and services, these limits are determined by the tolerances.

In general, knowing the *n* poles of Laplace transform of the transfer function, the inverse Laplace transform of it will be a time function (a wave form) and it will have a general shape expressed by the equation:

$$h(t) = A_1 \cdot e^{s_1 t} + A_2 \cdot e^{s_2 t} + \cdots + A_{n-1} \cdot e^{s_{n-1} t} + A_n \cdot e^{s_n t}$$

Please have in mind that although the transfer function is defined and calculated as ratio between the input and output signals, it is actually the characteristic of the system (the impulse response of the system) which will help you to understand what will be the output if a particular signal on the input is connected.

In Figure 4.5, two simple cases in time domain are presented: The first one for $\mathbf{Re}(s_n) = \sigma_n < 0$ (it belongs to the left side of the *s*-plane or to the Stable Region) and the second one for $\mathbf{Re}(s_n) = \sigma_n > 0$ (it belongs to the right side of the *s*-plane or to the Unstable Region!).

As it can be noticed on the left side of Figure 4.5, the system with such a transfer function $h(t)$, even if there is no input $x(t)$, will produce increasing output $y(t)$ in time. It means that after power supply is connected to the system (although there is no

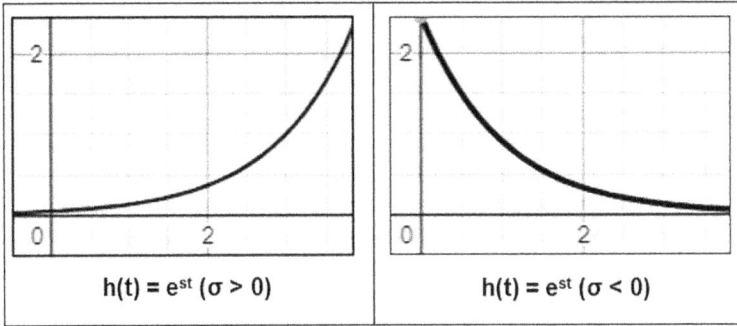

h(t) = est (σ > 0)   h(t) = est (σ < 0)

FIGURE 4.5 The two cases for $h(t)$ in time domain. *Left:* for real part of the pole, $s_n > 0$ ($\sigma > 0$). *Right:* for real part of the pole $s_n < 0$ ($\sigma < 0$).

signal on the input), the system will be unstable, which will result in the "damped" output (suppressed) due to the limitation posed by the power supply or in fault of the output devices of the system.

On the right side of Figure 4.5 is a case where the output will be limited, which means the system will be stable. Do not be confused that output will go to zero! It will happen only if there is no signal on the input. Usually, the transfer function will be designed as

$$h(t) = B + e^{st}$$

where B is the expression chosen to be a Working Point of the system and the transfer function will produce output which in short time will asymptotically move toward B.

This explanation is in accordance with the fact that the poles are those which are responsible for the unforced response[8] of the system, which means that even without the signal on the input, system could be stable or unstable.

The important thing is the time needed for the system to be in fault (in the case of $\mathbf{Re}(s_n) = \sigma_n > 0$) or to calm down (in the case of $\mathbf{Re}(s_n) = \sigma_n < 0$). It depends on the value of σ_n. If σ_n is much bigger than 0, then the fault will come soon and vice versa and if σ_n is much smaller than 0, the calming down will be faster. This time could be treated as "transition time" for the system to stabilize or to pass the limits, depending on the situation.

4.6 ADDITIONAL METHODS FOR CALCULATING STABILITY OF LINEAR SYSTEMS

The methods which will be briefly presented in this chapter are as follows:

1 Routh–Hurwitz criterion
2 Root Locus method
3 Bode plot (diagram)   4 Nyquist plot (diagram)

---

[8] "Unforced response" means there is a signal at the output of the system in situation when there is no signal at the input.

Do not forget that most of the circuits for monitoring and control of the Complex Systems are based on electronics and computers. The use of these methods is very much convenient for those electrical engineering systems, especially for those used for control of the systems or operations. The reason for that is that the Laplace transform is very much useful in solving linear differential equations used there.

I would like to add something very important here . . .

For all these methods above, I have presented only basics without many details in this chapter. I just tried to give short explanations how the methods are used. Do not forget that the book is about "forgotten knowledge" and the intention is to remind the engineers of these things. But for the readers who would like to go (again) into more details, there is bunch of literature on the Internet and there are many useful videos on YouTube. Regarding the YouTube videos, I strongly recommend the videos of Brian Douglas about control of the systems (Control System Lectures)!

4.6.1   ROUTH–HURWITZ CRITERION

In general, the Routh–Hurwitz criterion is known by many different names in the literature, one being "Routh–Hurwitz" which applies to continuous systems expressed by linear ordinary differential equations and the second criterion is "Schur-Hurwitz" which applies to discrete systems expressed with difference equations. Of course, you can find it also under the names as Routh criterion, Hurwitz criterion, and Schur criterion. All of these do not matter, if we are aware that these three mathematicians have made a huge contribution to the development of this method. So, please have in mind that I will use the name as given in the title of this section, but you may use whatever you like.

We can start from the characteristic polynomial presented in the previous section and its expression for the general purpose:

$$a_1 \cdot s^m + a_2 \cdot s^{m-1} + a_3 \cdot s^{m-2} + \cdots + a_{n-2} \cdot s^2 + a_{n-1} \cdot s + a_n = 0$$

The Routh–Hurwitz criterion actually consists of two criteria which are necessary, but they are not enough to state that the system is stable. It means that if the system is stable, these two criteria must be satisfied, but if these criteria are satisfied, it is not necessary that the system is stable. We need additional checking to prove that the system is stable.

These criteria are as follows:

1 All coefficients $a_1, a_2, \ldots, a_{n-1}, a_n$ should have the same sign.
2 All coefficients $a_1, a_2, \ldots, a_{n-1}, a_n$ should be present (none should be equal to 0).

Of course, if these two criteria are satisfied, they present the necessary and sufficient conditions for the poles of the characteristic polynomial to be on the left of the y-axis in the Stable Region in Figure 4.4.

Checking the stability of the system with integrity and finding sufficient condition by using the characteristic polynomial are through the creation of Routh–Hurwitz matrix.

   The equation that follows is a practical example of how it looks using the characteristic polynomial of the eighth order:

$$a_0 \cdot s^8 + a_1 \cdot s^7 + a_2 \cdot s^6 + \cdots + a_8 \cdot s + a_9 = 0$$

The Routh–Hurwitz matrix for polynomial above should look like this one:

s^6	a_0	a_2	a_4	a_6	a_8
s^5	a_1	a_3	a_5	a_7	a_9
s^4	b_1	b_2	b_3	b_4	0
s^3	c_1	c_2	c_3	0	0
s^2	d_1	d_2	0	0	0
s^1	e_1	0	0	0	0

The coefficients a_n are used from the equation above and the coefficients b_1, b_2, b_3, and b_4 are calculated by using the following formulas:

$$b_1 = \frac{-\det \begin{vmatrix} a_0 & a_2 \\ a_1 & a_3 \end{vmatrix}}{a_1} = \frac{a_1 a_2 - a_0 a_3}{a_1}$$

$$b_2 = \frac{-\det \begin{vmatrix} a_0 & a_4 \\ a_1 & a_5 \end{vmatrix}}{a_1} = \frac{a_1 a_4 - a_0 a_5}{a_1}$$

$$b_3 = \frac{-\det \begin{vmatrix} a_0 & a_6 \\ a_1 & a_7 \end{vmatrix}}{a_1} = \frac{a_1 a_6 - a_0 a_7}{a_1}$$

$$b_4 = \frac{-\det \begin{vmatrix} a_0 & a_8 \\ a_1 & a_9 \end{vmatrix}}{a_1} = \frac{a_1 a_8 - a_0 a_9}{a_1}$$

As it can be seen, for each b coefficient, we create a 2×2 matrix where the first column is the column of b_1 coefficient and the second column is the column of the next b_n coefficient. Above, where the Hurwitz matrix is presented, only the 2×2 matrix for b_1 coefficient (dashed circle) is presented by the dashed rectangle, but I do believe that the reader can understand how the things go on for other coefficients.

   Similarly, the coefficients c_1, c_2, and c_3 would be calculated as follows:

$$c_1 = \frac{-\det \begin{vmatrix} a_1 & a_3 \\ b_1 & b_2 \end{vmatrix}}{b_1} = \frac{b_1 a_3 - a_1 b_2}{b_1}$$

$$c_2 = \frac{-det \begin{vmatrix} a_1 & a_5 \\ b_1 & b_3 \end{vmatrix}}{b_1} = \frac{b_1 a_5 - a_1 b_3}{b_1}$$

$$c_3 = \frac{-det \begin{vmatrix} a_1 & a_7 \\ b_1 & b_4 \end{vmatrix}}{b_1} = \frac{b_1 a_7 - a_1 b_4}{b_1}$$

And so on . . .

At any place in the Routh–Horwitz matrix, where there is no coefficient or the coefficient is zero, we put zero.

For a system to be stable, it is necessary and sufficient that each element of the first column of Routh–Hurwitz matrix (b_1, c_1, d_1, and e_1) of its characteristic equation be positive if $a_0 > 0$ (negative if $a_0 < 0$). If this is not fulfilled, the system is unstable.

What is important is to pay attention: If there is a mixture of positive and negative elements in the first column of the matrix, then there is a need to count the number of sign changes of the elements of the first column of the Routh–Hurwitz matrix. This number corresponds to the number of poles of the characteristic equation which will be in the Unstable Region of the s-plane presented in Figure 4.4.

4.6.2 ROOT LOCUS METHOD

The Root Locus[9] is a method of writing a plot of the poles and zeros of the characteristic equation of the open-loop or closed-loop system[10] as a function of the change of one of the system parameters. The chosen parameter under consideration is usually the variable which is used to control the particular characteristic of interest of the system. Mostly, it is the gain in the system transfer function, but the method applies to any other parameter of interest (inertia, friction, damping, etc.).

Although it is a simple method, it provides information how the position of the poles and zeros in the s-plane changes for each value of the chosen control variable. So, it can be said that it is a graphical presentation of behavior of the poles and zeros of the transfer function and explains how they could affect the stability of the dynamic linear system.

I will use one example to explain how the Root Locus method works.

Figure 4.6 presents a system under consideration (above) and Root locus presentation for that system (below).

The Complex System under consideration is known as "unit negative feedback" system. It consists of Controller (K), System with transfer function $G(s)$, and Feedback with transfer function equal to 1 (feedback line). The Controller is characterized with parameter $K(s)$ and this parameter can go from 0 to ∞. This is a particular controlling configuration for closed-loop systems where the Feedback is equal to 1.

---

[9] The inventor of the root locus method was the American Walter Richard Evans (1920–1999). He was a renown theorist in engineering control.

[10] These types of Control Systems will be explained later in the book.

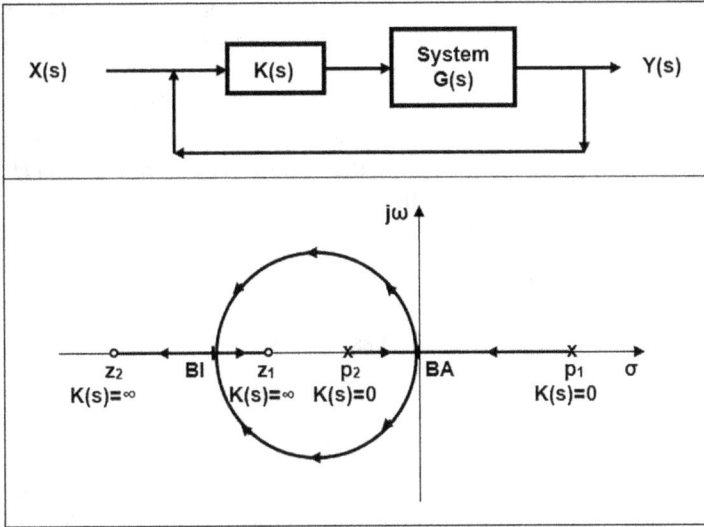

FIGURE 4.6 The system under consideration (above) and its Root Locus presentation (below).

It is a closed-loop system, but without any Feedback; it will be an open-loop system.[11]

The transfer function to overall system (under the assumption that transfer functions of Summator (**S**) and Divider (**D**) are equal to 1) will be

$$H(s) = \frac{Y(s)}{X(s)} = \frac{K(s) \cdot G(s)}{1 + K(s) \cdot G(s)}$$

Obviously, for purpose of the Root Locus, drawing the equation of the open-loop system (without feedback), the $K(s) \cdot G(s)$ will be used.

For the above system (as example),

$$K(s) \cdot G(s) = \frac{K \cdot (s+2) \cdot (s+4)}{(s+1) \cdot (s-2)}$$

The first step will be calculating the zeros and poles of the characteristic equation[12] above. Zeros are $z_1 = -2$ and $z_2 = -4$ and poles are $p_1 = 2$ and $p_2 = -1$.

The second step will be to identify the number of loci (lines) which will build the diagram. This is the maximum number of the zeros and number of poles. Having in mind that there are two poles and two zeros, the number of loci (lines) will be 2.

The next step is to identify the number of asymptotes. Asymptotes (mathematical term) are lines which are limits for the movement of the loci from the diagram.

---

[11] More about this in the next chapter.
[12] This is known as "factored form" of the characteristic equation. The numbers are actually the roots (with opposite signs) of the nominator and denominator.

In $t = \infty$, the one or few of the loci will reach the asymptote. The number of asymptotes is equal to the number of poles minus the number of zeros. In our case, it will be $2 - 2 = 0$ (no asymptotes in the diagram for this case).

The fourth step is to identify the *breakaway* and *break-in* points. At the breakaway point, the loci will leave the σ-axis and at the break-in point, the loci will join again the σ-axis. To calculate it, we need the characteristic equation from $H(s)$ equation[13] above which must be equal to 0:

$$1 + K(s) \cdot G(s) = 1 + \frac{K \cdot (s+2) \cdot (s+4)}{(s+1) \cdot (s-2)} = 0$$

From this equation, we can express $K(s)$ as follows:

$$K(s) = -\frac{(s+1) \cdot (s-2)}{(s+2) \cdot (s+4)} = -\frac{s^2 - s - 2}{s^2 + 6s + 8}$$

For breakaway points, we need $dK(s)/ds = 0$. Solving this, we obtain one breakaway approximately equal to -2.71 and another one approximately equal to 0.1. Which one will be breakaway and which one break-in, we will decide when we will draw the zeros and poles in the diagram.

This data is enough to draw loci, taking into account some important rules. The first rule is that loci always start at poles (there the $K(s) = 0$) and they finish at zeros (there the $K(s) = \infty$). From one pole, only one locus may get out. And in one zero, only one locus may come in.

Do not be confused by the lower part of Figure 4.6, where the Root Locus method is presented. At the point BA, two loci come from each pole, but one locus goes upward and one locus goes downward. The same thing happens also at BI point: One locus from one pole goes to one zero and another locus from another pole goes to other zero.

Maybe you will not be satisfied with the explanation given in this paragraph, but the point of Figure 4.6 and the short explanation here are just to show you approximately how the things are going. The point is that you will never use manual drawing of Root Locus method because there are plenty of Software applications and even the MATLAB can be used for such a purpose. There is command "*rlocus*" which will help you in calculations.

In addition, there are plenty of rules how to deal with loci drawing, but as I have said, it is not important to be presented here. If you want to know in more detail how it works, there is plenty of literature on the Internet and a lot of good videos on the YouTube channel.

4.6.3   BODE DIAGRAM

Bode diagram applies for oscillations or signals which contain different frequencies. Actually, mathematically, every signal (function) could be presented as a sum of many sinusoidal signals and as such it can be analyzed depending on the frequency.

---

[13] Please note that the characteristic equation (poles) for closed-loop system will be the same as the characteristic equation (poles) for open-loop system .

For such signals, the roots (zeros and poles) will always be in pairs and they will be a complex conjugate. These roots will produce the oscillatory response of the system and they can be expressed as $s_1 = \sigma - j\omega$ and $s_2 = \sigma + j\omega$. An important thing to understand is that frequency of the oscillations will be higher if the distance between the complex conjugate poles in s-plane is bigger.

The criterion for stability will be the same as previously mentioned: The real part of the poles (σ) must be negative or must be situated on the left part of the diagram (Stable Region in Figure 4.4). In this case, the characteristic equation with these poles can be expressed as follows:

$$(s - s_1) \cdot (s - s_2) = (s - \sigma - j\omega) \cdot (s - \sigma + j\omega) = s^2 + 2\zeta\omega_n + \omega_n^2$$

Here, ω_n is called *break point*[14] and ζ (zeta) is called *damping factor*.[15] We can express them as follows:

$$\omega_n = \sqrt{\sigma^2 + \omega^2} \quad \text{and} \quad \zeta = \frac{\sigma}{\omega_n} = \frac{\sigma}{\sqrt{\sigma^2 + \omega^2}}$$

In reality, for the stable system, the system will achieve its stability after a particular short time and it means that in this short time, $\Delta\sigma \to 0$, so σ can be neglected and for the frequency response of the system, we can assume that $s = \pm j\omega$.

Figure 4.7 presents a simple mechanical system which can be analyzed by using the Bode plot.

The Bode plot[16] is actually a measure of the bandwidth[17] and phase changes introduced by the system under consideration and it is presented by two diagrams: One diagram is for amplitude (magnitude) of the signal, depending on the frequency, and another one for the phase of the signal, depending on the frequency, too.

The sinusoidal response to the sinusoidal input in time domain could be presented as follows:

$$y(t) = h(t) * x(t) \quad \text{or} \quad A \cdot \sin(\omega t + \varphi_1) = h(t) * B \cdot \sin(\omega t + \varphi_2)$$

As it can be noticed, the transfer function of the system $h(t)$ changes the amplitude of the input signal ($A \to B$) and its phase ($\varphi_1 \to \varphi_2$). The shape and the frequency of the signal will not be changed.

From engineering point of view (in electronic industry especially), this is very much important. We can use very accurate and stable signal generators which can produce sinusoidal signals and we can use these signals to evaluate the design of our system. And this is done almost always in real life.

---

[14] Somewhere in the literature, you can also find it under the name "corner frequency" or "critical frequency".

[15] This type of damping factor (ζ) will also be mentioned in Chapter 10 as one of the essential tools for decreasing risks from vibrations.

[16] The main interest of the Bode plot is dependence on the frequency, so σ is mostly neglected ($s = j\omega$).

[17] The bandwidth for Control System is defined in Section 5.3 and this definition applies to any system: The range of frequencies of the input signal which can be processed by the System without disturbing the output signal is called *bandwidth*.

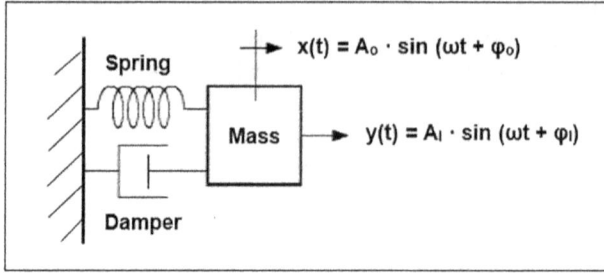

FIGURE 4.7 Simple example of mechanical system with damped oscillations.

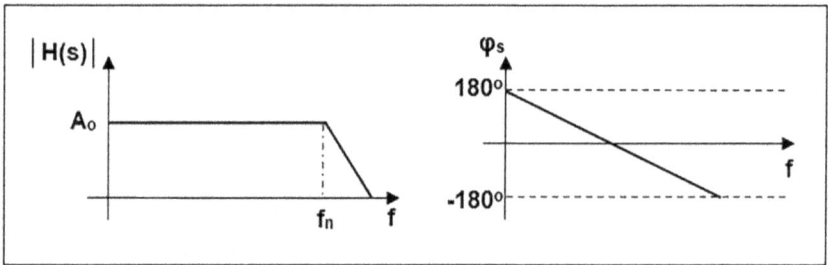

FIGURE 4.8 Non-disturbance (theoretical) criteria for module and the phase of the $h(t)$ in s-domain for ideal system.

Anyway, if there is a non-linearity of $h(t)$, then some other frequencies can be created and it could happen if the input signal consists of at least two different frequencies. It means that in the s-domain, the amplitude characteristic of $h(t)$ ($|H(s)|$) must be linear and constant up to f_n and the phase (φ_s) must be linear (Figure 4.8).

On the left side of Figure 4.8, the frequency $f_n = \omega_n/2\pi$ is the break point of $H(s) = H(j\omega)$. It is involved in the stability criterion also, because we may say that a closed-loop system is unstable if the frequency of the response of the open-loop transfer function $H(j\omega)$ has an amplitude ratio greater than 1 at the break point frequency. In all other cases, it is stable.

The plot on the left side of Figure 4.8 is known as magnitude plot and the plot on the right side is known as phase plot. Also, there is need to mention here that usually the frequency (x-axis) and the magnitude $|H(s)|$ are usually given in logarithmic scale and that is the reason that on the plot there are only straight lines. The phase φ (y-axis) is given in degrees (usually from $-180°$ to $+180°$).

From engineering point of view, this frequency (f_n) is a critical frequency when the transfer function starts to distort the transfer of the frequencies in the input signal bigger than f_n. Actually, after this frequency, the system damps the amplitudes of all frequencies higher than this one.

The Bode plot is very much important during the design of transmission systems in electronics and telecommunications where bunch of signals with different frequencies are transmitted by wire or wireless. It also applies to mechanical engineering, mostly in systems dealing with vibrations and resonance. If the criteria for $|H(s)|$ and for φ_s are not fulfilled, then the input signal will be distorted and the information can be lost (partially or fully).

During the design, the system is theoretically investigated through the Bode plot, but when the design is finished and the first prototype is produced, there is experimental validation of the Bode plot. The signals (one after one) with particular amplitude, frequency, and phase are brought on the input of the designed system and the measurement instrumentation on the output will measure the output amplitude, frequency, and phase. This is done repetitively for many frequencies from zero up to two times more than f_n.

I must state here that it happened mostly in the past. Today, in electronics and telecommunications, there are many dynamic Signal Analyzers for obtaining the Bode plot from an electrical engineered system. The displayed data on the screen (the Bode plot) can be used to analyze, design, or determine the mathematical model for the system.

From the point of stability of the system, the Bode plot is not so important, as it can be noticed by the stability criterion mentioned at the beginning of this section.

Anyway, one can also use the MATLAB to draw the Bode plot, but considering that it is a commercial application, there are many other cheaper applications which can be found on the Internet.

In general, let's mention here some characteristics for poles and zeros with regard to the Bode plots:

For the poles:

(a) In electrical engineering, they behave as low-pass filter.
(b) For real poles, the cutoff frequency (ω_n) of the system is the position of the pole.
(c) For complex conjugate poles, the cutoff frequency (ω_n) is the absolute value of the module of $H(s)$.
(d) Poles cause the phase lag (phase changes negatively).
(e) Etc.

For the zeros:

(a) In electrical engineering, they behave as high-pass filter.
(b) For real zeros, the cutoff frequency (ω_n) of the system is the position of the zero.
(c) For complex conjugate poles, the cutoff frequency (ω_n) is the absolute value of the module of $H(s)$.
(d) Poles cause the phase lead (phase changes positively).
(e) Etc.

4.6.4 Nyquist Diagram

It can be said that the Nyquist plot resembles the Bode plot, but for the Bode plot we use presentation in Cartesian coordinate system and for Nyquist plot we use a Polar coordinate system. Anyway, the Bode plot gives information about the frequency response (changes of magnitude and phase) of the system and the Nyquist plot gives information regarding the stability of the system. Of course, both the Bode and Nyquist plots are used for signals with sinusoidal forms and, having in mind that the sinusoidal signals are periodic, the Nyquist plot could be presented by only one period of the signal.

It is interesting to emphasize here that the magnitude and phase of the transfer function in the Bode plot are not independent, but rather they depend on the frequency. As such it can be presented as two curves in one diagram. The point is that presenting them in two diagrams contributes to the clarity of diagrams. With the Nyquist plot, we have only one diagram where the magnitude and phase of the transfer function are presented with one curve.

The Nyquist plot is presented by the Nyquist contour half-encircling points which are mapped onto ω-plane (Figure 4.9). The Nyquist contour contains the imaginary ($j\omega$) axis and it encloses the right half of the plane. These are the poles which are responsible for instability of the system. The movement of the system output (caused by the change of the parameter $j\omega$ (from $-\infty$ to ∞) is always clockwise and it always starts and finish at the origin (at 0). It is a closed-loop line and can have many encirclements around 0.

To write the Nyquist plot, as example, I will use the Bode plot.

Figure 4.10 presents one real Bode plot of electronic device with magnitude and phase changes. As it can be noticed, point **B** is the break point (cutoff, corner, or critical frequency f_n) by definition: It is the frequency where module of the transfer function $|H(s)|$ decreases to square root of 2 ($\sqrt{2}$ = 0.707) from its normal (constant) value.

By using the data in the table and the points from Figure 4.10, we can write the Nyquist plot for the above example of electronic device (Figure 4.11).

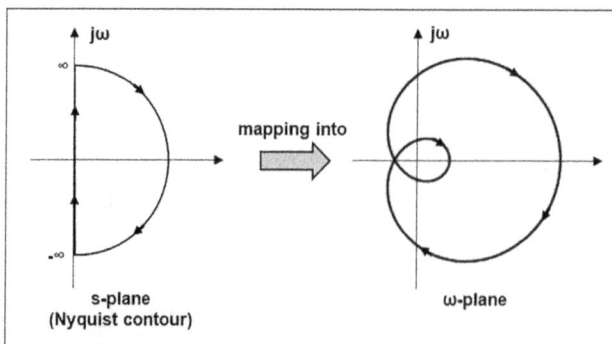

FIGURE 4.9   Example of the mapping from Nyquist contour in the s-plane into ω-plane.

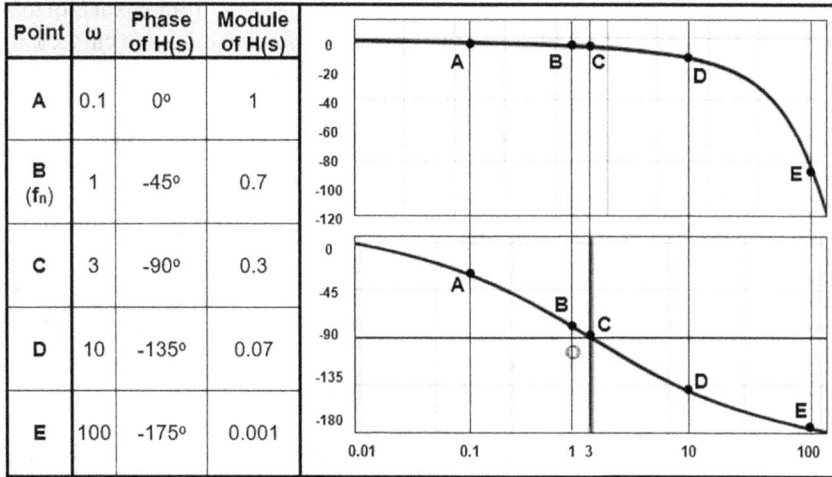

Point	ω	Phase of H(s)	Module of H(s)
A	0.1	0°	1
B (fn)	1	-45°	0.7
C	3	-90°	0.3
D	10	-135°	0.07
E	100	-175°	0.001

FIGURE 4.10 Example of the real Bode plot of simple electronic device with magnitude plot (above and right) and phase plot (below and right).

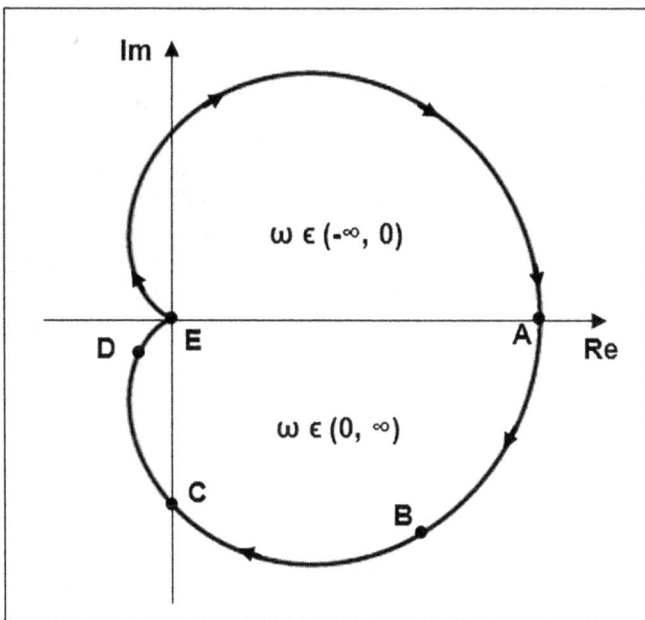

FIGURE 4.11 Example of the real Nyquist plot of a simple electronic device.

As it can be noticed from Figure 4.11, the Nyquist plot is drawn clockwise only for the lower part. The upper part is just added using the symmetry with respect to the **Re**-axis (real axis).

There are few rules for creating the Nyquist plot which are also supported also the Nyquist theorem, but I would not explain it here. It is beyond the purpose of the book.

Having in mind that the Nyquist plot is actually mapping from s-plane onto ω-plane where magnitude and phase of the signals are presented by one curve, I will present short procedure for drawing the Nyquist plot manually:

1 Find the poles and zeros of the transfer function of the system under consideration.
2 Draw those poles and zeros in s-plane.
3 Pick a point in the s-plane which you would like to transfer to the Nyquist plot.
4 Draw all the connections (phasors) from each pole and zero to the chosen point.
5 For the magnitude of the point in the Nyquist plot, multiply all the zero phasor magnitudes and divide each of them with the poles phasor magnitudes.
6 For the phase of the point in the Nyquist plot, add the zero phases and subtract the pole phases.
7 Connect the points (dots!) in the Nyquist plot (have in mind that the Nyquist plot must be symmetrical with respect to the real axis).

Of course, no one is doing that manually today. There are lot of applications for use with computer. Even the MATLAB have command *"Nyquist"* for drawing the Nyquist plot.

4.7   STABILITY AND RESILIENCE ENGINEERING

The newest movement in the quality and safety area are based on Resilience Engineering (RE). By implementing Resilience Engineering in our system design process, we provide tools and methods for the system's ability to sustain possible fault and to continue maintaining continuity of its operations. The RE accepts the reality that faults of any system could happen due to many reasons, but it tries to build a system to continue normal operation despite those faults.

It is understandable that the resilient system should maintain its stability, but it is little bit different in science. Scientifically, stability may be defined as the ability of any system to restore its equilibrium (stable) position when disturbed, and the stable system will have a bounded response for a bounded output.

From engineering point of view, it is about operation, not about stability.

For engineering system, we try not to keep the equilibrium of the system, but to keep the normal operation. The normal operation does not necessary mean to be the same as the system equilibrium. Let me remind you that engineering operation is determined by choosing the Working Point of the system and it is not necessarily the equilibrium point. For example, the equilibrium of the system can also be achieved when we switch off the power. The system then goes to rest and nothing except switching on the power would make it to leave this equilibrium position. But this is not why we have built this system. So, this is the main difference between the stability in science and (resilience) engineering.

Usually, keeping a normal operation is achieved by optimizing the parameters and variables to be sustainable to internal and external influence factors, but this is not always solution. To achieve total resilience of the engineering Complex Systems, there is need to provide capability of self-optimizability which can be achieved by automation (control). Choosing a proper building material could help, but the characteristics of the Complex Systems are not only based on its structure but also on the interdependencies and interactions of Subsystems inside the Complex System (which actually provide intended operation).

In its fundamentals, the RE is a "new" concept where designers build a capability of the system to be "tough" and "elastic" during its operational time, or better say to be dynamically adaptable regarding its "toughness" and "elasticity" not only by its structure, but also in its interactions and interdependencies. "Tough" and "elastic" are two words which look like they oppose each other, so I will offer additional explanations.

All systems experience a lot of "stress" during its functioning. We design systems to cope with most of the "known stresses". These "known stresses" could be caused by all known controllable internal and external influence factors, which are already considered and solved during the design of the system.

But there are also "unknown stresses" caused mostly by uncontrollable external factors. Some of them can be defined as "unknown unknowns" or Black Swan events (BSe). These are a category of "uncontrollable" factors which never happened before and the designer and user of the Complex System never assumed that they could happen.

As you can notice, the word "uncontrollable" is given in quotation marks. The reason is that we did not consider these factors to be uncontrollable, but not knowing them, we do not provide controls for them. So, during the operations of the system, they come as surprise. In general, "fighting" something which you are not aware of that it exists is not easy.

However, if we try to explain stability of the system through its "toughness" and we use everything which has been said in this chapter, we obviously need to provide systems with poles which are less than zero (to be on the left half of the s-plane). Also, these poles should be "very left" from zero ($\sigma \ll 0$), because in the case of small variations triggered by some influence factors, the poles could "transfer" itself into the right half of the s-plane and they will become unstable.

In the cases when there are no poles on the left side of the s-plane, some of the poles on the right side of the s-plane should be "removed" into left side by reconfiguring the system during the design. These two things are of utmost importance in the design.

But if you produce a system to be strong enough ("tough") to survive these BSe, then the problem is solved. The point is that not always the system can be built to be strong enough to survive all of the "stresses". It will need more material, more steel, more weight, more power to operate and to maintain it and as such it could be unsustainable.

In general, the "toughness" has an economic limit and further improvement of the endurance of the system could be achieved by "elasticity". "Elastic" system has the ability to survive the "stress" by changing its state, and it is also capable to resume its normal operation after the "stress" is removed. It is similar to a young tree which bends itself during the strong winds and when the wind stops, the tree resumes its normal position.

So, we can define a "tough" system as a system strong enough to have the "energy" inside to oppose every interaction with adverse "energies" from external or internal influence factors. In the cases when the adverse "energy" is too high, the "elasticity" of the system will transform this "energy" into something sterile (into heat, for example). It means that system will adapt itself to survive and later will "come back" into normal operation. What is very important to mention is that the "elasticity" hides itself into recovery. I do not mean recovery of the system when the damage is already done, but it is recovery from transformed "energy" into normal operation without any damage. The important thing about this "recoverability" is the time needed for the Complex System to recover normal operation: The sooner the better!

We can design "toughness" of the system having in mind that most of the "stresses" are known, so their adverse "energy", which could damage the system, can be calculated. Simply, during design we can embed preventive measures to handle the known "stresses". But unknown "stresses" can be handled only by "elasticity". The problem here is that we can just assume the quantity of adverse unknown "energies". That is the reason that "elasticity" of the system is not easy to produce with the same integrity as system "toughness".

There is another point which is also very important. There is a question how far away we need to go in our efforts to create stable and resilient system which would be "immune" on any disturbance (faults, failures, noise, etc.) during its normal operation. We shall be careful because extremely stable system could be hard to control through the changes of variables. Maybe the best solution is to provide "controlled instability" in our Complex System, which is another challenge.

So, be careful: Everything is about compromises and to know when to compromise, you need knowledge!

5 Control of the Complex Systems

5.1 INTRODUCTION

We struggle to have control over our lives and our destinies and this is something innate in the humans.[1] We design and use Control Systems with an intention to influence the "behavior" of the industrial systems. It has to be done by maintaining the order in their operation with intentions to provide particular benefit. There is nothing wrong with that, but every control needs more resources and, in industry, it means more knowledge, more skills, more equipment, more money, etc.

The control over linear systems is not so complex and it is achievable. Linear systems are mostly predictable (even when they are very much complicated and complex), so control over them is not a big issue, as it is with the non-linear systems.

I emphasize that there are two types of Control Systems: Passive and Active Control Systems.

Simple examples of Passive Control Systems are helmets on the head of bikers, harnesses for drivers and passengers in the cars, personal protective equipment (PPE) for the industry workers, etc. All these are Control Systems which do not use any energy or imply any process to stop bad things to happen. They are actually the "shields" which protect the humans from adverse energies around them.

The Active Control Systems are devices which are characterized by sensing (measurement) equipment, processing device, and actuators to change the situation within the system under control.

The Passive Control Systems are cheaper and they need almost no maintenance, but (to be honest) the Active Control Systems provide more flexibility, more control, and more benefit. However, the Active Control Systems affect the "safety paradox" by increasing the system's Complexity and as such make the job of Safety Professionals more demanding.

In this book, only Active Control Systems will be considered.

5.2 CONTROL SYSTEMS

The Control System is engineering mechanism (manual, electrical, or mechanical) introduced to the systems in use, with intention to keep (stabilize) or change (track) the present state of the system in use and to reject the internal or external disturbances.[2] Stabilization is actually a process of keeping the operation in the range of

---

[1] A particular level of self-control is innate for the Nature also, especially in the biological and chemical processes.

[2] The disturbance is any event (triggered by change of the external or internal factors) which can cause change of any of the variables and parameters in the system, so the system will keep its operation (process, activity, etc.) inside tolerances.

established or standardized tolerances. The unwanted changes are followed by the capability of Control System to provide "tracking" of operation based on Software algorithm and to conduct the operation as planned. Tracking is more difficult to achieve, because the Control Systems equipped with this capability must be capable to follow dynamics of the system and sometimes it could be very challenging for the fast operations. Also, in addition to the tracking, there is a need to implement requested corrections to keep the system stable.

The example for stabilizing capability of control system could be temperature control in the fridges or in the ovens, keeping steady velocity of the cars with Cruise Control and switching on/off the fan for cooling the liquid from the engine cooler in the car, etc.

For tracking and correction capabilities, good examples are FMS (Flight Management System, popularly known as auto-pilot) in the aircraft which could follow (predetermined by pilots) route of the flight, the automatic tracking of ordinary and radio telescopes in astronomy and programmed operation of CNC (Computer Numerical Controlled) machines, etc.

As mentioned before, the Control Systems are big contributors to the increased Complexity of the engineering systems. It is important to emphasize here that the Control Systems mostly impose some limitations (constraints) on the system in use or its operation and as such they keep the system inside the operational boundaries. In other words, the Control Systems force the systems in use to keep the previously adjusted Working Point within the tolerances.

I must mention here that the Control Systems are usually implemented as parts of Subsystems or parts of overall Complex Systems. As such, they will control only the performance of that particular Subsystem or the whole Complex System. Whatever might be the situation, the hierarchy level of the Control System is higher than the hierarchy level of the controlled Subsystem or the Complex System. It is also important to understand that system in use could be a linear system, but the introduction of Control System will provide considerable non-linearity in the Complex System.

There is another area of controlling and this applies when there is a non-linear system in use. The ways of non-linear control are little bit complex and not easy to achieve. That is the reason that in this book, there is considerable attention paid to non-linear systems and their behavior.

There is a discipline in mathematics (used mostly in science and engineering) known as "Control Theory" which deals with the strategy of design and operations of Control Systems. The main point is that the Complex System, if not controlled by a Control System, could endanger its internal and external boundaries. Also, during the design of the Control System, the designers may encounter (very much) a considerable uncertainty about the behavior of the system in use and the behavior of the environment, the external and internal influence factors (disturbances). The uncertainty is the reason to misunderstand how the disturbances will affect the system in use.

So, all constraints in the operation of the Complex System, which are imposed during the design with intention to maintain the operational boundaries, faults of equipment, failure of operations, or other safety issues, are done through Control

Systems. In the Risky Industries, these systems are duplicated and they usually control the main system, control themselves, and control the other (redundant one) Control System.

The main point is that all these control measures embodied through Control Systems are done automatically and as such they need no human presence or at least, minimal human presence. In industry and engineering, the area dealing with all these measures is known as Automation. However, the new movement in industry, expressed as Industry 4.0, prefer the use of the word Digitalization.

Please, have in mind that by controlling the operation of system in use, we actually control the process performed by that system. The control is usually provided by feedback of some of the signals in the systems in use and this principle is very much considered mostly with linear systems with many inputs and non-linear systems with one input. The control of the non-linear systems with many inputs is still under development.

However, it can be stated that the Feedback device is actually a complex device itself. It consists of monitoring device which, in conjunction with processing device (equipped with control algorithm) and appropriate actuators, could automatically control the Complex System (Figure 5.1).

As it can be seen in Figure 5.1, there is a Controlled System (which is some kind of Complex System) conducting or supporting some operation (process, activity, etc.) and there is a Control System.

The Control System is based on the Process Model which explains operation of the Controlled System. There is a particular Control Algorithm which tries to keep the Process Model constant. The Feedback is actually communication channel which provides data from the sensors embedded into Controlled System. These data are compared with the Process Model by Control Algorithm and if there are discrepancies, the particular Commands are issued and appropriate changes are made by actuators inside the Controlled System.

The operators monitor the overall operation. It is important to mention here that the operators and their influence on the Complex System are also based on feedback

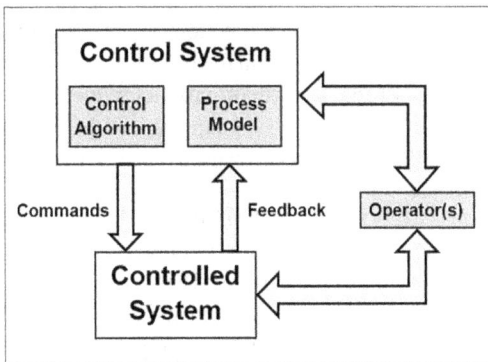

FIGURE 5.1 General model of Complex System based on control.

activity. They monitor the system, and based on gathered data, influence the system changes and adaptations (if needed) to keep the operation within the tolerances.

And it may be the most important thing which must not be missed: Writing this book I have gone through many books on Control Systems and no book was less than 400 pages, so please have in mind that in this chapter, only necessary basics about Control Theory are given!

5.3   CHARACTERISTICS OF CONTROL SYSTEMS

The main characteristics of the Control Systems are as follows:

(a) *Accuracy:* This explains how accurate is the Control System in providing control of the system in use. It can be expressed by the number expressed as quadratic sum of all errors which are noticed in the operation of the system in use. Those errors depend on the level of dynamics of the operation, influence of the internal and external factors, errors in sensor (measurement instruments), communicating the data used by the Control System, errors in processing the data, and errors in execution of the commands which are used to keep the operation (activity, process, etc.) in required tolerances.

(b) *Sensitivity (S):* The parameters of the system in use are prone to changes of environmental conditions that cause changes in external and internal influence factors, which can cause changes of the variables and parameters. The point is to build a Control System which will be insensitive to external and internal influence factors (resilient!), but sensitive only to changes of the controlled parameters. There are different definitions of sensitivity in literature, but I will define it by the following equation:

$$S = \frac{\frac{\Delta T(s)}{T(s)}}{\frac{\Delta G(s)}{G(s)}}$$

Here, $T(s) = Y(s)/X(s)$ is the transfer function of total system (system in use + Control System); $\Delta T(s)$ is the change in the transfer function of the total system caused by the change of the variable or parameter due to some disturbances; $G(s)$ is the transfer function of the system in use; and $\Delta G(s)$ is the change of the transfer function of the system in use caused by the Control System.

(c) *Noise Rejection:* Any kind of undesired signal from inside and outside, which is entering the Control System, is known as noise. A good Control System should ignore (reject) the noise. If not ignored, the noise can produce false operation of the Control System and the system in use.

(d) *Stability:* It is an important characteristic of any system, not only for Control System. More details will be presented in the next chapter.

(e) *Bandwidth:* The range of frequencies in the control signal which can be processed by the Control System without disturbing the output signal (command) is called *bandwidth* of the Control System. The bandwidth of the Control System should be bigger than the bandwidth of system in use. Bigger bandwidth, in general, provides better fidelity of the commands.

(f) *Speed:* The Control System must be fast in its operation. The speed of processing the sensed change (caused by the disturbance) of the parameter or variable by the Control System must be bigger than the speed of change of the parameter or variable (caused by the same disturbance) of the system in use.

(g) *Reliability:* Section 3.5 deals with the Reliability and I would just add here that it is calculated using data for Reliability of the components and Subsystems used to build a Complex System.

It is important to mention here that some of these characteristics are opposite to each other. Improving some of them can spoil others. However, it is important to understand that accuracy, sensitivity, and stability could decrease the Reliability of the Complex System.

5.4 LINEAR SYSTEMS IN MATHEMATICS

Using the name "linear system" in mathematics means that we would like to describe a system which can be analytically (mathematically) expressed as function where its graph is straight line. For simplification, I will use one-dimensional equation which can be presented by two-dimensional graphic. Such a linear system can be described, mathematically, by the general equation which can be written as follows:

$$f(x) = a \cdot x + b$$

Here $f(x)$ is the function which explains the changes of the state of the system for different values of variable x, $a = tg\ a$ and b are just rational numbers (constants).

As it can be seen, the linear system is a system where variable x in the formula must not be on the power different than 1. The graphic presentation of the linear system which corresponds to the equation above is given in Figure 5.2. As it can

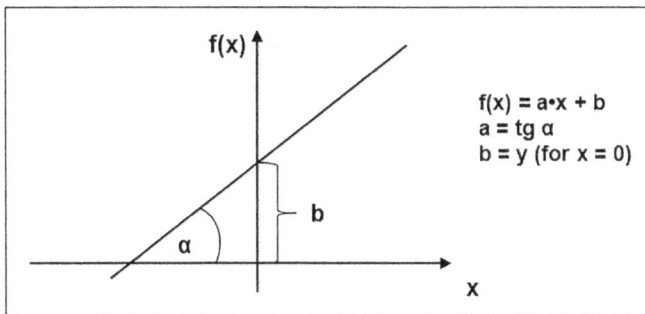

FIGURE 5.2 Graph of the function $f(x) = a \cdot x + b$.

be seen, the graph is a straight line, so this is the reason that the name "linear" is used.

Speaking in the scope of the Complex Systems, this case implies that the variable x is the input parameter which we use to change the output of our system $f(x)$ (which I will rename to y). Different values of x will provide different values of y. Having in mind that the input can be variable in time, so the output will also vary in time, I will use $x(t)$ for input and $y(t)$ for output.

The linear systems are easy to analyze mathematically. For them, the principle of superposition applies, which means that their solutions can be easily multiplied by constants and added to each other, and this new multiplication or addition will provide another (mathematical) solution for the system. I can say that for the linear systems, the "whole is just sum of its parts" or making connection with Section 1.4, the Complicated Systems can be described by linear equations.

This is actually the main difference between linear and non-linear systems: For non-linear systems, the principle of superposition does not apply!

This superposition principle can be used to provide another definition of linear and non-linear systems: If the superposition principle applies to the solutions of mathematical equation, the system is linear. If it does not apply, the system is non-linear.

The linear systems could be "simulated" inside the non-linear systems if we use small signals (small amplitudes) as inputs. You will see later that many of the problems in non-linear mathematics can be (approximately) solved by using Taylor series, which is actually linearization of the non-linear function in the vicinity of the point of interest.

But if you do not pay attention to two important things, there is a problem with this linearization from engineering point of view:

1 The linearization of function at one point of interest is just an approximation of what is going on in the neighborhood of that point of interest. So, the result of this operation will explain only the local behavior of the non-linear system in the vicinity of that point. We still do not know what will be the behavior of the system far away from that point. In the words of engineering, it means that we should strive to keep our system close to this "linearization" point, because if the system moves away, we will not know what will be its behavior. It can be said that the linearization can increase uncertainty about behavior of the system in areas far away from the point where linearization is executed.

2 The dynamic of overall non-linear system is much richer compared to the dynamics of any linear system. There are some non-linear phenomena which cannot be found in linear systems, so you need to be careful with linearization (if not done properly).

However, it is worth mentioning here that in the area of Control Systems (most of them are non-linear), linearization is a much-used process in the vicinity of stable and unstable fixed points. Usually, linearization could be applied in the areas between two such fixed points.

5.5 DIFFERENT CONTROL SYSTEM'S CONFIGURATIONS

The duty of the Control System can be expressed in two essential functions of interest regarding our system in use:

1 The Control System must shape the output of the system in use to give the desired (planned) behavior (activity, process, operation, etc.).
2 To maintain and to "stick" to this desired (planned) behavior (activity, process, operation, etc.) during operation of the system in use.

Do not forget that any system is endangered by the fluctuations caused by external and internal influence factors that could affect the operation of the system during its use. So, the main idea behind the use of Control Systems is to make them capable to generate control laws (constraints) that satisfy the two essential functions mentioned above.

I will present in the following sections few methods for achieving these two functions.

5.5.1 OPEN-LOOP CONTROL SYSTEM

The general system containing the open-loop control system is given in Figure 5.3.

The upper part of Figure 5.3 is a schematic of the open-loop control system consisting of System and Controller and the lower part is a schematic of System and Controller in Laplace transform. As it can be noticed, the input signal $x(t)$ is coming into Controller. The Controller actually manages the System responsible for the operation providing the output $y(t)$. In other words, the output is actually a variable which is controlled by the Controller.

The problem of the open-loop control is that it cannot handle noise (it is present inside and outside the system) which could affect the output signal and the operation. Another thing is that the open-loop control system cannot handle other disturbances, so if some parameter changes inside the System, the Controller will not know that and it will not react. So, the open-loop control system needs outside subject (human monitoring and control) to handle these possible changes.

There is one good example of the deficiency of the open-loop system, where the output signal is limited by the limit of power supply needed by the system to function:

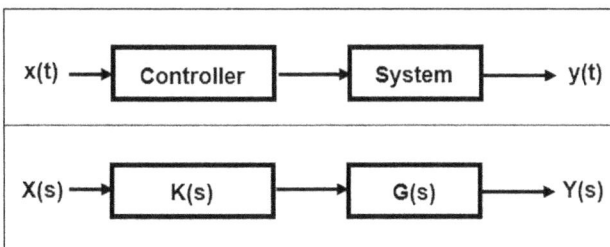

FIGURE 5.3   General system with open-loop control in time domain (above) and with Laplace transform (below).

That is amplifier in electronics. Imagine an amplifier with input voltage signal of 0.1 V[3] with constant amplification of 100 times. This amplifier with this input will produce output signal of 10 V. If the power supply of this amplifier is 12 V, then there is no problem with the output signal. If the input signal changes from 0.1 V to 0.2 V, then the output signal should be 20 V. But it is impossible to have output signal with amplitude of 20 V, simply because neither the controller nor amplifier could provide voltage bigger than the power supply of the amplifier, which is 12 V.

There is no system which can produce more energy than it is put inside. The phenomenon of producing more energy than it is put inside is known as *perpetuum mobile* and it does not exist. In our case of amplifier with 0.2 V input signal, the value of power supply will limit output signal of amplifier to less than 12 V[4] and it means that the amplification of the amplifier will fall to 60 times. Obviously, the output signal will be distorted and as such it will not be useful. So, to use this amplifier, we need to change the transfer function of the combination of Controller and System in this configuration or to limit the input signal to 0.1 V. Such a device in electronics is Automatic Gain Control (AGC) which can be put on the input of the receivers to limit the receiving signal. Actually, all our radios and TV sets use AGC.

In the context of the Laplace transform, explained in the previous chapter, under assumption that Laplace transforms are for Controller $K(s)$ and for the System $G(s)$, the Laplace transform to the transfer function of the open-loop control system will be given by

$$H(s) = K(s) \cdot G(s)$$

We may explain the open-loop control system through human body. The five senses of humans can be treated as "inputs", the brain in the human head would be the Controller (it processes all signals from the "inputs"), and the System would be all human body (it provides the operation).

In addition, most of the kitchen and home appliances are also open-loop systems. For example, the washing machine for clothes has a bunch of preset programs. When we choose some program and press the *on* button, the program will be executed from the beginning to the end without any possibility to intervene.

5.5.2 CLOSED-LOOP CONTROL SYSTEM (FEEDBACK)

The most used concept of control of any type of systems is the concept of closed-loop control, also known as Feedback (Figure 5.4).

This is a concept where another system is used as Controller, but the connection with the system in use makes a Closed Loop. This connection actually allows the Controller to receive the output signal, to compare it with the requested (reference) value, and, if there is an error, to give a signal to the input of the system to rectify the error in the output signal. It solves most of the problems of open-loop control system and it can handle all interactions inside and outside the system in use.

---

[3] V stands for volt (unit for measurement of voltage).
[4] "Less than 12 V" because there are always some losses of energy in the system due to different factors.

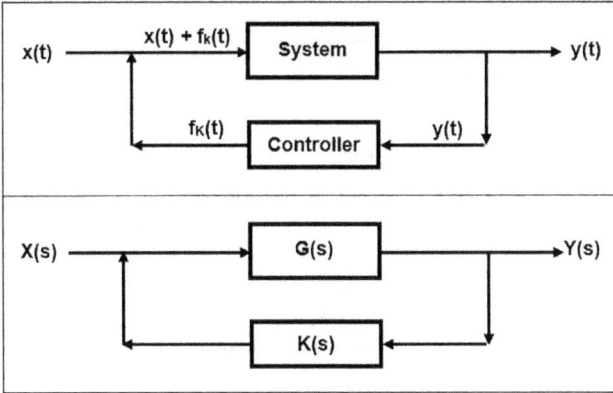

FIGURE 5.4   Circuit in time domain (above) and the same circuit with Laplace transform (below).

The upper part of Figure 5.4 shows a normal system with feedback in time domain and the lower part is the same system expressed by Laplace transform.

In general, the closed-loop control system (Feedback systems) can be with both positive and negative feedback. Positive feedback is the feedback when the signal $f_K(t)$ is in phase with the input signal $x(t)$ and negative feedback is the feedback when the feedback signal $f_K(t)$ is in anti-phase with the input signal $x(t)$.

If the feedback is positive, the controlled system in use may fall into oscillation. Very often, in electronics and telecommunications, the positive feedback is created intentionally (in the devices called oscillators) to provide signals which can be transmitted as radio waves.

The Laplace transform of the system $G(s)$ actually describes the nature and level of change of input signal which will produce output signal without the Controller. The transformation factor $G(s)$ is the reason to produce and use the system. It can be amplifier, transducer, transformer, filter, etc. and it can be constant (static) or variable (dynamic). If it is constant (static), to control the output $y(t)$, we must be capable to manually change the $G(s)$ to get the output signal as we like, but if it is variable (dynamic), a simple (human) change of the $G(s)$ will not work. As the input signal dynamically changes, there is also need to dynamically change $G(s)$ to provide control for the output $y(t)$.

This is a problem which can be solved, but it needs more resources.

The transfer function of closed-loop control system would be as follows:

$$H(s) = \frac{Y(s)}{X(s)} = \frac{G(s)}{1 + K(s) \cdot G(s)}$$

As it can be seen, $K(s)$ and $G(s)$ are variables in the equation above and, for the sake of truth, they, together, form a non-linear system. If you ask why then it is mentioned here, I will offer the explanations that there are methods based on Laplace transform which can be used to assess the stability of the closed-loop control system based

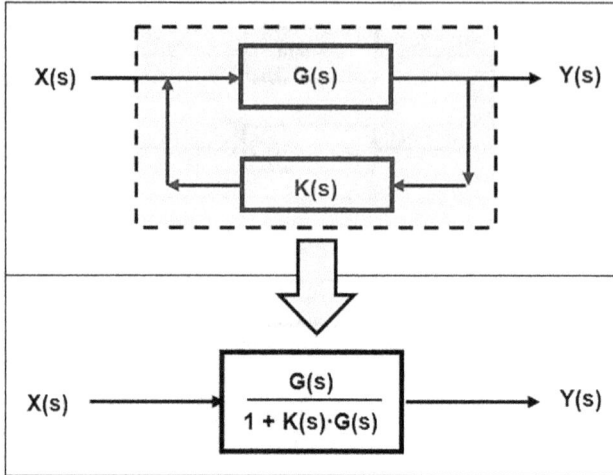

FIGURE 5.5   Transformation of closed-loop control system into open-loop control system.

on the stability of the open-loop system presented by $K(s) \cdot G(s)$, which is part of the equation for open-loop control system.

Refer to Figure 5.5.

If we design the System with $G(s)$ and Controller with $K(s)$ into one system (a Complex System), the transfer function of this system should be the same as closed-loop control system transfer function. Now imagine plenty of these controls in our system in use and you will understand that this is how we make the systems complex.

As mentioned earlier, the closed-loop control system is more complex than the open-loop control system because it has a sensor (at least one) or a measurement device at the output of the System (or at input of the Controller). The signal is compared with the referent signal stored in Controller. All these do not necessarily change a lot the system in use, but they could make the Controller very much complex. So, let me say here again that we can define the Complex Systems as systems designed with plenty of control devices providing different closed-loop controls.

The benefits of the closed-loop control system are many.

For example, it can notice deviation on the output and immediately it can react at the input to adjust it in accordance with the reference. This is actually achieving BIBO Principle's stability. Having in mind that it deals with the output and input in our system, it does not care about the system, so the system can be analyzed as "black box". The Controller used for feedback is usually the so-called *PID* (*Proportional Integral Derivative*) controller which is pretty much versatile and robust. It can be easily retuned (readjusted), if the conditions (disturbances or parameters) change.

The disadvantage could be that no control action is taken until the output did not change. As such, it can be good for the systems which last long and/or have long time delays. It cannot provide predictive corrective action to deal with the known disturbances.

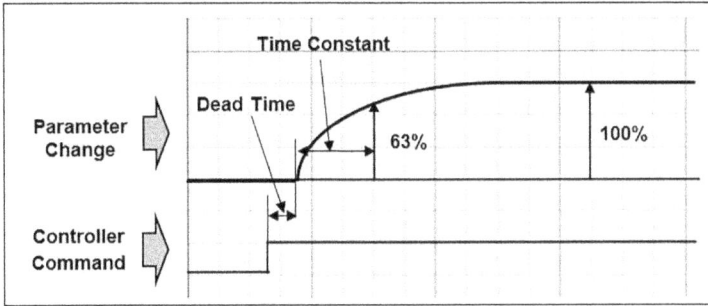

FIGURE 5.6   Dead time and time constant.

There is a criterion[5] to help decide when to use the Feedback controllers. It is the ratio of the system Dead Time[6] to the process Time Constant.[7] When the value of that ratio is equal to or greater than 1 (Dead Time bigger than Time Constant), the closed-loop control usually cannot help with the disturbance in the system.

This is shown in Figure 5.6.

5.5.3   FEEDFORWARD CONTROL SYSTEM

If we would like to generalize, the problem with the closed-loop control system (feed-back) is that the signal for control is taken from the output of the system. It looks like in industry we take a product (the output of production system) and we realize (after comparing it with the reference) that it is not good. So, we put into the production system resources (materials, power, time, human, etc.) and at the end of the production process, the product is not good. Obviously, the closed-loop control system cannot help with the problems inside the system.

Having this in mind, it will be good if we "sense" the problem before the product reaches the final phase (which is the output of the system). The Feedforward control can be explained as a concept of control which is trying to find and reject the persistent disturbances that cannot be rejected by the closed-loop control system.

The principle of Feedforward Control System is shown in Figure 5.7.

The Feedforward Controller takes the values of variable (parameter) of interest from inside the System, not from the output. It means that it has information how the parameter (affected by disturbance) behaves and if it deviates from the reference value (stored in the Feedforward Controller), the Feedforward Controller produces

---

[5] This criterion was first explained by Armando. B. Corripio and Michael Newel, in their book *Tuning of Industrial Control Systems* (issued by Instrumental Society of America, 2015, 3rd Edition).

[6] The dead time of the system can be defined as a delay in the response to a control action triggered by the Controller.

[7] The Time Constant shows how fast the process variable changes in response to a change triggered by the Controller. It is defined from the time when the Controller issues command until the process variable changes to 63% of the requested value of the change.

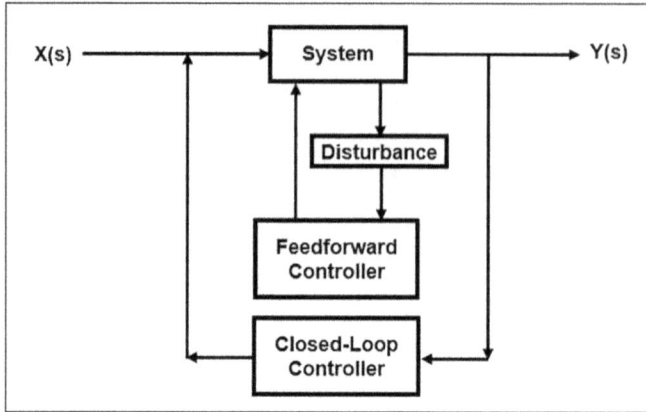

FIGURE 5.7   Feedforward control system in combination with closed-loop controller.

correction which changes the parameter adjustments in the System and rectify the irregularities. It means that the results of disturbances at the output are rejected in advance inside the System.

Usually, the Feedforward control is used together with closed-loop controller because the closed-loop control system is required to track changes of the System output. If there is inappropriate signal at the input, the Feedforward Controller alone cannot handle it, so the output will be distorted. For such cases, we need Closed-Loop Controller to monitor the output. That is the reason that combination of these two controllers is a very good tool to provide good control.

In the scope of this book, it is good here to mention that this concept is very much important for the operation of the Complex Systems. As mentioned earlier, one of the characteristics of Complex System is that the approach to understand it should be holistic, which assumes that we need to analyze the system as a whole (taking care of its emergent properties). It means that Feedforward Controller can deal with the functioning of the Subsystems inside, but the closed-loop control system will take care of the overall Complex System. This is because the closed-loop control system is controlling the reason for design of the Complex System: A system which transforms particular input into exactly determined output.

The Feedforward Controller can suppress internal or external disturbances of the parameters (undetected by the closed-loop control system) that are always present in any industrial system. The Feedforward Controller should be able to detect disturbance and as such it can take action immediately, but it will not be able to adjust the total system performance. Sometimes, it can also produce stability problems, especially if the system in use is also unstable. It is also (very often) susceptible to modeling errors.

That is the reason that we need combination of these two Control Systems. Usually, if the criterion mentioned in the previous section for usability of the closed-loop control system is not satisfied, the Feedforward Controller would be added to the system.

Under assumption that all blocks in Figure 5.7 have Laplace transform system—$G(s)$, Feedforward Controller—$F(s)$, Closed-loop Controller—$K(s)$, and Disturbance—$D(s)$, the transfer function of the total system $H(s)$ will be

$$H(s) = \frac{D(s) - F(s) \cdot G(s)}{1 + K(s) \cdot G(s)}$$

The Disturbance $D(s)$ must be eliminated, so the Feedforward Controller must have transfer function:

$$F(s) = \frac{D(s)}{G(s)}$$

This equation looks very simple, but its realization is not so easy. It is clear that for such a configuration, the Disturbance must be known accurately in advance or be measured in real time. If measured in real time, Feedforward Controller must be very fast.

In general, the Feedforward Control System is mostly used in industrial systems to keep the vital parameters of the system in use within the tolerances (SPC!) or to follow a particular algorithm (trajectory) in the process provided by the system in use.

For the second use, the computer system with stored algorithm in the internal memory is mostly used. Also, it is a very popular controller with Complex Systems characterized as MIMO (Multi-Inputs–Multi-Outputs) systems.[8] Sometimes, the Feedforward Controller is also used as Ratio Controller where instead of keeping steady some parameters, it will keep steady the ratio of two connected or disconnected parameters. A simple example of such a controller is the electronic fuel injection controller in the cars, where the ratio between the fuel and air is kept steady during driving.

It is important to emphasize that the Feedforward Control System deals mostly with dynamical performance of the system and it cannot change the stability of the system in use. So, if the system is (un)stable by itself, it will also be (un)stable with the Feedforward Controller. The reason is that it only deals with tolerances of one or few system parameters and it does not check the overall system stability.

5.5.4 CASCADE CONTROL SYSTEM

The Cascade control is used when there is more than one parameter for control in the system in use, but only one variable parameter for providing control is available. It is usually done through two controllers where output of the first controller (Outer Control Loop) is input in the second controller. The second controller (Inner Control Loop[9]) actually provides the control over requested parameter. It is important to state that the changes of the signal measured by the Inner Control Loop does not have to be faster than the changes of the signal from parameter in the Outer control Loop.

---

[8] There are also SISO (single-input–single-output) systems.
[9] You can find in the literature also names Master and Slave for Outer and Inner Control Systems, respectively.

As such, the Cascade Control System has two objectives:

(a) To suppress the effect of disturbances on the system (process) output via the action of a secondary controller
(b) To reduce the sensitivity of the changes of the primary parameter to gain variations of the part of the process in the inner control loop.

The upper part of Figure 5.8 shows a schematic of Cascade Control System and the lower part presents the real Cascade Control System for maintaining the level $y(t)$ of the water in the tank.

There is a pipe at the bottom of the tank which tries to empty the tank and there is a flow $x(t)$ of the water, which is filling the tank with water. There are two sensors—Level Sensor and Flow Sensor—which provide data to the Level Controller and Flow Controller, respectively. Level Controller provides signal $x_2(t)$ with information about the level of water in the tank and Flow Controller (which already have information about the flow of the water $x_1(t)$) will adjust the flow of the water $x(t)$ through the valve to maintain the necessary level $y(t)$ given by the Reference as soon as possible.

So, one controller provides information what needs to be done and another controller handles the problem as soon as possible, by changing the requested parameter.

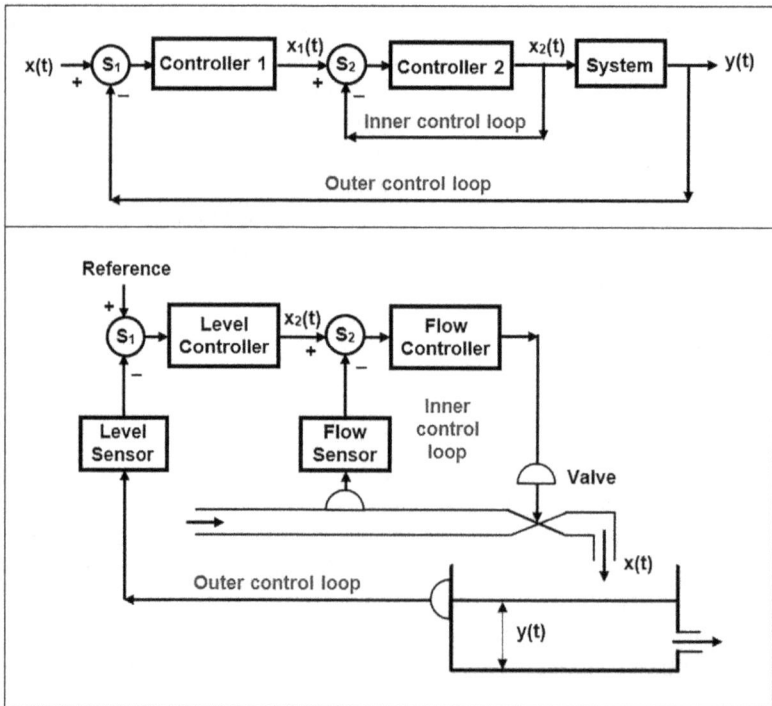

FIGURE 5.8   Cascade control system schematic (above) and real system for maintaining the level of water in the tank (below).

The Cascade Control System provides good control for non-linear systems where the law of changing the controllable influence factors (parameters!) is non-linear. Also, it can be successfully applied to the systems which are continuous and to the systems which are discrete.

5.5.5 ADAPTIVE CONTROL SYSTEM

The Adaptive Control is an approach in the design of Control Systems where the values of the variables which should be used for control are known with uncertainties. This happens when there are uncertainties regarding internal and external disturbances. And this is the main difference between Adaptive Controllers and other Controllers: The Adaptive Controllers have the ability to automatically adjust itself in real time to handle these uncertain changes of the disturbances.

During the operation of the system, the Controller collects data with information about the behavior of the system. These data help in reducing the level of uncertainty regarding the value of the disturbances and with its help, the responsible variable is adjusted to handle disturbances. It means during operations, as the level of uncertainty is reduced by the gathered data, the Controller is capable to adapt itself more accurately to the system disturbance and as such to provide better level of performance.

There is a difference between the conventional closed-loop control system and Adaptive Control System. The closed-loop control systems are beneficial in elimination of the effect of disturbances which affect the output of the system through changes of the variables and Adaptive Control System is dedicated to the elimination of the effect of unknown internal and external influence factors (disturbances, variations, etc.) upon the overall performance of the system in use.

Here, I must mention that the control provided by the Adaptive Control System looks more iterative because, as the new data are gathered, the adaptation need time to adjust and the control over the system also needs time to change. So, for optimal adjustment, these Controllers need more time.

Adaptive Control Systems are inherently non-linear and they can be divided into two categories based on the approach used to build them: Direct and Indirect. This division comes from the way how the knowledge about the changes of the Controllable influence factors is gathered.

Indirect methods estimate these disturbances in advance by data gathered through sensors and for control, they use the estimated values to adjust the controller. Also, the indirect methods cannot detect the fast-changing disturbances in operation, so for such fast disturbances, they can be unacceptable. However, they can make a better adjustment of the parameters over a particular time.

Direct methods are those where previously estimated (known) disturbances are directly used in adaptation of the controller.

Let's consider the example of Adaptive Control System shown in Figure 5.9.

As it can be seen in Figure 5.9, there are few more devices for the Adaptive Control System than for other Control Systems. There is a Regulator which provides the correct input signal $x(t)$ into the System through one of the outputs of the Divider D_1. The output of the System $y(t)$ goes to Data Sampler which is connected to the

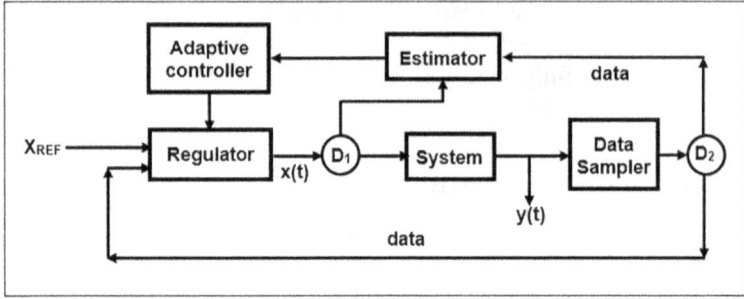

FIGURE 5.9 Adaptive control system.

Estimator and to the second input of Regulator (for comparison with the x_{REF}). The connection is done through Divider D_2 which provides necessary portion of data to both the Estimator and the Regulator. Data Sampler samples data periodically at a particular sampling time which should be shorter than time needed by the disturbances to spoil the output.

The correct input $x(t)$ into the System is the result of the comparison of data from the Data Sampler and signal x_{REF} in the Regulator. The second output of Divider D_2 is connected to the Estimator which obtains information about the behavior of the output of the System $y(t)$ and compared with the input signal $x(t)$ obtained from the Divider D_1. The Estimator provides data for Adaptive Controller which adapts itself and issues the command to the Regulator to adapt (change) the input signal $x(t)$ to the optimal requirements with regard to the output signal $y(t)$.

The Adaptive Control Systems are (roughly) divided into two categories: Model-Referenced Adaptive Control (MRAC) and Self-Tuning Controllers (STC).

The MRAC is based on reference model containing the reference value with tolerances of the controlled variable or parameter (as the reference stored in the Regulator in Figure 5.8). As such, it is a better controller for dealing with known disturbances and change of variables and parameters.

The STC is sensing the changes of the controlled variables or parameters and it adjusts itself to make them as small as possible. As such, it is a better controller for dealing with unknown disturbances and change of variables and parameters. I will not go into details in this book, but there is plenty of literature available on the Internet to find what you need.

Adaptive Control Systems are fully applicable today only for linear systems. For non-linear systems, they are based on ML (Machine Learning), Deep Learning (DL), and AI (Artificial Intelligence) and these three areas are still subject to considerable efforts in the research. Anyway, from the point of safety of Complex Systems, this is additional problem because there are no standards in these areas and hazards and risks are still not defined accurately. The Adaptive Control today is very much used in robotics (almost everywhere), aviation (control of the aircraft by auto-pilot, etc.), space exploration (control of rockets and spaceships), marine (control of ships movement, etc.), control of chemical and metallurgical processes, adaptation of big power distribution systems to change the loads in time, and medicine (wide area of usage).

5.6 PROVIDING RESILIENCE TO THE CONTROL SYSTEMS

All systems are prone to internal and external influence factors. Not always we can handle these factors, because whatever Control System[10] we use, it cannot cope with everything. There is always a lack of knowledge about the Complex System and all other physical factors (linear or nonlinear, static or dynamic, frictions, damping, hysteresis, etc.) connected with him. There is also a lack of knowledge how the environment (where the system will be used) will behave and lack of knowledge about the character, culture, religion, education, and social aspects of the humans who will use to operate and to maintain the Complex System.

All these things create a considerable uncertainty which can affect the design of the system in use and the design of its associated Control System. So, mostly, the design and use of any of the systems is based on compromise.

This compromise is more critical for the Control System, because if the system in use is inaccurate in maintaining operation in tolerances, the Control System must deal with these inaccuracies. So, the system in use can make errors, but the associated Control System must react to rectify these errors. If the Control System fails, then every fault of the system in use will be catastrophic for the operation. That is the reason that we try to put our emphasis on the design of the Control System (constraints!) and providing Resilient Control System is imperative in science and engineering. Resilience (as mentioned earlier) is actually the *robustness* introduced first by Genuchi Taguchi in the last century.

Under the word "*robust*" (or "*robustness*") is hidden a characteristic of the system which is capable to sustain all adverse internal or external influence factors without endangering its process, activity, operation, etc. It is important to understand that resilience (robustness) is a quality issue with very big impact on safety.

We can say that the Control System is resilient under the following conditions:

(a) It has low sensitivities, which means that the disturbances or uncontrolled changes of the parameters of the Control System will not affect the operation, no matter how big they are.
(b) It is stable over the range of any parameter variations, which means the system will stay in control, although there are uncontrolled changes in the system parameters and variables.
(c) Its operation (process, activity, etc.) continues to meet the required specifications in the presence of all disturbances, noise, and any change of the external and internal influence factors.

Resilience in Complex Systems is provided mostly by Adaptive Control Systems and through Robust Control Systems, The Robust Control Systems are not mentioned here because I do not consider it as a different class of Control Systems (although in Control Engineering it exists). In reality, the Robust Control System could be any of the previously mentioned Control Systems which use high-quality components

---

[10] The resilience also applies to the Complex Systems, but the point is that the capability of Control Systems to provide control is a base for providing resilience of Complex Systems.

characterized by very small changes of the variables and parameters in time due to different disturbances.

The difference between the Robust Control System and the Adaptive Control System is that the first one deals mostly with uncertainties caused by constant or slowly varying parameters. The Adaptive Control System needs time to learn about behavior of the disturbances, variables, and parameters. So, as the adaptation goes on, it becomes better in handling these changes. The Robust Control System is the one which simply attempts to provide consistent performance.

In addition, the Adaptive Control System requires little or no information about the unknown disturbances and uncertainties of variables and parameters, while a Robust Control System usually requires reasonable estimates of the changes which need to be kept within the tolerances.

So, the Robust Control System will provide "toughness" (built by its design) by dealing with disturbances, quickly varying parameters, and unmodeled dynamics and the Adaptive Control System will provide "elasticity" by dealing with disturbances, slowly varying parameters, and modeled dynamics. Very often, these two characteristics of each of the Control Systems are combined and the "toughness" and "elasticity" would provide resilience.

Here, it is important to mention that the existing adaptive techniques for nonlinear systems would require a linearization of the Complex System dynamics, which is actually the uncertainty expressed as linearity of a set of unknown disturbances and parameters. In some cases, full linearization could be achieved only by combination of the Adaptive Control System and Robust Control System.

I must emphasize that the resilient systems are needed for any jobs that need to be done. So, speaking about Resilience, I must extend it to all systems, not only to the Complex Systems. That is the reason that I will continue with the Resilience in next chapters of the book.

5.7 COMPUTER-BASED CONTROL SYSTEMS

Everything explained in this chapter is just the basics of the Control Theory. Today, these "basics" (many of them) are integrated with the system in use, into one big Complex System which contains everything needed to provide continuous and accurate operation (activity, process, service, etc.) for any purpose of the humans.

The point is that this integration is designed, built, monitored, overseen, commanded, and maintained by the so-called CPU[11] (Central Processing Unit) and/or by a simple processor or computer (COTS or specialized). It can be said that this CPU is the heart of the Computer-based Control System. Roughly speaking, this Computer-based Control System is congregation of Hardware (different electronic or mechanical devices) and Software (different application programs written separately or as one big holistic program).

---

[11] It means that almost every Control System today is a digital system, although there are still analog systems. However, everything is controlled by computers.

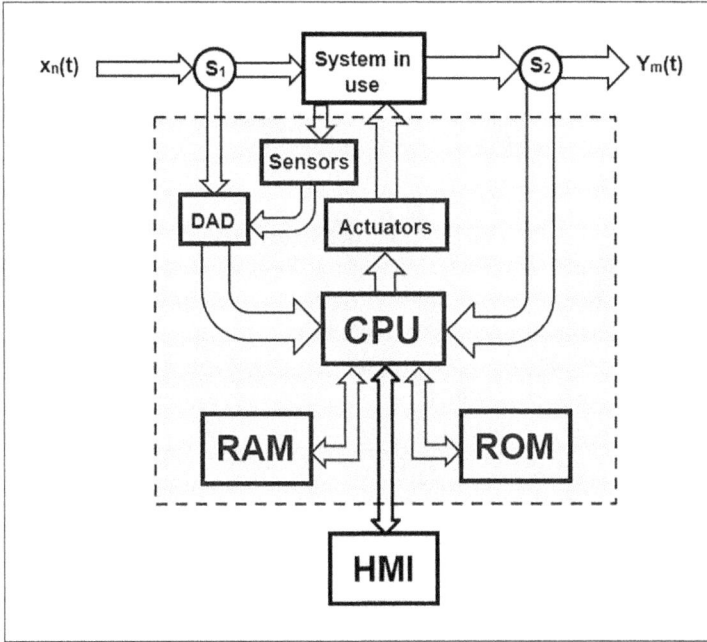

FIGURE 5.10   Computer-based control system.

One of the most important aspects of computers is the fact that they are digital machines. It means that any signal which is analog[12] would be digitalized into train of pulses. Digitalization means that the quantity of each signal would be sampled in periodic time, the sample will be coded, and the signal will be presented as train of electrical pulses. These pulses are very low in amplitude and as such they could be prone to noise. That is the reason that there are special techniques to recover "masked by noise" pulses. However, the digital signal could be only received or not received, it cannot be disturbed as analog signals. If you have received it, it is very certain that this is good signal (data, information). But if you do not receive it, it means that there are some disturbances, which may not be necessarily something catastrophic: For example, only small error in synchronization between devices could provide loss of digital signal.

The block diagram of one modern Computer-based Control System is presented in Figure 5.10. Please note that all blocks in the figure could vary and not all of them should be part of general system. Of course, for some systems in use, some other blocks could be added; so, Figure 5.10 presents only one simplified example for general application.

---

[12] All our sensors (eyes, nose, ears, tongues, skin) provide analog (continuous) signal to our brain which processes these signals and provides feelings as a result of our senses.

But I do not like to offend you by explaining in detail what is it about in Figure 5.10!

The point is that I do not believe that there is any engineer in this world who cannot understand what is this about. But, let me just remind you . . .

There is one or more systems in use (part of our Complex System!) which are designed to provide some benefit to humans or to finish some job (process, operation, activity, etc.). This system in use is characterized by its input(s) $x_n(t)$, its output(s) $y_m(t)$,[13] and its transfer function(s) $h(t)$.

To the system(s) in use is added Control System(s) (the gray area in Figure 5.10) which provides limitation (constraints) of the level of activities conducted by the system(s) in use. That is the way how it provides control. Both together, system(s) in use and Computer-based Control System(s) create our Complex System

As part of the Control System, there are many other devices (Sensors and Actuators) which can be separated from the system in use or could be part of it. These devices can be configured by any of the control methods explained in this chapter. Which of the methods would be used depends on the requirements posted for the system in use or by the creativity of the designer.

In general, to provide monitoring, overseeing, fault finding through BITE (Built-in Test Equipment) maintenance and control, we need data. The data is gathered through Sensors or measurement instruments connected to the point of interest in our system in use. Usually, there is DAD (Data Acquisition Device), which gather all these data and submit them to the CPU. SCADA is one well-known standardized data acquisition system, but there are many others as well. If the data gathered is analog, there must be ADC (Analog to Digital Convertor) as part of the sensors or DAD. In general, all communications and processing are digital.

The gathered data can be transferred through wires or wireless. The CPU collects these data and there are one or few powerful processors which decide (based on the Software algorithms) what to do after processing these data. There are few types of memories which are used, but in Figure 5.10 only RAM (Random Access Memory) and ROM (Read-Only Memory) are presented. The CPU provides processing of the data and, based on the criteria "stored" in the memories, it issues commands which activate the Actuators and particular parameters and variables of the system in use are changed. The intention of these imposed changes is to keep these variables and parameters within the tolerances and to provide correct operation (process, activity service, etc.) or tracking. There is also HMI (Human–Machine Interface) where employees (operators) can operate (set, adjust, monitor, maintain, test, etc.) the system. However, all this can be provided through one powerful computer with appropriate Software and additional systems/devices (the gray shaded area in Figure 5.10).

The point is that all these devices (or better say, thousands of them) together, shown in Figure 5.10, establish our Complex System. The thing which is missing from the figure is the Software. It can be capable to execute one or many thousand tasks, embedded as algorithms in one holistic "piece" of Software, which can also be very complex.

---

[13] In the case when there is single-input–single-output, the system is known as SISO and when there are multiple-inputs–multiple-outputs, the system is known as MIMO. Of course, there are also combinations: SIMO and MISO.

Speaking about the Complexity of the system, let's just mention here the Formula 1 car, one simple example of high-technology product. There are over 120 different sensors in all parts of the car tasked to monitor the performance of different parameters. The speed of collecting the data from the sensors could be at a rate of over 10 MB per second. That data is wirelessly send to the computerized system[14] into the garage via antennae in the front of the car. So, you can imagine the power of the Hardware and Software capable to process these data in real time and to undertake measures to protect the car and the driver.

This is a Formula 1 car, but let's move to aviation. There, the Airbus 350 model of aircraft has approximately 6,000 sensors across the entire plane and the Airbus 380 has approximately 10,000 sensors in only one wing.

If you think it is really huge, you are wrong. Imagine the Complexity of the space rocket together with the interplanetary sondes which are launched toward some of the planets of our Solar system. That is something really, really, really complex!

---

[14] Overall, supervisory system in Formula 1 car is known as Telemetry.

Part II

Non-linearity in Complex Systems

6 Introduction to Dynamics of Systems

6.1 INTRODUCTION

What is the connection of Complex Systems with dynamics?

The answer to this question is obvious. The dynamic systems change with time and as such, they add more Complexity to already Complex Systems. If we speak of non-linear dynamics, the things are all the more complex.

The main point is that systems change with time and space. These changes are intentional (caused by humans) and some of them are non-intentional (wearing, decay, etc.). In Part I of the book, I have provided explanations about linear systems and in Part II, I will speak about non-linear systems.

This and next few chapters are for purpose of gathering comprehensive knowledge regarding the reasons and features of Complex Systems and as such (although there is very big simplification) the intention is to provide scientific approach to the operations and challenges for the Complex Systems during their operations. This knowledge, about non-linearity of the structure and the operations of the Complex Systems, will help later with the Risk Assessment.

6.2 STATES OF SYSTEMS IN NATURE

System which can change its states[1] in time and in space is known as dynamical system. Usually, it is described by one or more equations (system of equations) where there could be one or few variables. The equations (formulas) describe changes of the states of the system in time or changes from one state to another (in space[2]). Also, there are plenty of variables in the system which can change with time and space, so system changes, depending on the number of variables, can be very simple or very complex.

There is a particular area of physics which deals with dynamical systems and it is known as Dynamics. There are two types of Dynamics: Linear and Non-linear. Both are gathering their name from the type of the graph produced by graphical presentation of the equations (formulas) which describe the variables in the system. If the variables, which are parameters of the system, are only those presented

(a) as power of 1 ($y = ax + b$);
(b) they are not expressed as a function (for example, not to be as $y = sin\ x$, $y = e^x$, etc.); or
(c) they do not multiply themselves ($z = ax + xy + by$);

---

[1] In the book, I will later use word "event" for the "state", because in this "context of the things" they are synonyms.

[2] It does not mean that the system changes its position in space (moving from one place to another). It could mean that it changes its configuration through movement of its parts (Subsystems).

DOI: 10.1201/9781003404811-8

then the system is linear. In all other cases, the system is non-linear.

There is another (rough) explanation of these rules: If the graph of graphic presentation of the equation is a straight line, the system is linear. And vice versa: If the graph is some other curved line, the system depicted by formulas used to produce the graph is non-linear.

The simple examples of linear systems are amplifiers, integrators, differentiators, etc. in electronics, simple harmonious oscillators, electromagnetic radio waves and their combinations (interference!), etc. Simple examples of non-linear systems are pendulum, weather, lasers, turbulence, fluids, etc.

In addition, the dynamics of the systems can be described by the following:

(a) Differential equations, where the changes of the system with time and position are explained by derivatives of the variables. It applies for systems which continuously change with time.
(b) Simple formulas (difference equations, iterative maps, mapping) where the changes of the system are described by the movement of the system's states from one state to another. It applies for the systems which discretely change in time.

A general form of differential equation of system with one variable and one parameter is presented by the following equation:

$$\frac{dx}{dt} = \dot{x} = f(x,a)$$

where $dx/dt = \dot{x}$ is the first derivation of variable x in time[3] and $f(x, a)$ is the function depending on variable x (depending on time) and parameter a[4] (which can also depend on time). The equation above is known as autonomous because time t is not explicitly included in function $f(x, a)$,[5] but it is "hidden" inside the variable x. It means that $f(x, a)$ changes in time only when changes of x (or a) in time happen. If x (and possibly a) does not change in time, $f(x, a)$ also does not change in time.

The general form of difference equation[6] is given by formula:

$$x_{n+1} = f(x_n)$$

It means that next values of variable x will be calculated putting into the formula the previously calculated values of $f(x)$. The next state of the system is dependent only on

---

[3] $\dot{x}$ is generally used in physics (science) only for derivations with respect to time.

[4] The parameter a can be expression of the influence of the Control System (as a Subsystem) to the Complex System.

[5] This is a strongly mathematical definition. In mathematics, systems expressed by function f(x, a, t) are known as non-autonomous systems. They are strongly deterministic and if something unanticipated happen, the system will not adapt (the system will experience fault).

In engineering, things are different: The car is a non-autonomous system, because the behavior of the car is controlled by the driver, but the car with the driver (together) is autonomous system because driver can adapt the car "behavior" to the unanticipated changes in the conditions of driving. For the sake of truth, the physical systems are not-autonomous.

[6] Due to used method of iterations, these difference equations can also be found under the names of mapping or iterative maps.

x and other parameter(s) are not included. However, x could be dependent on other parameters.

This is an iterative method and due to this operation, the graphs of difference equations are known as iterative maps. In words of physics, we calculate the particular state of the system and the result is used in the same equation to calculate the next state (iteration).

For better understanding, let's see Table 6.1 for the discrete function described by the following equation:

$$f(x) = 2x$$

The arrows in the table show values of $f(x)$ which are used for next iteration of x.

It is important to mention here that if we know that the change of the system happened by same law (expressed by equation), then we can describe the system by graphs known as network diagrams with nodes and arrows, or the so-called process maps. This presentation is known as iteration map or mapping (Figure 6.1).

As you can notice in Figure 6.1, in the dynamic system, very often, a decision needs to be made (rhombs with question marks) and depending on the decision, the present state of the system can change into different states. In addition, there is need for feedback (presented by dashed lines) and, as it has been explained previously in the book, the Subsystems providing feedback are the ones which provide non-linearity and increase the Complexity, in most of the cases.

TABLE 6.1

n	1	2	3	4	5	6	7	...
x_n	$x_1 = 1$	$x_2 = 2$	$x_3 = 4$	$x_4 = 8$	$x_5 = 16$	$x_6 = 32$	$x_7 = 64$	...
$f(x_n) = 2x_{n-1}$	2	4	6	16	32	64	128	...

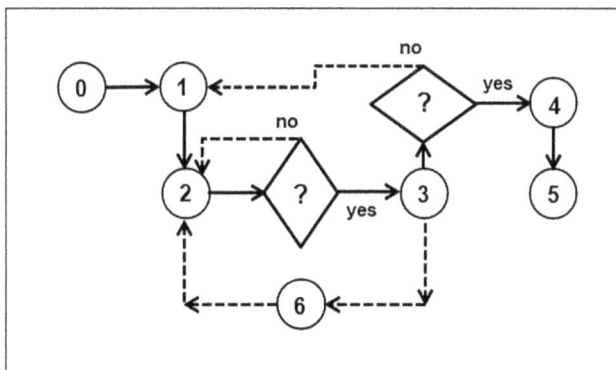

FIGURE 6.1 Example for iteration map which describes the changes of five states in the system during its activity (state "0" means the system is in rest [not working]).

The dynamic systems can be stable or unstable and they can change these states, alternatively, into one to another. It is important to understand when and why stable dynamic system will transfer itself into an unstable system. In general, the unstable system is a system when we do not have control over it.

The most important characteristics of dynamic systems are Phase space, Time, and Law of Evolution.

A set of all points (states) of the system is known as *phase space*.[7] The phase space is a set of all points from the diagram presentation (for differential equations) or iteration map (for difference equations) where the law of changing the states is presented. For diagram presentation, the state of the system is presented as point (you may assume it is a moving particle) where the coordinates and the velocity vector (direction of change) are initial conditions. For iterative map, the operating states of the system are presented with circles and direction of change with arrows (Figure 6.1).

Time implies the schedule of time needed for the system to change its states. In practical terms, this is the time for producing some product or providing some service. It can be continuous (if we are interested about the process) or discrete (if we are interested about the states).

The Law of Evolution can be described as a rule(s) which is(are) used by a system to change the states. It means that the Law of Evolution is usually a deterministic law: Knowing the previous state of the system, we can calculate next states. The Law of Evolution usually does not depend on time: There is a time needed to change the states, but in total, the next state does not depend on time.

As you can notice from your experience, in real world, not always the Law of Evolution will be the same to change the same system from one state to another.

---

[7] In some literature, you can find it as "state space".

7 Non-linear Dynamical Systems

7.1 INTRODUCTION

In Section 5.4, I have explained definition of the linear systems. Here I will pay attention to the non-linear systems presented in different ways.

The examples for non-linear systems are systems explained mathematically by functions when $f(x)$ is determined as fraction and variable x is in denominator, any power of x (different than 1), logarithmic and trigonometric functions, multiplication of variables (for few variables), etc. All of these functions have graphical presentation as curved line (curved surface in cases of two and curved volume in cases of three variables). Roughly, in general, if the line in graphical presentation of the function $f(x)$ is curved line (surface, volume, etc.), then the system is not linear.

Having this in mind, with the non-linear Complex System, the outputs of the operation are not proportional to the inputs. More characteristically, a small (or large) change in some variable and parameter would not necessarily result in a small (or large) change in the system. What our primary interest in this book is that it could be expected that a small (external or internal) intentional change of any of the variables and parameters may have large, unexpected outcomes.

Pay attention to this because I will come back again later in the book (Chapter 13).

7.2 INTRODUCTION TO DYNAMICS OF NON-LINEAR SYSTEMS

As I already explained in Section 6.2, the "system which can change its states in space and/or in time is known as dynamical system".

The way how we can present the dynamics of change will determine whether the system is linear or non-linear. Most of the systems in our real lives are non-linear, but most of the analyzing methods used in science are made for linear systems. This is an issue for exploring non-linear system, so it is common to "linearize" non-linear systems in the vicinity of states (subjects of exploration) and to use there some of the analyzing methods for linear system. Of course, this is connected by modeling and simulations. The power of computers (Hardware and Software) today helps in all these analyses.

However, this approach of analyzing non-linear system cannot be always used, because it is not applicable to many of the non-linear systems.

The point is that non-linearity is one of the most common contributors to the Complexity of the systems. From engineering point of view, the Complex System has many levels of control and automation and these two cannot be provided by linear systems. So, every system, subject to control and automation provided by feedback

DOI: 10.1201/9781003404811-9

is a non-linear system. To provide quality and safety of operations of these systems, we need to be familiar with non-linearity and its influence on the behavior of the Complex Systems.

The non-linear systems are usually described by differential equations or by difference equations. The non-linear system is a Complex System due to non-linearity of changes which this system experience in space and/or in time. Each non-linear dynamical system is complex, but not each Complex System is non-linear.

From the safety point of view, there are few important differences between the linear and non-linear systems:

(a) The states of linear systems will not go to infinity in finite time, which does not apply to non-linear systems. There, such abrupt and sudden changes could very much happen.

(b) The non-linear systems may have many stable or unstable states, but linear systems could have only one stable state which is usually used to adjust its Working Point there to provide good operation (process, activity, etc.).

(c) The linear systems cannot oscillate, have bifurcations, or Chaos. The non-linear systems are prone to all of these situations.

(d) A linear system can have only one fixed point.[1] It means in the linear system we can adjust only one Working Point and that point could attract (stable point) or repel (unstable point) the state of the system, ignoring its initial state. The non-linear system could have more than one fixed point which can be stable, unstable, or hybrid.

(e) The linear system will keep stability of its fixed point whatever it is. Non-linear system may bifurcate and change the stability (stable state to unstable state, or vice versa) to any of its fixed points.

(f) Linear systems cannot produce sudden jumps from one state to another (the so-called *catastrophe*) and they cannot produce hysteresis.

(g) A linear system with few periodic inputs with different frequencies will produce periodic output signals with the same frequencies as the input frequencies. A non-linear system with few periodic inputs with different frequencies will produce the output signals with the same frequencies. But it will also produce a bunch of signals with new frequencies which will be combinations of the input frequencies. These new "output signals" are known as harmonics (integer multiples of input frequencies), subharmonics (submultiples of input frequencies), and intermodulation[2] products (different combinations of input frequencies).

All these things will be explained later in this chapter, but to "arrive" there, let's first explain (mathematically) what the non-linear systems are . . .

---

[1] In the next few paragraphs, you will understand what is this about . . .

[2] Non-linearity in electronics is not necessarily bad. Non-linear systems are very much used in telecommunications as a source of modulated signals. Without non-linear systems, this type of telecommunication will not exist.

7.3   NON-LINEAR SYSTEMS

As I said in the previous chapter, the dynamical systems are systems which change their states in space and/or in time. Every state is determined by a particular set of variables and they can be generally described with equations for each state which is dependent on time. For non-linear dynamical system with n variables $(x_1, x_2, x_3, \ldots, x_n)$, we can explain the system behavior using these equations:

$$\frac{dx_1}{dt} = \dot{x}_1 = f_1\left(x_1, x_2, \ldots, x_n\right)$$

$$\frac{dx_2}{dt} = \dot{x}_2 = f_2\left(x_1, x_2, \ldots, x_n\right)$$

....

$$\frac{dx_n}{dt} = \dot{x}_n = f_n\left(x_1, x_2, \ldots, x_n\right)$$

Each of the equations f_n above describes the rule of change of the result by change of each variable with time. Please note that this is an autonomous system, which means that f_n does not depend on time. Anyway, having in mind that all variables x_n are dependent on time, the system obviously changes in time only with the changes of variables x_n. The point is that all functions $f_n(x_1, x_2, \ldots, x_n)$ are actually non-linear functions of x_n.

Another option is the functions f_n to be expressed as $f_n(x_1, x_2, \ldots, x_n, t)$ which means that the functions change with change of x_n in time and, in addition, they change with time (independent of x_n). This will be a non-autonomous system.

From engineering point of view, the expression dx_n/dt is an expression for the quantity of change (velocity) of the variable x_n which is caused by changes in time of the function f_n (which, again, depends on the changes of $x_1, x_2, \ldots, x_n$ in time).

Let's repeat again: "Non-linear" means that inside the functions f_n, the x's are connected by mathematical operations as multiplication, division, logarithm, and power (different from 1), or any other combinations between them. Any system of non-linear functions presented as above is very difficult (or better say, almost impossible in most cases) to solve analytically (mathematically), so the mathematicians are using mostly graphical method to understand what is hidden behind the equations.

Of course, there is need for initial conditions of the system described by the equations above, but we will just assume here that we know them. So, the system described by these equations (together with the initial conditions[3] of each variable) presents the states of the system in time. Having in mind that we have n variables, we need n equations, so the order of the system is n.

Although the "initial conditions" have clear meaning in mathematical theory of differential equations, it is worth mentioning here that these initial conditions are actually settings (adjustments) of the Working Point of our engineering Complex System done by the operator to execute a particular job. It is assumed that all these

---

[3] Initial conditions are values of each of the x_n at the beginning of our analysis of the state of the system.

different initial conditions are beginning of the operation for a particular product to be produced (service to be delivered) and they are well-known to the operator. There shall be finite number of these initial conditions (Working Points) in our system.

The variables x_n can represent velocity, force, location, temperature, volume, pressure, etc. Each derivation of variables includes a function which is dependent on all other variables in the result means that the variables are strongly interdependent on their values.

The solutions of the equations of the system describe a particular state of the system. It does not mean that system changes its states by jumping from one state to another, because we speak here (mathematically) for functions which have at least first derivate (continuous functions). The change moves continuously and speaking about states we are speaking about presentation of the system in a particular portion of time. All these states can be described as points in a graph which will have n dimensions and each state will have n coordinates depending on the value of each variable x_n.

In addition to the variables, in equations we can also put some parameters, which could or could not be independent in time. These parameters affect the level of change of variables. For example, the pressure in the car tire is a parameter which can affect the speed of driving the car or durability of the tire.

As I said before, the set of all points (states) of the system is known as phase space[4] and the first important thing is to know what are dimensions of it. The second thing which is important is we need to know whether the system of equations is linear or non-linear. Linear systems of equations can be easily solved, but for non-linear systems, finding solutions is a problem. Anyway, we can use graphical approach[5] which (to be honest) does not provide accurate solutions, but it provides visual information about the future behavior of the system.

Solving graphically the system of non-linear equations will provide different solutions depending on the number of variables included into the system. For $n = 1$, it could produce fixed points and bifurcations; for $n = 2$, it could additionally produce oscillations;[6] and for $n \geq 3$, it could additionally produce a Chaos. In this case, n is the number of variables included into the system and we can notice that Chaos will arise only where the system is built with three or more variables and parameters. It is important to understand it, because in industry, there are plenty of processes which are dependent on three or more variables and parameters and each of these processes (under particular initial conditions) can produce Chaos as behavior (if it lasts long enough).

I will not go into mathematical or engineering details in this area, but I will try to explain it as simple as it is possible . . .

Let's see . . .

7.4   DESCRIPTION OF NON-LINEAR DYNAMICAL SYSTEMS

As I said in the previous sections, non-linear dynamical systems can be described by differential equations and difference equations. Differential equations are not easy to solve, so other methods are used to explore non-linear dynamical systems.

---

[4] In some literature, you can find it as "state space".

[5] Graphical approach was first proposed by Poincare when he tried to solve the "three body" problem.

[6] To be scientifically correct, the oscillations are possible even for $n = 1$, but it is possible only if the line is a circle or an ellipse.

The point is that, having in mind that we are speaking about dynamical systems, the equations present the law of changing the state of the system and the solutions of the equations (associated by particular initial conditions) present the state of the system when changed from its initial conditions in new state at a particular time.

If the industrial process is continuous, then the changes of situations are continuous in time and change of the variable is following dynamics of the process described by differential equation. These differential equations can produce Chaos[7] in time.

If we have a process where the states jump from one position (step, node, etc.) to another and, if the time of jumping is not important, then the process can be described by difference equation. These difference equations can produce Chaos in space, known as fractals.

Let's see how the use of these descriptions could help us to analyze them . . .

7.4.1   NON-LINEAR DYNAMICAL SYSTEM DESCRIBED BY DIFFERENTIAL EQUATIONS

Let's try to analyze continuous dynamical system described by simple non-linear differential equation. Let's say that the equation is

$$\frac{dx}{dt} = \dot{x} = x^2 - 4$$

This is a simple differential equation, giving function of change of the velocity of some particles of interest[8] depending on their position in the space. So, x is the position of the particle and $\dot{x}$ is the velocity of the particle for this position. If it is autonomous ($f(x) \neq f(x, t)$), then it can be solved analytically. If we know the dependence of x from t and initial conditions, the solution will be found by separation of $\dot{x}$ and x and by integration of both sides of the formula:

$$\int \frac{dx}{x^2 - 4} = \int dt$$

Here, the solution of the right integral will be equal to $t + constant$. But we need to put more efforts to handle integral on the left side. Although it is not so complex integral, it will not be easy to solve it, so it is wise to use graphical approach[9] which is usually used for complex non-linear systems.

Maybe you will ask yourself, what is the point in this chapter?

Unfortunately, you must remind yourself that Laplace (and Fourier) transforms are valid only for linear systems. Or it can be used for non-linear system which are linearized in the vicinity of some of the stable states (poles in left part of s-plane). So, dealing with non-linear systems, we need something more and something qualitatively different.

---

[7] The mathematics behind the Chaos is known as Chaos theory.

[8] I will use term "particle" because I think this could be easier for the reader to understand how the things change. For our purposes, this "particle" (any point on the $\dot{x}x$-plane) could be also a state of the system which continuously changes with time.

[9] Very often, in the mathematical literature, you can find the name "differential topology" (more often in its short form as "topology") for this graphical way of dealing with non-linear differential equations. And there is nothing wrong with that!

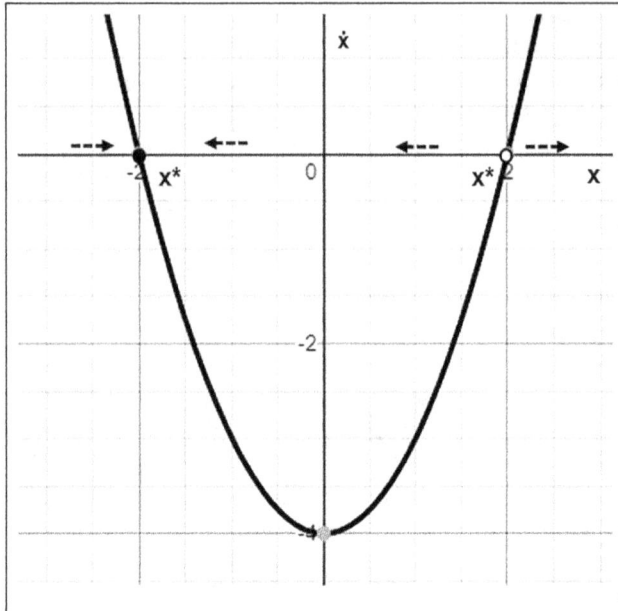

FIGURE 7.1   Graphical presentation of function $dx/dt = \dot{x} = x^2 - 4$.

In Figure 7.1 shows a diagram on function $dx/dt = \dot{x}$ (vertical axis) from above depending on the values of x (horizontal axis).

In Figure 7.1, we can notice that function (black line) intersects the horizontal axis at two places (for $x = -2$ and $x = 2$). These are places where $\dot{x} = 0$, which (physically) means that the velocity of change in these points is 0 (no change). These two points are known as **fixed** *points*.[10] They will be presented by the symbol x^*. Much of the exploration of non-linear systems is focused on the behavior of the system around these fixed points.

As it can be seen on the graphic, for the values $x < -2$, $\dot{x}$ is bigger than 0 which means that particle there will have positive (increasing) velocity. It means that the velocity of the imaginary particle changes (progresses) to the right (particle speed up). Physically, it means the distance of the particle changes in one direction in time.

For the values of x between -2 and 2, the value of $\dot{x}$ is smaller than 0 (negative velocity means speeding down), so in this area the velocity of the particle changes to left (decreases). For the values $x > 2$, $\dot{x}$ is bigger than 0 (positive velocity, increase), so the velocity of the particle moves to the right (increase).

It should be kept in mind that, in general, by using graphical presentation, the positive velocity causes the particle to move right (shown by black dashed arrows), which means that every particle which is left of $x = -2$ and right of $x = 2$ will move from left to right in the diagram. And, vice versa, the negative velocity (opposite

---

[10] In the strongly dedicated mathematical literature, the fixed point is also known as "equilibrium point" or "singular point".

direction) moves the particle left, so every point in areas defined by $-2 < x < 2$ will move from right to left.

Please note that for this presentation we have on x-axis the position of particle and on y-axis the change of the velocity of the particle. So, position and velocity are connected on the graph.

From engineering point of view, it means that in the areas where the particle which is left of $x = -2$ and right of $x = 2$, the process speeds up and in between ($-2 < x < 2$) the process slows down.

The point $x_1^* = -2$ is called *stable fixed point*[11] because all particles with positions close to this point (from both sides of this point) will tend to finish there[12] (reach stable position). The point $x_2^* = 2$ is called *unstable fixed point* because all particles with positions close to this point (from both sides) will tend to escape from there (will run away from that point).

Here I would just express simple rule: If the slope $f'(x^*)$ of the curve of line $\dot{x} = f(x)$ is negative near the fixed point ($f'(x^*) < 0$), then the point is stable; if it is positive near the fixed point ($f'(x^*) > 0$), then the point is unstable.

Stable fixed points of the non-linear systems, mathematically and physically speaking, are characteristics of "dissipative systems". The dissipative systems are those which gradually use the energy[13] and, in time, they spend its energy, so eventually, their dynamics ends in one position without moving.

From engineering point of view, the stable fixed points could be Controlled or Non-controlled.

The controlled stable fixed points are points which we would like to choose during our design of the system (adjust the variables and parameters) to be a Working Point of the system. These are values (coordinates) of the system which define the beneficial operation of the system. And it is very good if we succeed, because any small unwanted variation of any of the parameters or variables will be forced by the system to come back to the same stable fixed point chosen to be the Working Point of the system. It will be the characteristic of the Resilient (Robust) Complex System.

Non-controlled stable fixed points are those which do not provide any benefit to the system and if the system goes in any of them, the system will be faulty with high possibility to be damaged. It does not mean that this situation always result in any damage of the system or the environment around, but at least the system will be idle (of no use).

But it is not so bad . . .

It is wise during the design of the system to provide situation where, if there is fault of the system or failure of the operation due to some reasons, the system would move in any of these non-controlled stable fixed points, so that the damage of the system and operation could be avoided.

The unstable fixed points cause the dynamics of the system to change and if not prevented on time, they can result in bigger damage of the system or the environment around.

---

[11] As we can see later, the stable fixed points are also known as attractors.
[12] To be mathematically correct, they will asymptotically approach this point and they will never finish there.
[13] Loss of energy can be caused by uncontrolled friction or by some other reason of "leaking" the energy out of the system. In this category, we can put wearing of parts in mechanical systems.

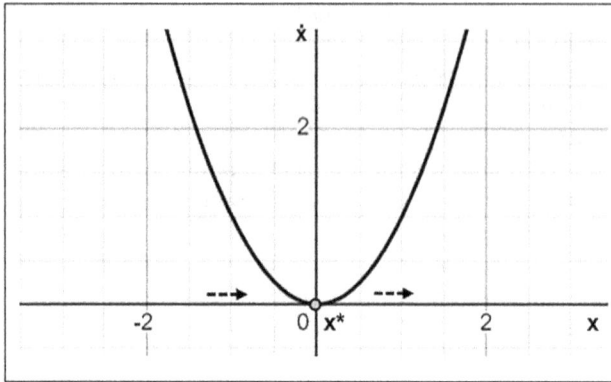

FIGURE 7.2   Graphical presentation of function $dx/dt = \dot{x} = x^2$ with its hybrid point.

Going further, using the wording of Resilience Engineering, the stable point means that Working Point of the system (chosen at this point) would not change for all disturbances applied by the change of variables and parameters, as soon as they stay in the vicinity of this point. On the contrary, the unstable fixed point chosen to be a Working Point for the system means that any change of the variables and parameters will provide failure of operation of this system. So, it is wise to use always the stable fixed points far away from the unstable fixed points as a Working Point of the system.

But there could also be the so-called *"hybrid" fixed points*, where $f'(x^*) = 0$.[14] Please note that if I have used in my example function $dx/dt = x^2$ (Figure 7.2), it has fixed point in $x = 0$ and this is a hybrid fixed point (marked as gray dot in Figure 7.2) because its derivative is 0 ($f'(x^*) = 0$). For particles which are left to $x = 0$, it is stable (they move toward $x = 0$) and for particles which are right to $x = 0$, it is unstable (they move away from $x = 0$). So, you need to be careful when you determine the stability of fixed points in the cases where $f'(x^*) = 0$.

For our case explained in Figure 7.1, these are actually points where the velocity ($\dot{x}$) is 0, which means that the position of the point (x) is not changed (the point does not move). Mathematically speaking, for case above, the fixed points of function $\dot{x} = f(x) = 0$ are extremes of the function $x = f(t)$ based on the position of the particles x in the time t.

Do not forget, $\dot{x}$ is the time derivative (dx/dt) of the function of variable x and variable x is just the function of t.

7.4.2   Dynamics of the States of the Non-linear Systems

Figure 7.3 shows the phase space with the movement of the particles (states) for the dynamic system presented by function $dx/dt = \dot{x} = x^2-4$. As it can be noticed, the curve presented by this function is actually a border between the directions of movement of the particles (states) of the system. All movements (directions of changes), depending on their initial conditions, are presented with dashed arrows. The dashed

---

[14] Please keep in mind that $\dot{x} = dx/dt = f(x)$ and $x = f(t)$, so $f'(x) = df(x)/dx$.

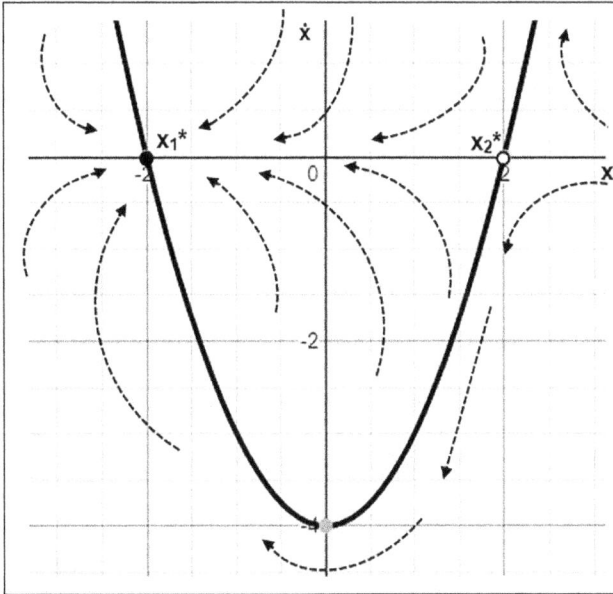

FIGURE 7.3 Graphical presentation of movement[15] of the particles in the phase space presented for the function $dx/dt = \dot{x} = x^2-4$.

arrows (vectors) are actually the speed trajectories of the movements given by equation $dx/dt = \dot{x} = x^2-4$ at these particular points. Having in mind that they change in time, we can call them "time series".

Another thing is that the negative axis where $\dot{x}$ is presented means that the speed of this movement decreases approaching $\dot{x} = 0$ and increases $\dot{x} < 0$. The point is that negative y-axis presents opposite direction from positive y-axis.

The point is that, as you can notice in Figure 7.3, the movement of the particles in non-linear system depends on their initial position. Some of the particles (depending on its initial position) will move toward x_1^* (stable fixed point) and some of them will move away from x_2^* (unstable fixed point) without finishing in x_1^*.

On the contrary, the movement of the particles in linear system do not depend on their initial position and there is only one fixed point, where the movement of the particles will finish. What will be this point depends on the equation which describe the linear system.

To better explain this feature of non-linear systems, I would use function $\dot{x} = f(x) = x^3-1$ graphically presented in Figure 7.4. This function is describing shape where the ball is moving and it is used just as example to clarify what are the fixed points in real systems.

In the two-dimensional system shown in Figure 7.4, the shape function $f(x)$ is a curve and there are points A and B marked on the graph. Obviously, locally, A is the maximum of the shape function $f(x)$ and B (locally) is its minimum.

Pay attention: In Figure 7.4, the function $f(x)$ is not a function $dx/dt = \dot{x}$!!!!

---

[15] In the literature, you can find it also under the name "vector field".

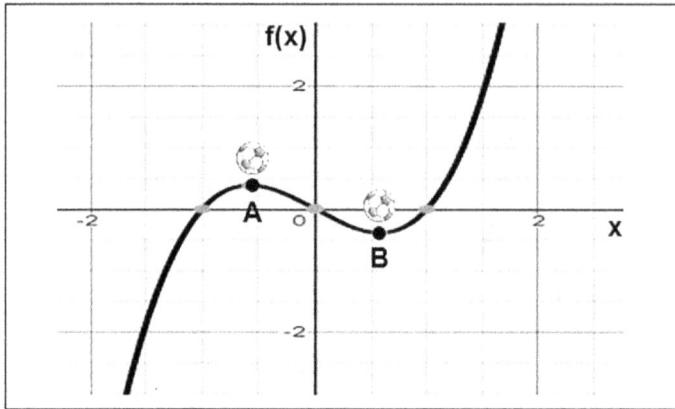

FIGURE 7.4 Shape of surface where the ball can move.

Imagine a ball placed in *A*. This ball could be carefully adjusted to stay at *A* (which is maximum of the trajectory), but any small deviations (caused by imperfection of the ball or the surface or maybe by the wind) will move it to left (to the negative infinity) or to the right (to *B*). Regarding the ball in the point *B* (minimum of the trajectory), any small deviation of the position will just temporarily remove the ball from *B*. When the reason for deviation ceases to exist, the ball will come again into *B*. So, we can conclude the following:

(a) *A* is an unstable point: Ball will stay in *A*, if there is no external force to change it. If there is such a force to provide even small deviation, then the ball will always be removed from the point *A*.

(b) *B* is a stable point: Ball will always stay in *B* in the case of deviations (if the deviation is not big enough to put the ball far away to the left of *A*).

In industry, the fixed points could be critical points when the system (operation, process, activity, etc.) changes. Now I will present an example (Figure 7.5) where, nevertheless the reason for change could be very small, the change itself could be sudden and huge.

Figure 7.5 presents a part of the graph of dynamical system determined by equation $dx/dt = \dot{x} = x^2-4$. This is the same system presented in full in Figure 7.1. There are two fixed points, point at $x_1^* = -2$ is a stable fixed point and point at $x_2^* = 2$ is an unstable fixed point. In addition, there are two chosen states, presented as point *A* and point *B*.

Let's see what dynamics of the system will do with these points.

Point *A* has initial velocity v_{iA} to the left, The combination of this initial velocity and attraction of the fixed point x_1^* will result in *change A* of its velocity. The point is that this *change A* will be negative, which means that the total velocity of *A* will decrease as the point *A* comes close and close to x_1^*. Sometimes, in the far future, the point *A* will reach the stable state x_1^*. As it can be noticed, the change of velocity of

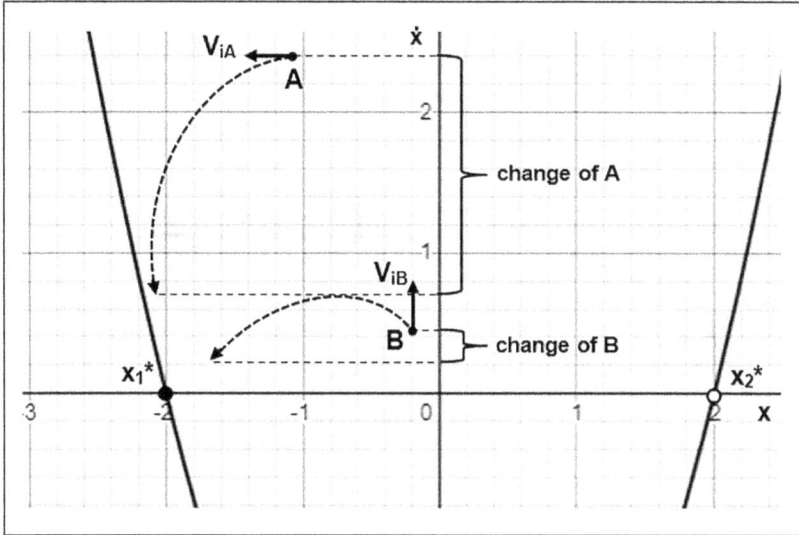

FIGURE 7.5 Change of velocity based on different positions.

A is very big. It starts with high velocity which gradually decrease, so it could be an explosion which happened in *A* and after that it settles in $x_1^* = -2$ (rest).

Point *B* has initial velocity v_{iB} directed up. We assume that v_{iB} is small. If it is big enough, the point *B* will go up to infinity.

This point is repelled from x_2^* (unstable fixed point!) and at the same time, it is attracted by x_1^* (stable fixed point!). So, its velocity (which is small) will be a combination of its initial velocity v_{iB}, attraction from x_1^*, and repelling from x_2^*. So, at the beginning, the velocity will increase and after that it will gradually decrease. In the far future, it will reach state x_1^*. This is a more probable situation which happens to *B*. The *change B* is not big and it is reversable: The velocity increases and later, it decreases.

This example could be valid for every point in the phase space which are close to the fixed points. I have assumed initial velocities v_{iA} directed left and v_{iB} as small and directed up, but any other choice will result in similar combination of resultant velocities.

You need to understand that the points which are far away from fixed points will not necessarily be attracted or repelled. So, in each case, there is a particular area around fixed points which is known as "basin of attraction" or "basin of repelling", depending on the nature of the fixed point.

There is another important rule: Two stable or two unstable points cannot be neighbors on the curve where three or more fixed points exist. There must be an unstable fixed point between stable fixed points or a stable fixed point between unstable fixed points. Another option is to have a stable or an unstable point with hybrid fixed point in vicinity.

Here you should note that only states close to the fixed points will behave like I explained above. You may not expect that stable fixed point $x = -2$ will attract states which are in $x \ll -2$ or $x \gg -2$ (Figure 7.6).

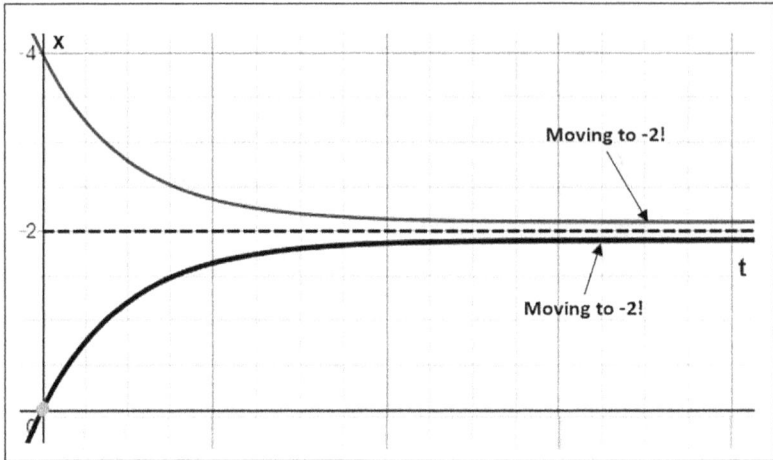

FIGURE 7.6   Change of the position with time for the particles close to fixed point $x^* = -2$.

On the left side of Figure 7.6, you can see the movements in time of the particles (states) close to $x^* = -2$ are with bigger velocity than on the right side of the graphic. An important thing to notice in Figure 7.6 is that x is presented on y-axis and the time t is present on x-axis, but there are no numbers on x-axis. The point is that we know that situations explained by the diagram will happen, but we do not know when. So, this is a qualitative approach and not a quantitative approach for dealing with non-linear dynamical systems.

Another thing to emphasize here is that, in Figure 7.6, the particles will asymptotically travel (converge) toward the value $x^* = -2$ in time, but it will never reach it (or to be mathematically correct, they will reach it in $t = \infty$).

What is the meaning of this example from the point of view of the safety?

If we assume that we have the operation described by the equation from the example above, it means that designers of operation (in this case the operation is presented by x_1^*) must strive to limit the operation to the values x of the system which are smaller than 2. In such a case, the operation itself, although the influence will cause the change, in the future will end into stable operation determined by $x_1^* = -2$.

If the system enters states for values of x bigger than 2, the system could lose control, because the operation (x_2^*) of the system (in time, in the future) will move to infinity: The operation will be out of control and as such it will produce incident or accident.

As it can be noticed from Figure 7.6, the graphical solution to the non-linear differential equation does not exactly provide quantitative solution, but it provides tool to understand how the situation changes depending on the position of the chosen state. Actually, we are looking on qualitative changes of the velocity and changes in trajectory of the state. I said "qualitatively" because we are not solving equation analytically, so the exact values of velocity expressed as numbers are not known.

Having in mind that solutions for complex non-linear equations cannot be found, this method, applied to such equations, will help us only to understand possible changes in states of the system.

Let's see how we can find a time dynamic in the phase-plane analysis. As you have already noticed, in all graphs for phase-plane, the time is not present, so we can only assume how it can be. So, finding time can be done in two ways.

The first one is by using the following equation:

$$\frac{dx}{dt} = \dot{x} = \frac{\Delta x}{\Delta t} \Rightarrow \Delta t \approx \frac{\Delta x}{\dot{x}}$$

The sign "approximately" applies because this formula is only for small periods of time due to non-linearity which is present. The next way is through the following equation:

$$\frac{dx}{dt} = \dot{x} \Rightarrow dt = \frac{dx}{\dot{x}} \Rightarrow t - t_0 = \int_{x_0}^{x} \left(\frac{1}{x}\right) dx$$

Here, state x_0 corresponds to t_0 (initial state) and x corresponds to t (final state). Graphically, this is a solution presented in Figure 7.6.

7.4.3 Non-linear Dynamical System Described by Difference Equations

To analyze the non-linear dynamical system described by simple difference equation, I will use the similar graphical method, but in this case the method will apply to the following equation:

$$f(x) = 3 + \ln(x)$$

In this case, I will use iterative process[16] which will be expressed by the following equation:

$$f(x_n) = x_{n+1}$$

It means that future state of the system will be determined when I apply function of change $f(x)$ to the previous state x_n to get the value of next state x_{n+1}. Analytically, I can do calculations step by step and results can be presented by table where all solutions are filled,[17] but for presentation, I will use graphical method. The point is that drawing a graphical presentation will be quite different from that in the previous chapter. For drawing, I will use combination of two functions. The first one is given by the equation above and second function will be as follows:

$$g(x) = x$$

Iterative process by 16 steps for three different initial vales of x is given in Table 7.1. The bolded values show asymptotic attraction of all iteration for different values of

---

[16] Be careful! Using differential equations means the system is continuous and using iterative maps for difference equations means the system is discrete. Discrete systems are presented by maps, not by diagrams! Anyways, I am using the same graphical method here to provide continuity with the previous paragraph, so the diagrams in Figures 7.7–7.9 are built using Table 7.1.

[17] As it is done in Table 6.1 in Section 6.2 in this book.

TABLE 7.1

Iterative Process by Steps for Three Different Initial Values of x

Steps		$f(x) = 3 + \ln(x)$	
(n)	$x_{01} = 0.5$	$x_{02} = 3$	$x_{03} = 6$
1	0,5	3	6
2	2,30685282	4,0986123	4,791759469
3	3,83588418	4,4106485	4,566897665
4	4,34439996	4,4840217	4,518834127
5	4,46888765	4,5005203	4,508254024
6	4,49713953	4,5041930	4,505909944
7	4,50344153	4,5050087	4,505389856
8	4,50484189	4,5051898	4,505274426
9	4,50515279	4,5052300	4,505248805
10	4,50522181	4,5052390	4,505243118
11	4,50523713	4,5052409	4,505241856
12	4,50524053	4,5052414	4,505241576
13	4,50524128	4,5052415	4,505241514
14	4,50524145	**4,5052415**	**4,50524150**
15	**4,50524149**	**4,5052415**	**4,50524150**
16	**4,50524149**	**4,5052415**	**4,50524150**
...	...	...	...

x (0.5, 3, and 6) to $x_2^* \approx 4.5$ which is a stable fixed point. Asymptotically, all movements of different states in time $t = \infty$ will finish in the stable state of x_2^*.

It is important to mention here that using difference equations, we put the time into the equations, although it is not explicitly seen there. It is self-understanding that transformation of the system described by difference equation will need time to go from one state to another state. This is a more realistic presentation compared to those where differential equations are used to describe the non-linear processes.

Values from the table are given just to show the analytics, but the real movement can be found by a diagram. I will choose one initial value for x and it will be used to calculate $f(x)$. After that I will just project the point from curve $f(x)$ to curve $g(x)$ and this will satisfy the iteration map.

Figures 7.7–7.9 show how it is done (respectively) for three initial values of x: $x_{01} = 0.5$, $x_{02} = 3$, and $x_{03} = 6$, respectively.

Following the arrows in Figures 7.7–7.9, you may notice that despite the initial conditions (x_{01}, x_{02}, and x_{03}) being different, the iteration is going from left to the right point where $f(x) = g(x)$. The reason for this is that for iteration maps the fixed points can be found by a different rule, i.e., when $f(x) = g(x)$. It means that solutions to the following equation

$$f(x) = g(x) \leftrightarrow 3 + \ln(x) = x$$

will give me the fixed points for function $f(x) = 3 + \ln(x)$. Analytically, solutions are given as $x_1^* = 0$ and $x_2^* \approx 4.5$.

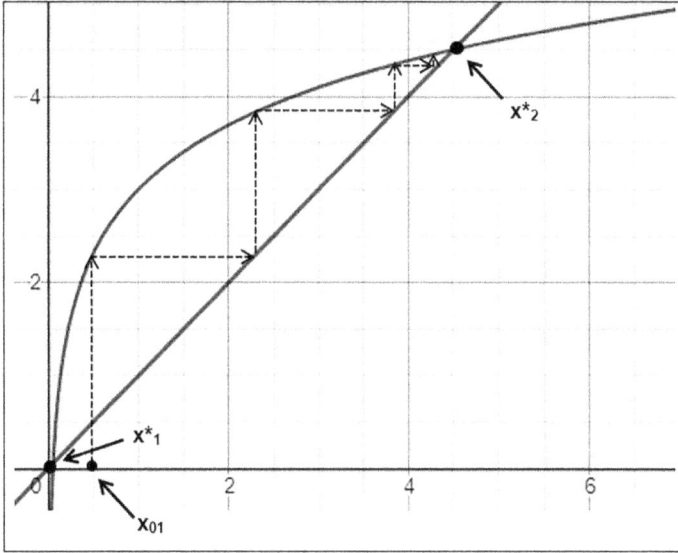

FIGURE 7.7   Iterative map for function $f(x) = 3 + \ln(x)$ for $x_{01} = 0.5$.

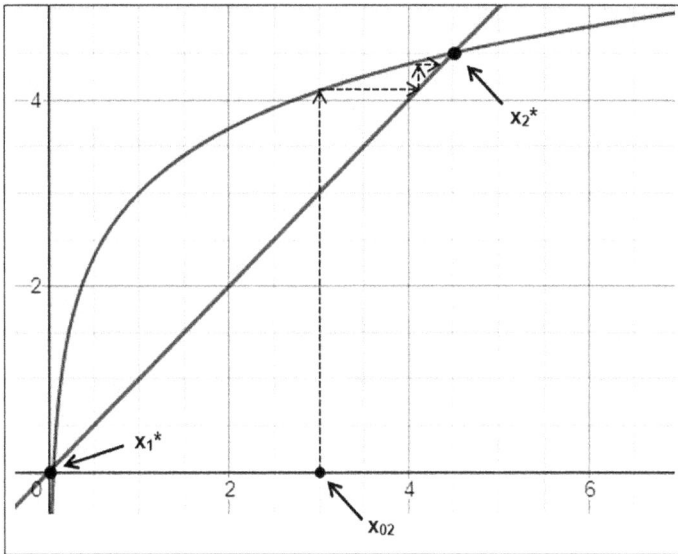

FIGURE 7.8   Iterative map for function $f(x) = 3 + \ln(x)$ for $x_{02} = 3$.

Let's be more detailed with explanation of this situation . . .

In Figures 7.7–7.9, you can see that all initial points which are in the interval between $x_1{}^*$ and $x_2{}^*$ are moving opposite from $x_1{}^*$ and in the direction of $x_2{}^*$. It means that $x^*{}_1$ is an unstable fixed point and $x^*{}_2$ is a stable fixed point.

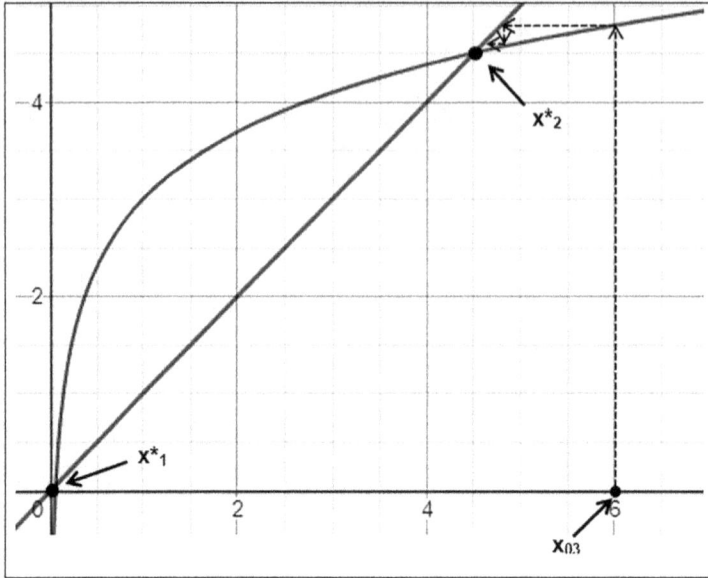

FIGURE 7.9 Iterative map for function $f(x) = 3 + \ln(x)$ for $x_{03} = 6$.

For difference equations (iterative maps), there are other rules for stability of fixed points:

$$\frac{df\left(x^{*}\right)}{dx} > 1 \quad \text{The fixed point is unstable.}$$

$$\frac{df\left(x^{*}\right)}{dx} < 1 \quad \text{The fixed point is stable.}$$

$$\frac{df\left(x^{*}\right)}{dx} = 1 \quad \text{We do not know what is the stability of this point.}$$

So, if I find derivative of $f(x)$,

$$\frac{df\left(x\right)}{dx} = \frac{d(3 + \ln\left(x\right))}{dx} = \frac{1}{x}$$

and I put the values of the fixed points $x_1^* = 0$ and $x_2^* \approx 4.5$, I will get the following values for derivatives:

$$\frac{df\left(x_1^*\right)}{dx} = \infty$$

$$\frac{df\left(x_2^*\right)}{dx} \approx \frac{1}{4.5} \approx 0.22$$

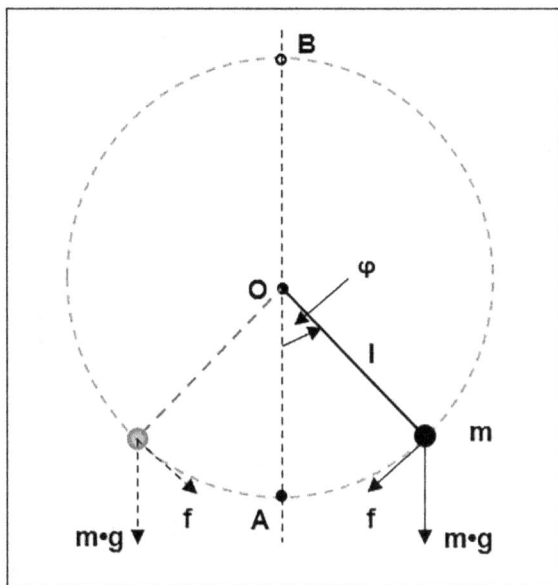

FIGURE 7.10   Pendulum.

You can notice that solution for $df(x_1^*)/dx$ has no sense, so it can be associated with unstable fixed point. From another side, the second solution $df(x_2^*)/dx = 0.22$ is smaller than 1 and it means this fixed point is stable.

Let's repeat again the rule about the neighboring fixed points which applies also for non-linear systems descried by difference equations: Neighboring fixed points must always be different. If one is stable, the next one must be unstable, and vice versa. So, the series of few fixed points (more than two) should alternate their stability (stable, unstable, stable, unstable, etc.). Two neighboring fixed points with same stability may not exist!

7.5   PENDULUM AS SIMPLE EXAMPLE OF STABLE AND UNSTABLE FIXED POINTS

Let's use simple machine called *pendulum* to explain the terms of stable and unstable fixed points. I will use Figure 7.10 to explain the pendulum behavior.[18]

Figure 7.10 shows a simple pendulum.

A bead with mass m is connected to point O by strong non-bending wire with length l and the mass m is displaced (by some other force, maybe a human intervention) in angle φ from vertical imaginary line.

---

[18] As it can be seen in this section, the pendulum is mathematically expressed by two differential equations, which means it is a two-dimensional system. You can find more about two-dimensional systems in the next sections in this chapter.

Under force of gravity **m·g**, pendulum will move toward point *A* as a result of force *f* = **m·g·sin** φ. Reaching point *A*, by inertia, the pendulum will continue its movement until force of inertia does not equalize gravitational force. At this point, pendulum will start to move in opposite direction and it will repeat its movement called *oscillation*.

The equation which describes the movement of the pendulum can be written as follows:

$$m \cdot l \cdot \ddot{\varphi} = -m \cdot g \cdot \sin\varphi - k \cdot l \cdot \dot{\varphi}$$

Or, after few transformations and introducing new variable $\dot{x}_1 = x_2$, mathematically, pendulum can be presented as a system of two non-linear differential equations:

$$\dot{x}_1 = x_2$$

$$\dot{x}_2 = -\frac{g}{l} \cdot \sin x_1 - \frac{k}{m} \cdot x_2$$

I will not use mathematics for this example, but I will use a common sense.

As it can be noticed, there are two marked points in Figure 7.10, point *A* and point *B*. Whatever angle φ the pendulum is deflected, after a particular time oscillating around point *A*, it will stop (due to friction) at point *A*.

If we put the pendulum at point *B*, smallest deviations of its position (due to any reason) will make the pendulum to move left or right and, after a particular number of (damped) oscillations around point *A*, it will finish again at point *A*. It is known in science as "criticality". Some of the systems are called "critical" if its state changes dramatically under the effect of some small disturbances. More details about this will be presented in Section 7.12.

So, *B* is an unstable fixed point and *A* is a stable fixed point. It means that whatever be the position of pendulum in the vicinity of point *B*, the pendulum will move away from point *B* (unstable) and whatever be the next position of the pendulum, it will move toward point *A* (stable).

If you disagree, you my use the system of non-linear equation from above and you may try to calculate fixed points yourself.

This is a good place to remind yourself of Section 4.1 and Figure 4.1. As you can notice there, with pendulum only indifferent situation is missing.

7.6 BIFURCATIONS

Physically, the bifurcation is actually a sudden change of the state of the system. Mathematically, it means that stable fixed point(s) is(are) destroyed and/or new one(s) is(are) created. From engineering point of view, bifurcation (if not intentionally imposed) means that equipment failed or at least, it is outside of tolerance limit, so the product (service) is not anymore as expected. From the point of view of Safety, bifurcation could be a sudden change of operation when incident or accident can happen. Actually, bifurcation does not determine whether it will be an accident or an incident, because they are actually based on consequences of sudden change in the operation when adverse event happens.

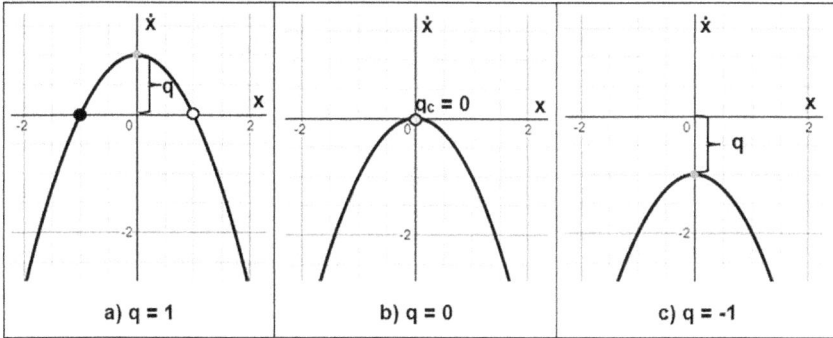

FIGURE 7.11   Situation with function $\dot{x} = q - x^2$ for different values of parameter q.

Three different situations are shown in Figure 7.11 with the parametric function $\dot{x} = q - x^2$ due to change of the parameter q.

On the y-axis is presented $\dot{x}$ and on the x-axis is presented x. So, let's clarify one more time: x is a variable in our system (factor used to control the system) and q is a parameter (influence of environment or something else, which we need to keep steady not to damage our operation).

In the cases of bifurcation, the parameter can be internal (part of operation) or external (part of the environment or human operator). Simple example for external parameter is rain during driving a car. Parameter "braking" in driving a car is drastically changed, so to stop bifurcation (sliding of the car) during braking, driver must adjust the speed of the car (which is variable in the process of driving of the car).

As you can notice in Figure 7.11 (based on the explanations in previous sections), the case (a) have one stable fixed point in $x = -1$ and one unstable fixed point in $x = 1$, the case (b) has one "hybrid" fixed point (gray fill, from left unstable and from right stable) in $x = 0$, and the case (c) has no fixed points.

So, this kind of situation, where the operation experiences sudden (undesired) change of parameters, expressed by the words of non-linear dynamics, is called *bifurcation*. This situation is associated with the process of creation or destruction of the fixed points or their transformation from one type to another. It means that the bifurcation can affect the stability of fixed points. Value of the parameter q when bifurcation starts is called *critical value of parameter q* (marked as q_c) and the parameter q (which change is cause for bifurcation) is called *bifurcation parameter*.

From the example above (Figure 7.11), you can notice that bifurcation spontaneously (even without our control) can

- produce a new fixed point(s); or
- merge the existing points, producing hybrid fixed points; or
- destroy the already existing fixed points; or
- change the stability of existing points.

Example presented in Figure 7.11 is the so-called *Saddle–Node bifurcation*. There are plenty other types of bifurcations, but I will mention here only two more: *Transcritical* and *Pitchfork bifurcations*.

The bifurcations happen when two conditions are satisfied:

1 The parameter change (presented as constant or function on the diagram) and function $\dot{x}$ have one common point. This common point is tangent on $f(x)$ made by line presenting the parameter.
2 At this point of tangent, the derivative of the parameter is the same as the derivative of function.

Let's see it in one "complicated" case . . .

I have explained emergence of bifurcations in Figure 7.11 with the parameter q which actually add or subtract each value of function $f(x) = \dot{x} = x^2$. But let's say that the parameter q changes as addition to variable of x following the rule $g(x) = q + x$. In such a case, Figure 7.11 will transfer into Figure 7.12.

Look at Figure 7.12.

The similar case of saddle–node bifurcation is presented, but the point of importance and cause for the dynamics of change are different!

In Figure 7.12a, there are no fixed points dependable on parameter q (no intersection between the line q and the parabola $-x^2$). In Figure 7.12b, the function of dependence of line q touches at one point $\dot{x}$ and at this point the bifurcation starts. In Figure 7.12c, the function of dependence of q intersects at two points $\dot{x} = -x^2$ and these two points are fixed points.

In Figure 7.12b, the fixed point is marked as black dot and for this value, $\dot{x}$ must satisfy the rules given in items (1) and (2) above:

$$f\left(x^*\right) = g\left(x^*\right) \qquad \text{and} \qquad \frac{df\left(x^*\right)}{dx} = \frac{dg\left(x^*\right)}{dx}$$

Actually, the bifurcation from Figure 7.11 is caused by the (controlled or uncontrolled) change of the variable x and the bifurcation from Figure 7.12 is caused by

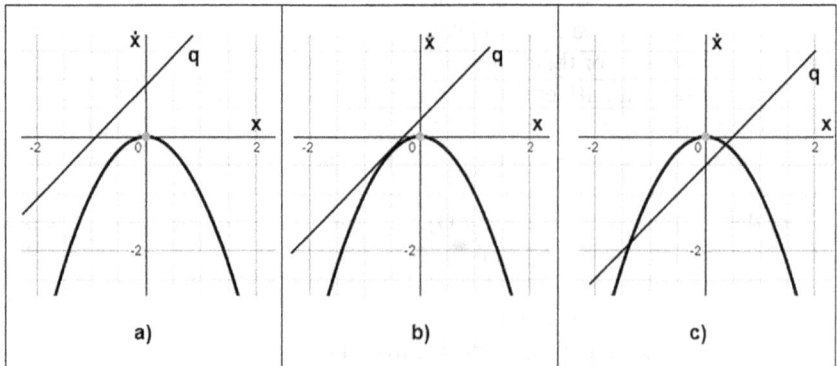

FIGURE 7.12   Change of parameter q and function of x and $f(x)$.

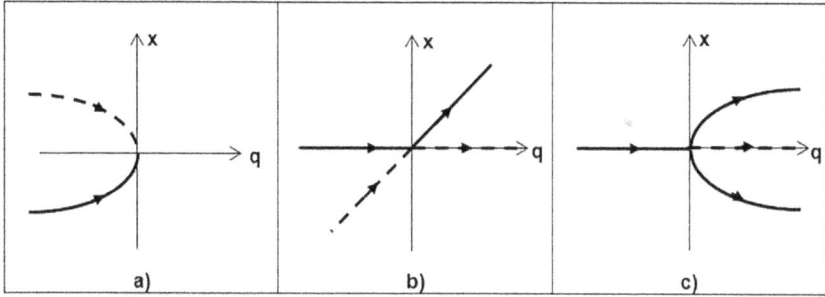

FIGURE 7.13   (a) Saddle–node bifurcation. (b) Transcritical bifurcation. (c) Pitchfork bifurcation.

(controlled or uncontrolled) change of the parameter q (which could present a change of the external influence factor(s) which are indirectly part of the system).

Figure 7.13 presents graphs of the functions of these three types of bifurcations depending on the change of parameter q. Full lines mark the moving of stable fixed points of variable x when the parameter q changes and dashed lines mark the movement of unstable fixed points. You can notice that the names[19] are given by the shape of the curves how x changes when q changes.

Saddle–node bifurcation[20] (Figure 7.13a) is common for the functions which are similar to function presented by Figure 7.11. For negative values of q, we have two fixed points (one stable and one unstable) and their change is presented by horizontal parabola (full and dashed lines!). As the q is approaching 0, the fixed points approach each other and in $q = 0$, they merge into a "hybrid" fixed point. For positive values of q, this "hybrid" point disappears,[21] so there are no fixed points at all.

In Figure 7.13b presents an example with two fixed points (one horizontal and another with 45° inclination to q-axis). Horizontal stable point moves from $q \ll 0$ to $q < 0$ and in $q = 0$, it changes its stability from stable to unstable. The same thing happens to "inclined by 45°" fixed point. At this point of bifurcation ($q_c = 0$), the fixed points do not disappear. Instead, their stability transit from stable to unstable (and vice versa) and that is the reason for the name. It is valid for functions which have combination of x plus (or minus) some number to power of x. Logical map for animal population also belongs to this kind of bifurcations.

If you are skeptical about the impact of bifurcations on safety of equipment in industry, then let me provide you with example regarding one of the biggest mysteries in few aircraft incidents and accidents which happened in the 1990s in the United

---

[19] Please note that in the literature, same type of bifurcations have different names. I am using the names which Stephen Strogatz used in his book *Non-linear Dynamics and Chaos* which is one of the best books on this problematic (by my humble opinion!).

[20] To be exactly correct, this type of bifurcation exists in non-linear dynamical systems with two variables and parameters, but just pure simplification is presented in the diagram. The name is coming from the surface which looks like horse saddle.

[21] Also, there are examples where the opposite could happen: For small q, there is no fixed point and later (as q increase) there is one which (as q further increase!) will split (bifurcate!) into two fixed points.

States. In this decade, a few Boeing 737 aircraft have experienced sudden rudder[22] movement in unacceptable position during their flights and in few of them, pilots did not recover the control of the aircraft finishing with crash. This happened to United Flight 585 (1991), USAir Flight 427 (1994) and Eastwind Flight 517 (1996). The first two finished with crash, killing 157 persons in total. Regarding the third one, the pilots were successful to gain control again. After the first two crashes, NTSB has done a lot of investigations without success. The third flight gave more details to the NTSB investigators, having in mind that the aircraft did not crash. After almost two years and millions of dollars' investigation, NTSB investigators realized first that in some situations there is possibility of rudder reversal. Rudder reversal means the pilot presses pedal of the rudder for left turn, but the rudder provides right turn. These situations could be at extreme cold temperatures (which could happen around height of 10,000 meters during the flights) or in situation when the push of the rudder pedal by the pilot is not so strong (then hydraulics responds with more strength in opposite position).

From the point of Transcritical bifurcations, this is actually what happened to these three aircraft: The parameters change caused the change of stable fixed point of non-linear dynamical system (rudder) into unstable fixed point.

Last example in Figure 7.13c is for Pitchfork[23] bifurcation. This bifurcation happens to the systems which can be described by symmetrical function. We can see that for $q < 0$, there is only one stable fixed point. But in $q \approx 0$, these points bifurcate into three points—two stable and one unstable. Operations (processes) expressed by cubic function are good example for this type of bifurcation.

In Figure 7.14 is given cubic function $\dot{x} = x^3 + qx$ in association with Pitchfork bifurcation which should clarify development of this bifurcation depending on the change of parameter q.

Don't be confused by the look of Pitchfork bifurcation in Figure 7.14 compared to the bifurcation in Figure 7.13c!

These two bifurcations are for different differential equations: The one in Figure 7.14 is for function $\dot{x} = x^3 + qx$ and the one in Figure 7.13c is for function $\dot{x} = qx - x^3$!

As it can be noticed from Figure 7.14, for $q = 1$, there is only one unstable fixed point; for $q = 0$, there is undefined mixed fixed point; and for $q = -1$, there are three fixed points (one stable and two unstable).

In engineering, for more-than-one-dimensional systems, these transformations of fixed points from stable to unstable (or to hybrid point) and vice versa happen continuously[24] and they can produce problems in the functioning of the Equipment. Sometimes, change of stable fixed point into unstable could also produce oscillations (only for two-dimensional systems) and this is known as "mild" loss of stability. In

---

[22] The rudder is part of a vertical stabilizer of an aircraft which provides lateral (right and left) movement of the aircraft.

[23] There are two types of Pitchfork bifurcations: Supercritical and Subcritical. Bifurcation example in this section is Supercritical. For Subcritical bifurcation, the same diagram is valid, but only stable situations are changed by the unstable situations, and vice versa.

[24] This loss of stability can also happen suddenly (abruptly) and it will be considered in the section dealing with Catastrophe theory.

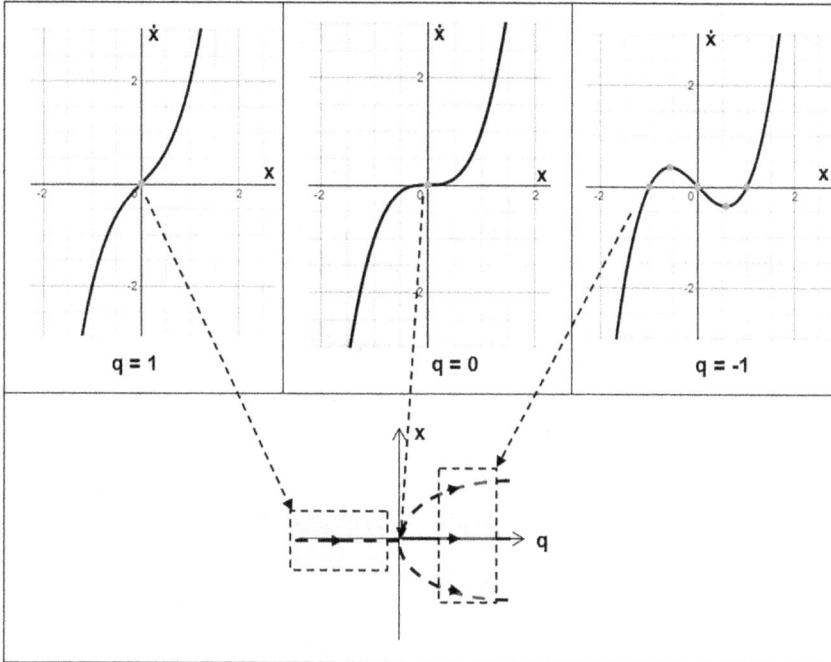

FIGURE 7.14 Example of Pitchfork bifurcation for different values of q in the system described by formula $\dot{x} = x^3 + qx$.

such cases, the stability of fixed points is transferred to the oscillations (the oscillations are usually stable) and the fixed point itself becomes unstable.

A simple example of what can happen with bifurcations in the real world is the subsonic and supersonic dynamics during the flight of a high-performance aircraft. These two modes of flight are radically different and as such they need different building strategies. The aircraft which is stable in subsonic flight will obviously be unstable in supersonic flight and vice versa.

The case where the unstable oscillations could transfer themselves into unstable fixed points is known as "hard" loss of stability and it can produce the so-called *attractors*. This is the area which will be considered later as Chaos. In producing attractors, the system changes the behavior (dynamics) into a complex movement which "attracts" neighboring conditions. Eventually, the attractors will settle in one place which would be a stable fixed point (equilibrium state). The attractors which are not settled in equilibrium states or have slightly aperiodic movement are known as *strange attractors*.

The movement of the fluid through pipe can produce strange attractor. On particular increasing of the speed of the fluid, the friction with the walls of the pipe will produce suddenly a turbulent movement which can be described as chaotic with emergence of strange attractor.

In general, the strange attractor is actually a type of bifurcation which happens when a fixed point changes into an unstable periodic orbit. This orbit is bounded and it continuously moves over a bounded region in the phase space. However, it will never transfer itself into a periodic limit cycle and will never stop at a fixed point (if there is any). This aperiodic movement (due to its boundness) is globally stable, but locally unstable. It means that the attractor contains (locally) plenty of aperiodic orbits which differ from each other, but they are very close to each other. Due to this local instability of the orbits, their frequencies are different and it results in movement which is similar to the fractals in space.

To be clearer, these are strange attractors in time and there are also strange attractors in space which are known as fractals (later in the book they will be explained).

7.7 NON-LINEAR DYNAMICS OF TWO-DIMENSIONAL SYSTEMS

What were mentioned in Sections 7.4–7.6 was a graphical way of dealing with non-linear differential and difference equations with one variable. These one-dimensional non-linear dynamic (differential and difference) equations actually present non-linear change of the state of the system, or better say, the velocity of change of the system (operation, process, activity, etc.) by the change of one or more variables and parameters.

Difference equations, nevertheless, can be expressed by continuous equations, they are discrete by nature. These equations actually present changes of different states in the system (process, operation, activity, etc.) which happen by "jumping" from one state to another and they will be treated by graphs and they are known as "mappings". I will not deal with them here, so for the time being, I will stay with differential equations.

The differential equations considered in Sections 7.4.1 and 7.6 can be expressed by the general formula:

$$\frac{dx}{dt} = \dot{x} = f(x)$$

As it can also be noticed, they can be presented by line in the diagram where ordinate[25] (y-axis) is $\dot{x}$ (graphical solution) and abscissa (x-axis) is x (variable of interest). Nevertheless, there are two coordinates, the graph is line which means the system is one-dimensional because it presents change of the system with the change of one variable.

In reality, we have systems with thousands of variables, which means that if the number of variables is n, the graphical presentation will be n-dimensional. It means that change of the state of the system could be explained for each n-variable by n diagrams in two-dimensions or it can be presented by one diagram with $n + 1$ dimension (n for variables and one for system state). Having in mind that for four variables, graphical presentation will be five-dimensional, it is really hard to draw this graphic. That is the reason why in the area of two-dimensions, the so-called *phase-plane analysis* is used.

---

[25] For non-mathematicians, in two-dimensional presentation (diagram!), the vertical axis (y-axis) is called *ordinate* and the horizontal axis (x-axis) is called *abscissa*.

You must have noticed that in the one-dimensional system, the graphic presentation of the changes of the states of the system were presented by line, but in two-
dimensional system they will be presented by surface.

To extend my explanation, I will mention here the two-dimensional system of non-
linear dynamical equations which can be presented by the general formula:

$$\frac{dx_1}{dt} = \dot{x}_1 = f(x_1, x_2)$$

$$\frac{dx_2}{dt} = \dot{x}_2 = g(x_1, x_2)$$

The graphical method for presentation in such cases is producing surface called
phase-plane. The phase-plane is actually the track where the point describing behavior of the states of the system moves. Any point of this surface can be a state of the
system which continuously changes with a particular velocity depending on the equations which describe the non-linear dynamical system.

The point with the phase-planes is that they may have different shapes; for example, they can be plane, cylinder, sphere, saddle, torus, or any other surface which
can be described by two variables. So, whatever be the phase-plane diagram you see
regarding non-linear systems in two-dimensions, it is just simplification of the reality
which is actually three-dimensional.

The simplification projection of the phase-plane in two-dimensional coordinates
can be seen in Figure 7.15.

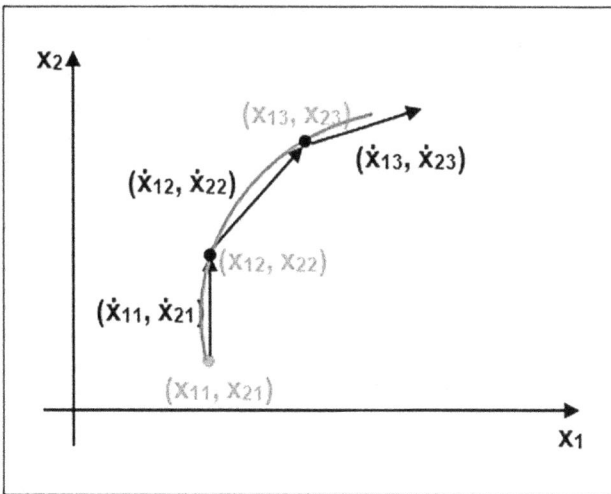

FIGURE 7.15 State of the system, trajectory of the system, and its vectors of velocity in
each state.[26]

---

[26] Nevertheless, this section is dedicated to readers good in mathematics, this graph is explaining states
of the systems for non-mathematicians. The good mathematicians will notice that the velocity vector should be tangential to the trajectory, but if I draw it correctly, it will (maybe!) confuse non-
mathematicians who have decided to read this section.

The explanation of the figure is as follows: If we have state of our system depicted with (x_{11}, x_{21}), then at this point, the system state is associated by the vector $(\dot{x}_{12}, \dot{x}_{21})$ which is actually the velocity of change (the arrow) of this state expressed as vector. This vector will bring our state into state (x_{12}, x_{22}) (gray dot) where the vector of velocity (the arrow) will be $(\dot{x}_{12} \dot{x}_{22})$. In this state, there will be another vector which will bring the system into the state (x_{13}, x_{23}) and so on.

Actually, the solution of this system with two variables (two-dimensional system) will produce the trajectory (gray line) of the changes of the system.

But, be careful!

Changes of the systems in the phase-plane are presented as changes of states, not changes in time!

Looking at Figure 7.15, we can say that the state of the system will move from point (x_{11}, x_{21}) to the point (x_{12}, x_{22}), but we do not know when it will happen; because on the phase-plane, the time is not shown.

I am using word "velocity", but this is not the velocity of movement of the bodies in physics. It is actually the speed of change of the states in the system. The system can be Complex System or complex activity (production process, population in the state, activity, etc.).

There should be few trajectories in the system which depend on the parameters, but whatever the parameters and trajectories may be, the trajectories will never cross each other! This is the most important thing to remember about trajectories in more-than-one-dimensional systems. The reason is that the point where these trajectories cross each other is the state of the system. Being in this state of the system where two trajectories cross each other means that there will be one state of the system, but it will be defined by two different trajectories coming or going away from this state. Of course, both of them cannot be good for our system. Mathematically, it will destroy uniqueness of our system. If such a thing happens, something is considerably wrong with such a system or something was pretty much neglected during design of the system (design flaw).

But the trajectories may approach the same point, if this is a stable fixed point. From the point of view of engineering, it means that there are few ways (trajectories) to achieve a particular state and the system will stay in this state (stable fixed point).

From the point of Safety, it means that there are few ways to provide safety to your system and all of them are good, if they finish in stable fixed points. But be careful: There could be always many stable points and not each of them is always desirable, so a designer should always take care while designing a system to provide different trajectories which will finish at desired stable fixed points called Working Point.

7.8   CALCULATION OF FIXED POINTS IN TWO-DIMENSIONAL SYSTEMS

For the two-dimensional system, the definition of fixed "point" will be given in the same way as for one-dimensional systems, but now we will have a system of equations:

$$\frac{dx_1}{dt} = \dot{x}_1 = f(x_1, x_2) = 0 = \dot{x}_1^*$$

$$\frac{dx_2}{dt} = \dot{x}_2 = g(x_1, x_2) = 0 = \dot{x}_2^*$$

Having a system of equations is plausible because we have two variables and parameters which need to be taken into account during consideration of the behavior of the system. Now, we will not have a line on the diagram, but it will be a surface.

So, the "fixed" points in two-dimensional systems will be the values for x_1 and x_2 where, at the same time on the surface described by equations above, $\dot{x}_1$ and $\dot{x}_2$ are equal to 0. You must have noticed above that I am using the phrase *fixed "point"*, where the word *point* is in quotation marks. The reason is that these will not be any more fixed points, but because of the two-dimensions of the system, they can be named as *fixed lines*. It means that the trajectories (states of the system) moving in this case will approach asymptotically the *lines* which will be *fixed* in the graphs. That is the reason that I will not explain everything in detail due to Complexity associated with that.

To produce graphs (diagrams) in two-dimensions for non-linear dynamical equations is a complex process, so I will use the process of linearization which will apply to the non-linear equations, but only in the limited areas around fixed "points". This linearization allows us to use Taylor series and Jacobian matrices and the behavior of the system around fixed points is found from the solutions of these Jacobian matrices in the forms of eigenvectors and eigenvalues.

The reason is that by using the linearization of non-linear equations around the fixed points, I will simply linearize the non-linearity of the system only for the fixed "points". This will not always work, but in most of the cases, these solutions will present the approximate behavior of the system only in this area (around fixed "points"). The point to remember is that they will not be correct for all other areas where the rest of non-linearity is present. Of course, that solutions will be only qualitatively correct, but not quantitatively.

I will not present here all the methods of calculation, simply because it is too complex and needs excellent knowledge of mathematics. The point is that it does not matter from applied engineering point of view, but for those who need more details, there is a bunch of literature on the Internet where you can find good explanations how the things go on. I, personally, used books *Nonlinear Dynamics and Chaos: With Applications to Physics, Biology, Chemistry and Engineering* by Steven Strogatz and *An Introduction to Dynamical Systems and Chaos* by G.C. Layek.

Dealing with those methods (linearization, Taylor series, and Jacobian matrices) for analysis, the solutions for bifurcations will actually depend on the solutions of determinant Δ of the two-dimensional system developed into Taylor series at one of the fixed points (subject of analysis!). The Taylor series ($P(x_1, x_2)$ and $Q(x_0, x_2)$) for the two-dimensional system from above will look as follows:

$$\frac{dx_1}{dt} = \dot{x}_1 = P(x_1, x_2) = P(x_{10}, x_{20}) + x_1\left(\frac{\partial P}{\partial x_1}\right)_0 + x_2\left(\frac{\partial P}{\partial x_2}\right)_0 + \frac{1}{2!}\left(x_1\frac{\partial}{\partial x_1} + x_2\frac{\partial}{\partial x_2}\right)_0 P + \cdots$$

$$\frac{dx_2}{dt} = \dot{x}_2 = Q(x_1, x_2) = Q(x_{10}, x_{20}) + x_1\left(\frac{\partial Q}{\partial x_1}\right)_0 + x_2\left(\frac{\partial Q}{\partial x_2}\right)_0 + \frac{1}{2!}\left(x_1\frac{\partial}{\partial x_1} + x_2\frac{\partial}{\partial x_2}\right)_0 Q + \cdots$$

Or, after neglecting some of the elements in the series (those with higher-than-1 order!), it can be presented as follows:

$$P(x_1,x_2)=ax_1+bx_2$$
$$Q(x_1,x_2)=cx_1+dx_2$$

where $P(x_0,y_0)=Q(x_0,y_0)=0$ (by definition, x_0 and y_0 are fixed "points"!) and only the first two elements of the Taylor series are kept. Having this in mind, we can say that coefficients a, b, c, and d are as follows:

$$a=\left(\frac{\partial P}{\partial x_1}\right)_0; \quad b=\left(\frac{\partial P}{\partial x_2}\right)_0; \quad c=\left(\frac{\partial Q}{\partial x_1}\right)_0; \quad d=\left(\frac{\partial Q}{\partial x_2}\right)_0$$

Having made these changes, the final equations will be as follows:

$$P(x,y)=\frac{dx_1}{dt}=\dot{x}_1=ax+by$$
$$Q(x,y)=\frac{dx_2}{dt}=\dot{x}_2=cx+dy$$

Or, in matrix form, it will look as follows:

$$\dot{X}=A\cdot X \Rightarrow \begin{vmatrix} \dot{x}_1 \\ \dot{x}_2 \end{vmatrix}=\begin{vmatrix} a & b \\ c & d \end{vmatrix}\cdot\begin{vmatrix} x \\ y \end{vmatrix}$$

The matrix A is known as Jacobian matrix.

Applying matrices for the equations above and using method with eigenvectors and eigenvalues, to find solutions for the system of two equations above, we need to solve the equation for eigenvalues λ:

$$\lambda^2-\tau\lambda+\Delta=0$$

Here, $A=$ (trace of the matrix of the simplified system) $=(a+d)$ and $\Delta=$ (determinant of the matrix of the simplified system) $=(ad-bc)$.

Solutions, with regard to λ, can be found by

$$\lambda_{1,2}=\frac{\tau\pm\sqrt{\tau^2-4\Delta}}{2}$$

Be careful, these solutions λ [27] are complex solutions ($\lambda=a+j\omega$)!

All possible solutions for λ will depend on expression $\tau^2-4\Delta=0$ and this is presented in Figure 7.16.

As it can be noticed, there are plenty of solutions and depending on the position in the graph in Figure 7.16, we have fixed "points" as spirals, nodes, saddles, centers, etc.

---

[27] This not a λ (Failure Rate) from Reliability paragraphs!

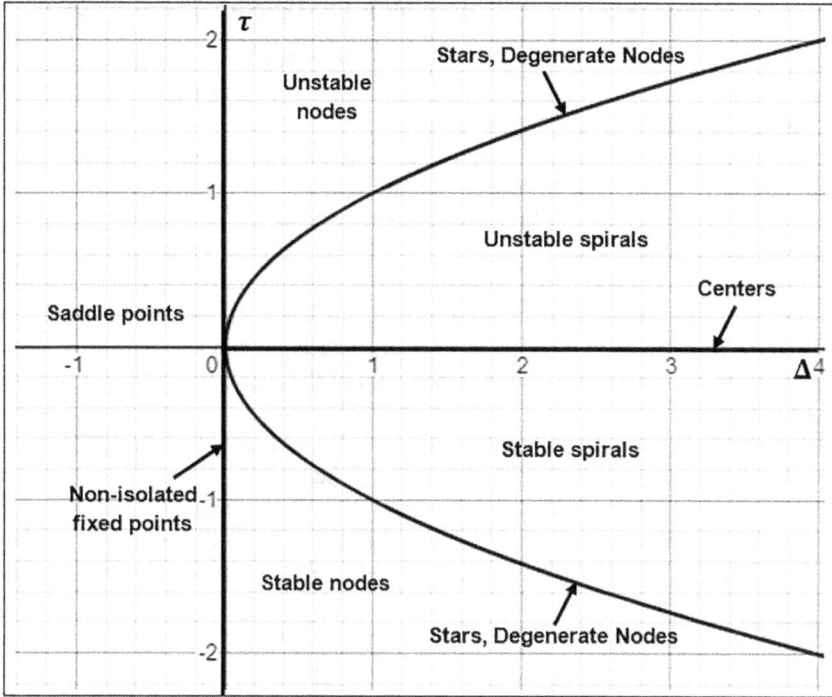

FIGURE 7.16 Different situations with bifurcations in two-dimensional system (the parabola is determined by equation $\tau^2 - 4\Delta = 0$).

There is no reason to go in details, because all the explanations are pure mathematics and this is not the intention of this book. The intention was only to show to the reader that as the number of variables and parameters increase, the Complexity of the situations dramatically increase. So, do not bother with this, just continue with the next section.

From engineering point of view, this linearization is OK, because we design the system to be stable at fixed points and we provide proofs that the design is such that it will not allow bigger deviations around these fixed points. This is something which is generally the aim of Resilience Engineering.

While speaking about stability (see Chapter 4) of the system, it is good to mention here few important things regarding the eigenvalues λ of the system calculated from the Jacobian matrix:

(a) If all eigenvalues λ have real parts less than zero ($a < 0$), then the system under consideration is stable.
(b) If at least one of the eigenvalues λ has real part greater than zero ($a > 0$), then the system under consideration is unstable.
(c) There is no other way to obtain conclusion (it could be a borderline case between stable and unstable lines).

7.9 TYPES OF FIXED POINTS IN TWO-DIMENSIONAL SYSTEMS

The fixed "points" in two-dimensional systems, as it can be noticed in Figure 7.16, have different names (saddles, nodes [normal or degenerated], spirals, Centers, etc.). Regarding their stability, all of them could be stable, unstable, and/or hybrid. Stability is determined from the trajectories in the neighborhood of fixed "points", but analytically, it can be determined from the values of eigenvalues.

Some of the trajectories in two-dimensional systems for particular fixed "point" are described in Figure 7.17.

I would like to mention here Centers (as particular type of two-dimensional fixed "points") to rectify the use of quotations marks for fixed "points". Centers are actually lines of fixed points, so, as you can imagine, each point of the line is fixed point by itself. They are called *Centers*, because each of these points could be a center of closed trajectory.

Figure 7.17 shows just few basic fixed "points" which are obtained by linearization. In non-linear systems, the fixed points will be similar to these fixed "points". Usually, the systems have more than one fixed "point", especially systems which depend on many parameters, so there is a possibility to have different fixed "points" in one system and combinations of these fixed "points" makes the analysis pretty much complex.

To get a real picture of the situation with fixed "points" and trajectories in two-dimensional system, the three-dimensional graph of the phase-plane of Saddle is presented in Figure 7.18 as a clarification of the Saddle fixed "point" in Figure 7.17.

Looking at the three-dimensional graphic situation in Figure 7.18, you can understand the two-dimensional behavior of Saddle in Figure 7.17.

But be careful: Figure 7.18 is just a three-dimensional presentation of two-dimensional system from Figure 7.17. It means that the arrows on Saddle in Figure 7.17 actually lie on the Saddle surface. All movements (trajectories) of the state of the system presented by the surface are limited to the movement only on the surface; there is no movement outside the Saddle surface.

From the viewpoint of effect on engineering of these fixed "points", I will give a simple example of spiral fixed points in the form of speed of the cars when they drive over spiral roads which exist somewhere.

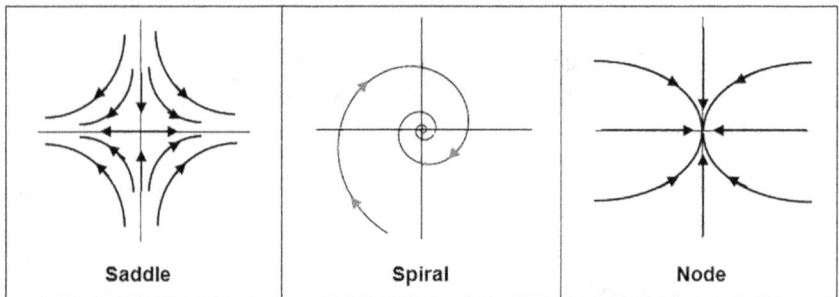

FIGURE 7.17 Few types of fixed "points" for two-dimensional system.

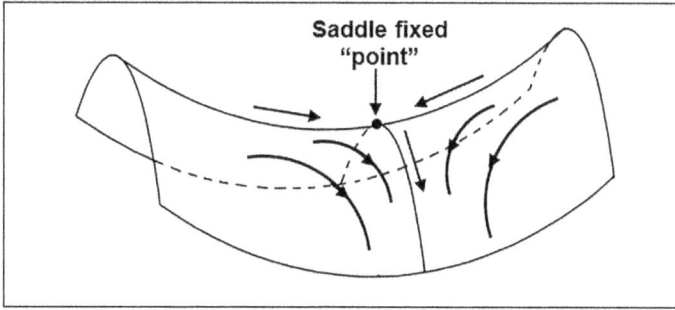

FIGURE 7.18   Saddle in three-dimensional system.

There is one example on the junction A4 of E-75 highway in North Macedonia which goes from Skopje city toward Skopje International Airport. It is a small traffic junction, but it is spiral in shape which is an issue for the drivers. The spiral shape of junction is presented as continually collapsing spiral which requests to decrease the speed of car to 30 km/h when approaching. Having in mind that the speed limit at the E-75 highway is 120 km/h and (although there is pay toll there), it is a real mental problem for drivers to decrease the speed from 120 to 30 km/h. This problem has caused a lot of "flight-outs" from the road of the drivers who overestimated themselves.

This is something which is already described by spiral in two-dimensional non-linear systems: As spiral is approaching the stable fixed point, the velocity of particle is decreasing as the curvature of spiral is increasing. This fact was realized by Traffic Authorities in North Macedonia and they put speed limit to 30 km/h on this sequence of the highway.

OK, let's be honest: I am not sure that they have investigated mathematically the spiral road and the consequences if drivers do not decrease the speed or they just realized that car "flight outs" in this location were too many. Probably, counting the car "flight outs", the sign for obligatory 30 km/h speed in this sequence was put as preventive action.

Maybe this simple example in this place can explain why I wrote this book: "Prevention" is better than "Correction".

There is also another very important characteristic of two-dimensional system regarding their change of states: The trajectories could close into their selves!

This could happen in the cases where there is fixed point inside and could happen without fixed point outside. Such a closing trajectory (known also as periodic orbit) is shown in Figure 7.19.

From the viewpoint of engineering, it means that the system will start from one state and, following the trajectory, will come again to the same state. In other words, the system will periodically move through the trajectory and this movement is called *oscillation*. This is something well-known in mechanical and electrical engineering. Some of the oscillations are good and some of oscillations are bad, so we always need to investigate them. The unwanted oscillations in the mechanical systems are called *vibrations*.

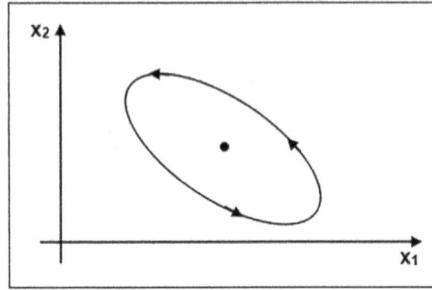

FIGURE 7.19   Closed trajectory (a limit cycle with periodic orbit) and the corresponding
center.

The closed trajectories are important due to *limit cycles*. These are isolated closed
trajectories where the neighboring trajectories are not closed. These "not closed" trajectories are mostly spirals going toward or away from the limit cycles. It means that
these trajectories (limit cycles) are limits for other trajectories: The other trajectories
may not cut limit cycle, but they may go infinitesimally close to them or infinitely far
away from them.

If they come close, they are "stable limit cycles" and if they go far away from
them, they are "unstable limit cycles". There are also "half-stable (semi-stable or
hybrid!) limit cycles" which are stable from inside and unstable from outside, or vice
versa. As you can notice, what is a fixed point for differential equations with one
variable, that is the limit cycle for differential equations with two variables. The limit
cycles could very often happen during bifurcations in two-dimensional systems of
non-linear dynamics.

The main point with the limit cycles is that the points which lay on the limit cycle
will always stay on the limit cycle, periodically moving on the trajectory defined by
the limit cycle. The points outside and inside the limit cycles could provide different
behaviors (movement, converging or diverging from the limit cycle) depending on
the type of the limit cycle (stable, unstable or hybrid).

7.10   BIFURCATIONS IN TWO-DIMENSIONAL
AND MULTIDIMENSIONAL SYSTEMS

In accordance with previously mentioned things regarding the two-dimensional
systems, the bifurcations will exist, but they will be different from those in one-
dimensional systems. They will be different, because the diagrams for changes in
one-dimensional systems are presented as lines and in two-dimensional systems,
the diagrams for changes are presented as surfaces. These specifics also change the
nature of fixed points and that is the reason that I have used them as fixed "points"
(in quotation marks).

The bifurcations in two-dimensional system can happen very often. Usually there
is bifurcation when Saddle coalesces with Node, limit cycle with Center, etc. But
I will not explain them into details here, because they are pretty much complex and

the material needed to deal with this area is beyond the scope of this book. Anyway, I will mention few bifurcations which illustrate the Complexity of bifurcations in two-dimensional systems.

Speaking about bifurcations in three- and more-dimensional systems, I can just mention that their Complexity is still a matter of consideration for the mathematicians and physicist and as such, there is no need to mention them here. Do not forget, this book is dedicated to the Safety Professionals in industry and their knowledge is far below the mathematics and physics levels needed to deal with these systems in a scientific way.[28]

So, there are many bifurcations in two-dimensional and more-dimensional systems which could happen frequently in reality and just with intention to emphasize the importance of bifurcations in "wrong-doing" of systems, I will present one example for bifurcations in two-dimensional system.

Let's consider the two-dimensional system presented by the following two formulas (which apply not to Cartesian coordinate system, but to Polar coordinate system):

$$\dot{r} = a \times r + r^3 - r^5$$

$$\phi = 1 + b \cdot r$$

This is a typical two-dimensional system[29] (very often) used to explain the bifurcations in two-dimensional non-linear systems. Variables are r (radius) and ϕ[30] (angle) and, as it can be seen, $\dot{r}$ and ϕ are independent of each other: $\dot{r}$ does not depend on ϕ and ϕ does not depend on r.

Actually, we have movement with angular velocity $\omega = \phi$ which depends on the distance r from the center (origin) of the used Polar coordinating system. a and b are parameters used to adjust the Working Point of the system. Depending on the value of a and b, system changes drastically and behave not always as we wish. That is the reason that parameters must be kept constant or allow just small (controlled!) changes of them.

One of the bifurcations (somewhere known as Saddle–Node coalescence) in this system (for $\dot{r}$ and ϕ) is shown in Figure 7.20 (graphically presented in the phase-plane) and this bifurcation is explained only for changes of the parameter a.

Please note that there are particular values of parameters a and b (critical values) and when these two parameters reach these values, system changes abruptly (bifurcations starts). This value for a on the diagram is marked as a-critical (a_c).

As it can be noticed, there is origin O and there are three different areas. O is the starting point for changes of $\dot{r}$ depending on the change of a. As it can be noticed,

---

[28] To be honest, my knowledge in this area is also below the level need to be familiar with bifurcation analysis . . .

[29] More thorough analysis of this two-dimensional system can be found in the book *Non-Linear Systems and Chaos* by Steven Strogatz (published by Perseus Books, Reading, MA).

[30] Do not get confused with the use of r and ϕ instead of x and y. r and ϕ belong to the so-called Polar coordinated system and x and y belong to the so-called Cartesian coordinated system. There is a mathematical possibility to move from one to another using the following formulas: $x = r \cdot \cos \phi$ and $y = r \cdot \sin \phi$ or $r = (x^2 + y^2)^{0.5}$ and $\phi = \text{arc } tg(y/x)$. Using Polar coordinates in this case, extremely simplify the calculations for behavior of this non-linear system. Anyways, the pictures of change of derivations of r and ϕ are presented in Cartesian coordinated system, again for simplification.

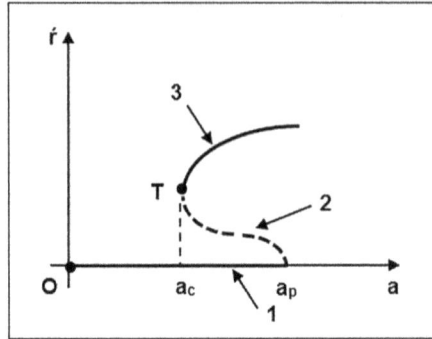

FIGURE 7.20   Bifurcation of *ṙ* depending on the values of *a*.

the system is stable in *O* for all values of $a = a_p$. This applies to area **1**. Then *a* starts to decrease ($a < a_p$), but *ṙ* increases. It can be noticed that at this point, dependence becomes unstable (dashed line in area **2**) and it goes backward (*a* decreases). This is actually a point where Hopf Subcritical bifurcation happens.

This movement continues until *a* reaches the value of a_c (point *T*). The bifurcation which happens at point *T* is known as Saddle–Node bifurcation. There instability of Saddle changes into stability of Node (now it is stable). From this point, change of *ṙ* continues by change of *a*, as a curve in area **3**.

Important is the event when the system is in area **3** and it changes its state as *a* is decreasing (going left!). When *a* reaches the value of a_c, then it "jumps" to another stability presented by a_c on the *a*-axis.

Figure 7.21 provides information on three possible situations for *ṙ* and φ̇ depending on the value of *a* with regard to a_c. All dynamics are presented by arrows.

In the first case ($a < a_c$), there are no fixed points for *ṙ* and the change of φ̇ depends on the change of *φ* in spiral.

In the second case ($a = a_c$), there is one single fixed point (*A*) which is popping up and this is the hybrid fixed point (stable from right and unstable from left) for *r*. This is actually the moment where bifurcation starts. At this point *A*, the change of φ̇ depending on the change of *φ* is divided into two areas by creating limit cycle. This limit cycle is hybrid: Inside the limit cycle, the spiral is moving all points far away from the limit cycle to the origin (stable fixed point) and outside the limit cycle, all the points are moving toward the limit cycle.

The third case ($a > a_c$) is the case where we have bifurcation formed and there are two fixed points (*B* and *C*) for *ṙ*. *B* is unstable and *C* is stable. The changes of φ̇ depending on the change of *φ* are affected by these two fixed points for *ṙ*. Two limit cycles are formed from the limit cycle of case with $a = a_c$. One of these cycles is stable (full line) and one is unstable (dashed line). Inside unstable limit cycle, there is spiral moving all points toward stable point in the origin.

I would like to mention (again!) that Figure 7.21 is shown with an intention to just emphasize the Complexity of the situation with bifurcations in two-dimensional system. For those Safety Professionals who would like to be more involved in the area of non-linear dynamics, there is plenty of literature available on the Internet.

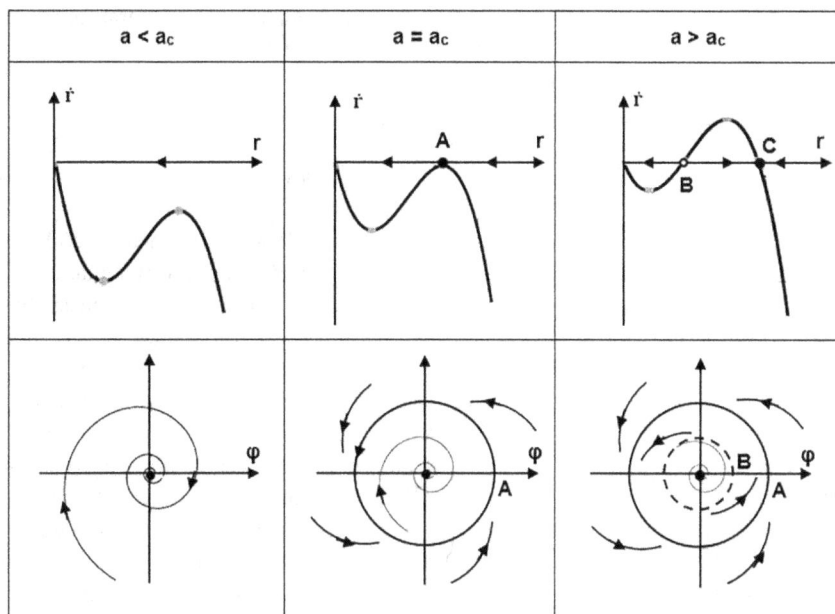

FIGURE 7.21   Some bifurcations in two-dimensional system.

7.11   STABILITY OF NON-LINEAR SYSTEMS

I spoke about the general stability of Complex Systems in Chapter 4, but let's give here a short insight into the stability of non-linear systems. This chapter is more mathematical and for all these stable, unstable, and hybrid points, there is also a criterion how to determine their stability. In the mathematical world, the Lyapunov Theorem is well-known which deals with this area. Unfortunately, it is very much complex and abstract to be presented here.

In simple words, there is need to find (create) the so-called Lyapunov function (usually presented as $V(x)$) to investigate in the vicinity of the fixed point which stability should be investigated. If that function is differentiable in the vicinity of the fixed point and its derivative in the fixed point is negative along the trajectories around, then the fixed point is stable.

However, the things are not so simple . . .

The Lyapunov Theorem for Stability is the basis for all calculations. This theory produces few methods (direct and indirect) for calculating stability of any dynamic system. There are also differences in application of this theory, depending on the fact whether the system presented with differential equations (continuous systems) or with the difference equations (discrete systems). Another problem is that Lyapunov function $V(x)$ is not easy to find, so it needs to be created. Sometimes, there is need to try few different functions. However, I do believe that level of mathematics presented in this chapter is enough, so I will not go into details with Lyapunov Stabilities.

For those who are interested about this, there is huge amount of literature (books and articles) on the Internet which can be downloaded free of charge.

7.12   CATASTROPHE THEORY

The bifurcations are very much important because they are basis (together with singularity theory) for the so-called Catastrophe theory. The Catastrophe theory originated from the work of the French mathematician René Thom in the 1960s and became very popular due to the efforts of Christopher Zeeman (British mathematician) in the 1970s. Both of them dealt with topology and, to be more precise, Thom produced the term "catastrophe". Later, Zeeman produced the phrase "Catastrophe theory" by unifying singularity theory and bifurcation theory.

The Catastrophe theory studies bifurcation phenomena characterized by sudden shifts in behavior of fixed points caused by small and smooth changes of parameters. These small and smooth changes of parameters are contributing to the transformation or disappearance of the stable Working Points of the systems (operation, processes, activities, etc.). The particular analysis of how the qualitative nature of equation's solutions depends on the change of parameters will provide hints how the system will behave. Having in mind that those small and smooth changes may lead to sudden (very dramatic) events (incidents and accidents), the Catastrophe theory can be used to investigate the events. Or better say, the Catastrophe theory can predict that such events could happen, but it cannot predict when they will happen.

Thom (and other scientists later) has not looked to the Catastrophe theory as a scientific theory having in mind that it is mostly qualitative and not quantitative. He looked at it more as a method, methodology, or an application how to qualitatively investigate or explain natural phenomenon using available experimental data. In general, this theory did not innovate anything else, but it helped to explain better what was already known. The Catastrophe theory had considerable effect not only in mathematics, biology, social sciences, and engineering, but also in philosophy of science.

As said before, even small changes in values of critical variables and parameters of a non-linear dynamical system can cause change of fixed points from stable to unstable. These changes can produce large and sudden changes of the stability of system under consideration. Catastrophe theory reveals that such changes tend to occur as part of qualitative geometrical structures. In reality, the Catastrophe theory deals with structural stability of the systems providing mathematical description how and when this stability can fail, changing the structure of the objects.

Thom was working mostly on topology of the states of the systems and based on changes of topology, he has produced seven classes of elementary catastrophes which are connected with the type of graphics for non-linear dynamics of the system under consideration. He has proved in his papers that any discontinuous (abnormal) behavior in any system controlled by not more than four variables (parameters) is one of these seven elementary catastrophes.

These elementary catastrophes and some of their characteristics are given in Table 7.2.

TABLE 7.2

Different Types of Elementary Catastrophes by Rene Thom

No.	Name	Formula of the Line (Surface)	No. of Behavior Axis	No. of Control Parameters
1	Fold	$x^3 + ux$	1	1
2	Cusp	$x^4 + ux^2 + vx$	1	2
3	Swallowtail	$x^5 + ux^3 + vx^2 + wx$	1	3
4	Butterfly	$x^6 + ux^4 + vx^3 + wx^2 + tx$	1	4
5	Hyperbolic umbilic	$x^3 + y^3 + uxy + vx + wy$	2	3
6	Elliptic umbilic	$x^3/3 - xy^2 + u(x^2 + y^2) + vx + wy$	2	3
7	Parabolic umbilic	$x^2y + y^4 + ux^2 + vy^2 + wx + ty$	2	4

In Table 7,2, x and y are variables for control of the operations (or the so-called *behavior axis*) and u, v, w, and t are parameters ("control factors")[31] for adjusting the Working Point of the system. During operation, we control variables and we do not touch parameters, but they could deviate from their already adjusted values, if there is internal and/or external interactions. When it happens, the system becomes unstable.

Speaking about geometrical structures, please note that in one-dimension, we were working with lines in the plane and in two-dimensions, we need to work with *phase-plane* in the space (volume). The shape of the plane in three-dimensional coordinated system is actually used to define the elementary catastrophes.

I will use Cusp catastrophe example (Figure 7.22) to provide clarification of catastrophes in a simplified way.

The phase-plane for cusp catastrophe is presented in Figure 7.22a and it is actually plane where our "Working Points" of the system may be found. Its projections in the plane (x_1, x_2) are presented in Figure 7.22a and in the plane $(\dot{x}_{1(2)}, x_1)$ in Figure 7.22b. The Cusp catastrophe could happen when the trajectory of the changes of the system is actually traveling through *phase-plane* where the folded shape is.[32] Looking at the projection on the plane (x_1, x_2), the catastrophe can happen when the trajectory cuts Catastrophe curve shape.

Assume that our system is in position A (on the *phase-plane*) and due to change of some parameters, it moves in the direction of dashed arrow. When it reaches the folded border, it will "catastrophically" fall over the folded surface. This is actually the situation when stability of our system is abruptly damaged: There is sudden change of the state of the system. The "fall" can be catastrophic itself, but this also changed the state of dynamics of the system which may cause defects in the system itself or in the environment around.

---

[31] These are names used in Wolfram Mathematica. Maybe you can find different names in other literature . . .

[32] This folding is characteristic for non-linear systems with Hopf Subcritical bifurcation which is part of non-linear dynamics in two-dimensional systems. A simple example of Hopf Subcritical bifurcation is vibration of the wings on the aircraft during the flight. If not controlled (damped), the wing could fall down from the fuselage of the aircraft.

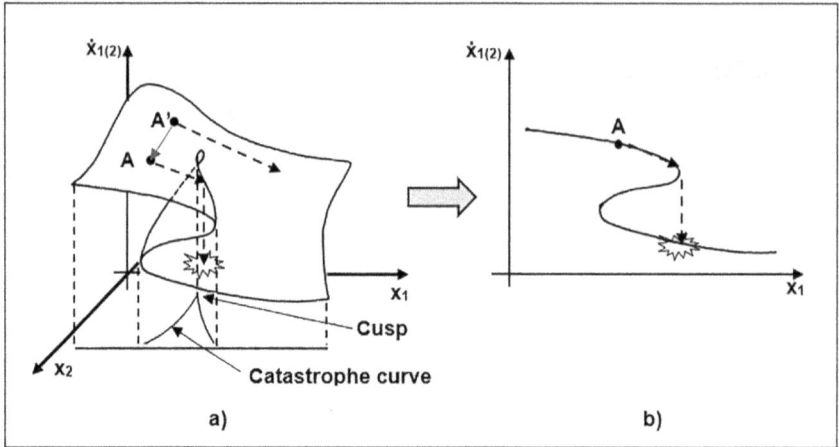

FIGURE 7.22 Folded surface (a phase-plane) for cusp catastrophe.[33]

But, look at the point A' ...

It is in a different position where its "dynamics" will provide movement, as shown by the dashed arrow. As it can be noticed, the trajectory of this point will not experience "catastrophe" (it is not cutting the Catastrophe curve) and actually, we may say that this could be a valid "Working Point"[34] of the system. In this particular case for the Cusp catastrophe, we can say that "catastrophic fall" is caused by change of parameter x_2 assuming that change of x_1 is not critical for particular values of x_2. But in its movement from A' to A (gray arrow), x_2 will endanger stability of the system. Being close to A will provide more damage and being close to A' will not produce any damage. As it can be noticed, it is everything about choosing appropriate Working Point for the system and keep it stable.

To be more precise, the "fall" of A in Figure 7.22a is more physical than a mathematical phenomenon: As mentioned before, all the systems with particular potential energy "strive", as soon as possible, to move themselves into a position with smaller potential energy. Keeping this in mind, we can say that movement of A' is also a natural phenomenon toward a state with less potential energy.

Here I would present another example for catastrophic event: Dropping of grains of sand onto a pile below. As we continue to pour the grains, the pile will grow into a cone-shaped form. But, from time to time, the addition of only one grain of sand will trigger an avalanche that will change the shape of the pile. This event is called *catastrophe*.

This will happen unpredictably, which means there is an uncertainty about the time when it will happen and about the amount of the pile which will redistribute.

[33] The surface for Fold catastrophe and Cusp catastrophe are same, but the Fold catastrophe depends on one parameter and the Cusp catastrophe depends on two parameters.

[34] Working Point in this case can be defined as a requested output of operation of Complex System based on the operator's settings of Controllable influence factors (adjustable parameters of Complex System).

The physical explanation is very simple: Every next grain of sand will change the potential energy of the grains at the top of the pile and there will be one grain which will pass the critical balance and the grains at the top of the pile will adjust themselves finding a new balance. This will continue to happen, if we continue to pour more grains of sand.

Having in mind this example, we can define the "catastrophe" as movement from one stable state to another stable state. Although both states could be beneficial or idle for functioning of the system, our interest is in the areas where the first stable state of the Complex System is the state determined with the Working Point of the system and next one is a fault of the system. This fault could be an incident or accident.

So, the movement of A can be explained as "revolution" and movement of A' as "evolution". Maybe these are valuable terms in biology and social sciences, but in engineering, we do not like "revolutions".

But there is more . . .

If the "fall" could happen, the "jump" also may happen. Or in other words, depending on the situation, if the Working Point is designed to be on the lower surface, small and smooth movement of parameters could make it to "jump" on the upper surface. So, it is a reverse movement from position of lower potential energy to position with higher potential energy.

Of course, this could not happen to engineering systems (Equipment), because it will violate the rule that bodies always strive to be in places where they have minimum potential energy (see the explanation of Figure 7.4). So, this will be very uncommon in engineering because the potential energy of the system will increase only if the energy is added to the system from outside. Anyway, such qualitative "jumps" could (more often) happen in biology (evolution), social sciences, economy (industrial revolutions), etc.

In biology, the mutations can also be associated with the Catastrophe theory. Some of the mutations contributed to the evolution, but some of them could be attributed to sudden changes in the genes of living subjects (humans, animals, plants, bacteria, etc.) which drastically change the state of some "Subsystems" in the newly born subjects. Sometimes they can be good (beneficial) and sometimes they can be bad (catastrophic).[35] The cancer is actually mutation of the normal living cells in the human body.

The Catastrophe theory is a very complex mathematical area of study and you do not think that you can easily get familiar with it. Again, I am just trying here to provide some simplified explanations what is this about. In maintaining safety in industry, all this mathematics applies during the design of the products and as such, it is a primary tool for modeling. Later, this theory can be used again in safety areas, during investigation of already happened incidents or accidents[36] to provide reasons why it happened.

---

[35] To be scientifically correct about beneficial mutations, I must clarify that these mutations are called *beneficial* (in the animals, for example) due to one characteristic which they have. But, in general, the mutation does not affect only one characteristic, so other mutated characteristics usually produce problems in the life of these animals, so they die soon.

[36] Using the Catastrophe theory in mathematical calculations for explaining how the ships capsize or how buckling of elastic structures happens has given very good results!

I have explained connections of Equipment (products) with Catastrophe theory, but Zeeman has used it to explain different phenomena in biology and behavioral sciences, so it makes it applicable also to offered services.

The Catastrophe theory is close to Theory of Chaos and they share a lot of common things—graphical presentations, non-linear dynamics, modeling practices, modes of explanations, etc. But here I must explain that there is another type of "catastrophe" which can happen inside the non-linear dynamic systems and this is when the state of the system is close to unstable fixed point and it moves toward infinity (unbounded system). This is the situation when our System will be destroyed (the explosion is a good example of unbounded system).

7.13   EPILOGUE TO THE CHAPTER . . .

I am almost certain that many of the readers will ask why I needed this chapter with so much mathematics and so many "hardly understandable for ordinary humans" points?

The reason is simple: I do believe that understanding the situation fully and gathering good knowledge what is going on will help you to cope with it! But there is also another thing: The phase-plane analysis presented in this chapter is one of the used methods for analysis of second-order non-linear systems with differential equations. Other methods are Lyapunov theory method (good, but complex) and describing functions method (approximate method).

All these methods are used as part of the design process of any Complex System which contain many non-linear Subsystems, at least in the form of Control Systems for providing control. In this chapter, the simplest points were explained and I hope many of you get a picture of what is going on.

There is another reason why this chapter received due attention. I explained here only invariant (autonomous) systems. These are systems where the state of the system does not depend on time. So, all complexity explained here would be worst if the system is variant (non-autonomous) or if its state is dependent on time. Keep in mind that most of the systems in our industries and our ordinary lives are dependent on time (non-autonomous systems)!

As the medical doctors must have excellent understanding of normal functioning of the human body, they also need to have good understanding of the abnormal functioning of human body (what happens when the humans are ill). This is imperative for the doctors to be effective and efficient (the first step) to diagnose the disease and (the second step) to find the good cure. I hope that you all agree that healing a human is not an easy job. There are plenty of pills and other methods of medicine (surgery, radiation, etc.) which can help in healing, but by providing a wrong diagnosis, doctors will use wrong methods and they could even produce bigger damage. Unfortunately, these things happen all the time . . .

So, trying to stop "bad things to happen to good people", in my humble opinion, is of utmost importance to gather the necessary knowledge and to understand what is going on with Complex Systems. And when I am saying "Complex Systems", I am thinking on Equipment as congregation of Hardware and Software in operational systems (as tools for "maintaining" operations).

My almost 20 years' experience in Quality and Safety areas assured me that, in maintaining the Quality and Safety, most of the companies and humans there are too bureaucratic.

Yes! I agree!

Today's achievements, especially in safety areas, are extraordinary compared with those in last century, but let's be honest: We still miss elementary things in maintaining the Safety. So, as I have mentioned at the beginning of the book and at the beginning of this chapter, I do believe that knowing all the things mentioned here will improve our attitude and knowledge in Safety.

Thes are the reasons that I think that any Safety Professionals should read this chapter with due attention. Again: Forgive me if it was too "scientific"!

8 Theory of Chaos

8.1 INTRODUCTION

The Chaos, as a word, is actually used very often in our everyday lives to explain something which is utterly confused or disordered. It is a synonym for total lack of organizations during incidents and accidents, especially in situations where people do not expect such a thing to happen. Long time ago, the humans understood that the Universe is a chaotic place and to explain it, more knowledge is needed. Later they realized that turbulent movement of the molecules in gases is a synonym for chaotic movement, but they assumed the mechanism to be very complex, so only the "collective" behavior expressed through the temperature, pressure, and volume can be deterministic. And this "collective" behavior is enough to understand the situation on macroscopic scale[1] using statistical tools.

In science (physics and mathematics), it is a movement presented by nonlinear dynamics which is deterministic, but it is very much unpredictable. It depends on the type of non-linear function which describes the behavior of the systems and its predictability is limited to the high sensitivity of initial conditions.

The Theory of Chaos is a part of science which describes complex dynamics of the systems in Nature, especially paying attention to the dynamics of "sensitive to initial conditions" systems. The point is that the Chaotic systems are intrinsically unstable, and as such having Chaos in your system (operation, process, activity, etc.) means trouble.

First, the Chaos appeared as a mathematical model, but later it attracted the attention of physicist and was accepted as a new concept of non-linear dynamics. We are trying to explain behavior of our systems by equations and when these equations are non-linear, we say that the solutions are complex and it is not easy to calculate them.

Although it showed as pure mathematical model, today it finds its applicability in areas of meteorology, biology, chemistry, cardiology, economy, etc. In general, the Chaos is part of Applied Mathematics which is used to study Complex Systems, especially those where Complexity is created by non-linearity of the operations. If the Catastrophe theory deals with changes from stable toward stable states, the Theory of Chaos deals with unstable states.

8.2 A SHORT HISTORY OF CHAOS

The Chaos is connected by dynamics of the systems and dynamics started with Sir Isaac Newton. Anyway, this is a book about Chaos in Safety, so I would not go into historical details about dynamics, but I will jump directly to Chaos.

---

[1] Robert Boyle (an Irish scientist) was probably the first one who started to take into consideration the macroscopic effects over microscopic movements of molecules in the gas.

DOI: 10.1201/9781003404811-10

The first scientist who noticed the Chaos was one of the greatest mathematicians in history, Henri Poincare. At that time, dynamics of movement of two bodies was already explained by Newton's law of gravitation using Sun and one planet and neglecting influence of all other space objects. Poincare went forward and tried to solve the gravitational problem of three bodies close to each other. He experienced big problems using the ordinary calculus, so he decides to use another approach by using the geometry (geometric approach). He noticed that the orbits are very chaotic, so the formulas can be produced only if some of the parameters are neglected.

Some of the historians of science stated that Poincare actually discovered the Chaos, but he could provide neither a good theoretical nor a practical explanation of that. The reason for this was that although Poincare was using a geometric approach, he was extremely bad in drawings. That is the reason that, even later, other scientist could not understand the meaning of Poincare's explanations.

The geometric approach of Poincare was dealing with initial conditions of the three bodies and he was trying to assume movement as trajectories of these bodies graphically presented, without solving equations analytically. And this approach is also used today in determining the Chaos. The only problem here is that this approach is qualitative and not quantitative. So, we cannot get numbers as solution, but we can get ideas what will happen.

At the beginning of the 20th century, the physicist were overwhelmed by efforts to advance Quantum Mechanics and Theory of Relativity, so Newton and Poincare were forgotten. At the beginning of the 1920s, the electricity had big development. Until the 1950s, the vacuum tubes, radio, radar, lasers, and semiconductors were invented and electronics started to shape itself as new areas of interest. Although all these inventions were directed in different areas, each of them was dealing with non-linear systems, oscillations, and transformations, so they gave good basis for future advance in non-linear dynamics.

Edward Norton Lorenz joined the US Army Air Corps in 1942 as meteorologist. He already had studied mathematics at Dartmouth College (B.Sc.) and he had continued to M.Sc. at Harvard University (Cambridge, Massachusetts, USA). During his army serving from 1942 to 1946, he fell in love with meteorology and after War World II, he decided to study it at Massachusetts Institute of Technology.

He noticed that prediction of weather was a big problem due to Complexity, interdependencies, and interactions of factors involved in the weather creation and started to think how to solve this problem. During these efforts, he tried to construct a mathematical weather model which would explain the dynamical movement of air due to the difference of temperature, humidity, and pressure in the Earth's atmosphere. His first model was built by 12 non-linear equations and he run it through (rudimentary) computer many times. Later, looking for other such non-linear systems, Lorenz found simplified model with only three equations.

His simplified model of convection[2] of the air in the atmosphere with three differential equations is given as follows:

---

[2] Convection is a physical event of movement of fluids and gases due to different density caused by different temperature of different layers.

$$\frac{dx}{dt} = \dot{x} = a \cdot (y - x)$$

$$\frac{dy}{dt} = \dot{y} = b \cdot x - y - x \cdot z$$

$$\frac{dz}{dt} = \dot{z} = x \cdot y - c \cdot z$$

Here, x, y, and z are 3D coordinates for the space and a, b, and c are parameters[3] bigger than 0.

These three differential equations are a system which describes the air movement necessary to produce a particular weather pattern. As you can notice, this was a non-linear system for modeling due to $x \cdot y$ and $x \cdot z$ terms in the equations. For calculation, he used the new LGP-30 desktop computer and he did the calculations for a wide range of parameters a, b, and c. He noticed that for chosen $a = 10$, $b = 28$, and $c = 8/3$, the calculations produce pretty much complex behavior which was deterministic, but so erratic that it makes it totally unpredictable and it depends on the parameter values. When he plotted the air movement in three dimensions, he discovered the so-called Strange Attractor (presented in Figure 8.1). This Strange Attractor actually proved that non-linear dynamical systems show order, but they never repeat themselves.

Figure 8.1 presents few pictures of Lorenz's Strange Attractor[4] for weather model in different planes and in time. The pictures are undertaken from the website:

https://media.pearsoncmg.com/aw/ide/index.html[5]

As it can be seen in Figure 8.1, the overall dynamics of Lorenz's Strange Attractor is aperiodic. Aperiodic movement can be explained as "oscillatory" movement where every next "oscillation" is not the same as previous one changing the trajectory.

Lorenz was not so sure about his results, so he tried another experiment in order to support his previous conclusions. He established an experiment that was based on water. Today it is known as the Lorentzian Waterwheel. He built a waterwheel with eight buckets, each of them spaced evenly around the common rim and each of them had a small hole at the bottom. The buckets were mounted on swivels and they always point upward.

The entire system was placed under a faucet, so, the slow and constant stream of water was falling on the buckets. Under the stream of water, they started to spin at a constant rate. When he changed (increased) the amount of water in the stream,

---

[3] In reality, **a** is the Prandtl number (coming from heat transfer physics), **b** is the Rayleigh number (coming from fluid mechanics), and **c** has no name (explains some physical dimensions of air layer in the atmosphere).

[4] For readers who are more interested about these attractors, I can mention another one which is similar to the Lorenz attractor: the Rössler system. The equations for the Rössler system are $dx/dt = -y - z$; $dy/dt = x + ay$; $dz/dt = b + z(x - c)$

[5] It is a very old website, so if you cannot open it with Chrome (at least I had problems!), then try with some other browser. I tried with Internet Explorer and, after few adjustments on Security Exceptions on Java, the site was fully capable to provide the pictures which are presented here. I strongly recommend to try to reach this site, because it offers very good insight (through animations) into the behavior of Lorenz's Strange Attractor.

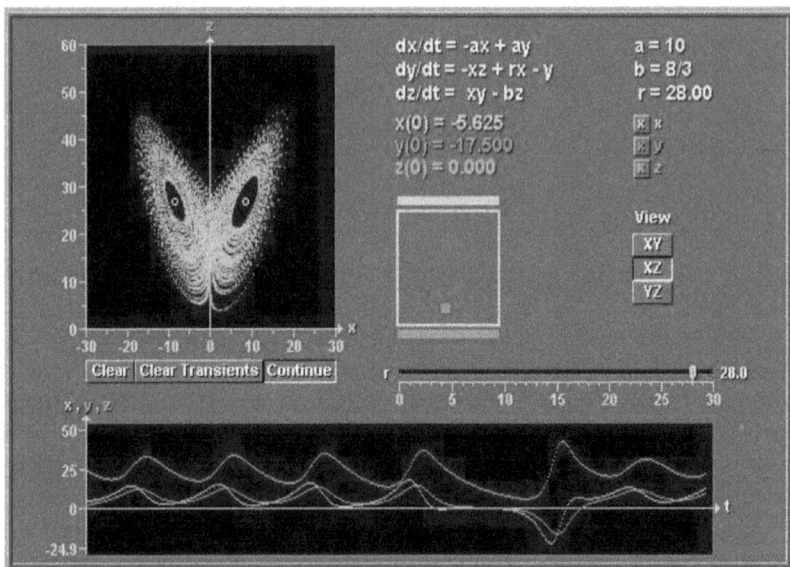

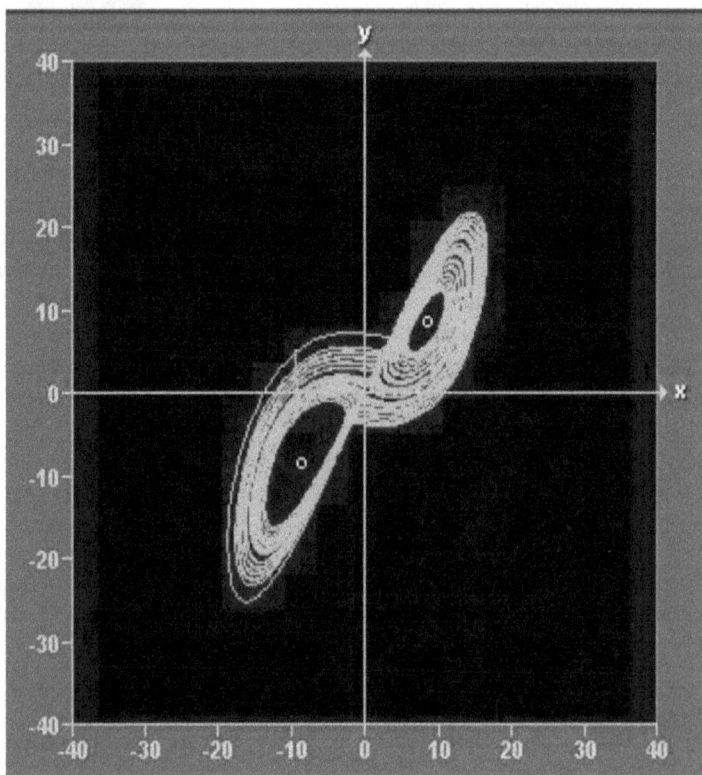

FIGURE 8.1 Lorenz strange attractor.

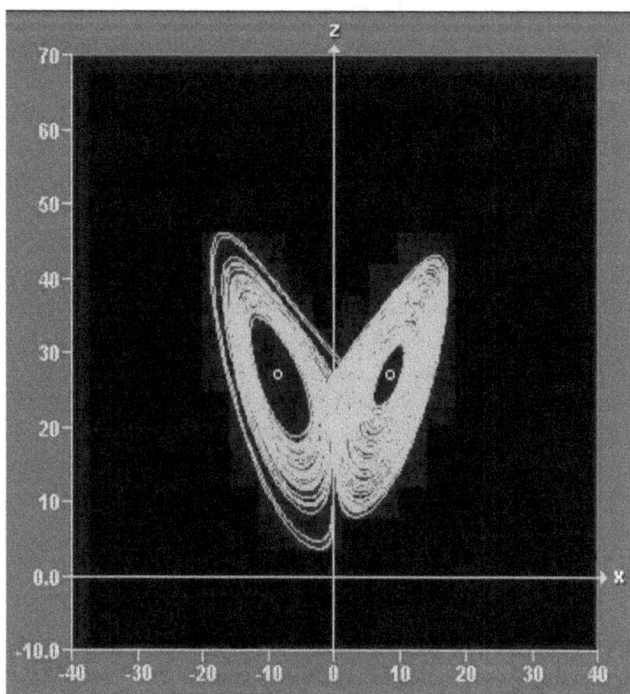

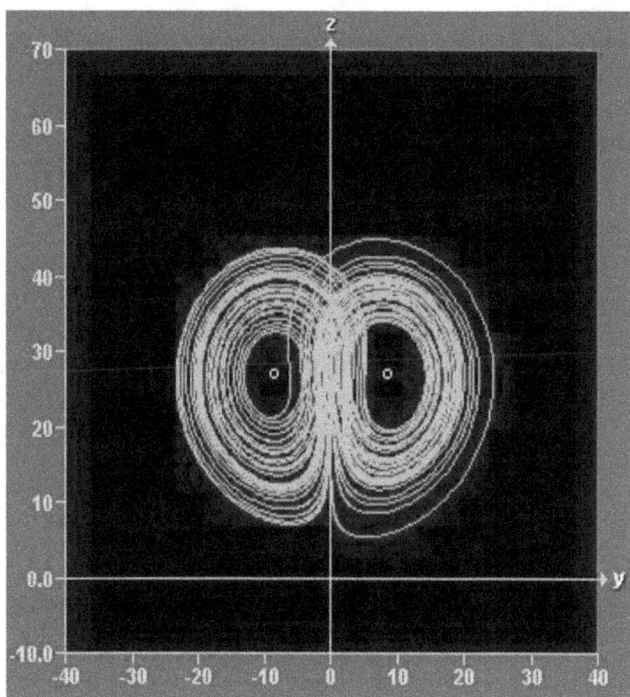

FIGURE 8.1 (Continued)

the same interesting phenomena arose. The increased flow of the stream of water resulted in a chaotic motion for the waterwheel. The waterwheel changed the speed and direction of spinning totally chaotic, without any rule. When he later (after many hours of monitoring) produced the graph of the waterwheel, it was very much resembling to the previously determined Lorenz Attractor.

The Strange Attractors and their aperiodic dynamics have one very important feature: Two trajectories produced from different locations (different initial conditions) will never recur to the same initial location, nor they will intersect each other.

The paper written by Lorenz did not attract any particular attention in science community and was soon forgotten, but the Chaos was later noticed by other scientists. Among the pioneers, I would like to mention Stephen Smale, Andrey Kolmogorov, Vladimir Arnold, and Jürgen Moser.[6]

In the beginning, the biologist was very much entitled to Chaos in 1975. Robert May published an article in *Nature* and it was read with interest not only by the biologists, but also by mathematicians. He described the chaotic changes of population of animals which arise if we are using logistic map given by the following mathematical formula:

$$x_{n+1} = a \cdot x_n \cdot \left(1 - \frac{x_n}{b}\right)$$

Here, x_n is the population previously expressed by a number. x_{n+1} is the next generation of population expressed by a number, a is the parameter which describes growth rate (how fast the number of animals will increase), and b is the parameter describing carrying capacity which depends on the environment around (is there enough food or good environmental conditions to support the animal population). As you can notice, it is a non-linear iterative map which behave very strange for values between 0 and 4.

In the area of Chaos in biology, a big contributor was Arthur Winfree who examined a periodic emergence of fruit flies by using short pulses of light. He found that these pulses actually can change the internal biological clock in the flies. His job triggered examinations in cardiology when doctors noticed that atrial fibrillation[7] also behave in the same way.

Benoit Mandelbrot[8] is also well-known for his contribution in the area of Chaos in space. He was dealing with prices of cotton in 1963 and had noticed that there was a periodical recurrence of the patterns at every scale.

Mitchell Feigenbaum realized that there is universality[9] of bifurcation in non-linear dynamics. Universality is a term used to explain that a totally different system may achieve Chaos in same way and it can be presented by a simple formula:

$$\sigma = \frac{x_{n+2} - x_{n+1}}{x_n - x_{n-1}} = 4.6692\ldots$$

---

[6] These three were mathematicians and they are known by their KAM theory about dynamic systems with quasiperiodic motions under small irregularities.

[7] Atrial fibrillation is heart disease caused by heart rhythm abnormality.

[8] In 1982, Mandelbrot published *The Fractal Geometry of Nature* which is today an anthology in space Chaos.

[9] He published a paper "Quantitative Universality for a Class of Nonlinear Transformations" in 1978.

This value is a "magic number" and it is valid before non-linear systems move to Chaos. It is also valid not only for particular dynamical systems, but also for systems from different disciplines which have nothing in common. The interesting point here is that Feigenbaum used only simple HP-65 handheld calculator to find this value.

A year later, in 1979, Albert Libchaber proved experimentally the bifurcation cascade that leads to Chaos and turbulence in Rayleigh–Bénard systems.

Well-known scientists in the area of Chaos are mathematicians David Ruelle and Floris Takens, known by their work in the area of turbulence.

The Chaos (together with non-linear dynamics and its fractals) became fashionable between scientists in the 1980s. In 1987, James Gleick published his book *Chaos: Making a new Science* and this book had tremendous impact on the ordinary public, accelerating the interest for Chaos, not only in physics, but also in other natural sciences.

At the end of last century, Chaos was accepted as an important explanation of events in many scientific disciplines.

8.3   FRACTALS OF CHAOS

Chaos (roughly) is an irregular behavior of non-linear deterministic systems. As I already mentioned in Chapter 7, the Chaos can happen only in non-linear dynamical systems with more than three variables and parameters and those are Complex Systems which we have all around us today.

"Deterministic system"[10] means that having the present or past knowledge about the behavior of the system can help us to describe future behavior of the system with a considerable accuracy. Deterministic, in this situation, could mean regular, periodic, predictable, and/or expected.

The interesting thing in the definition of Chaos is the word *irregular*. It means that the Chaos in deterministic systems (by its nature), under particular conditions, can move out from the regular behavior and start to behave irregularly which would bring a lot of unpredictability. Actually, I can say that the fine changes in details of the system performance with Chaos can evolve into big changes in large-scale behavior.

The Chaos can be seen in space and in time.

In the space, the Chaos is presented by *fractals*. Fractals are geometric shapes which can be constructed starting from bigger to smaller (or vice versa). However, the fractals can be explained as *strange attractors* in space. We can find in Nature such an effect when we change progressively the scale of looking at Nature. The simplest examples in Nature are snowflakes, fern leaves (Figure 8.2), sea and ocean shores, clouds, mountain ridges, etc.

Figure 8.2 presents fractals in fern leaves. You can notice that similarity is very high between the overall fern leaf and the constituents of this fern leaf. Actually, in Figure 8.2, I have changed the scale of looking at the fern leaf and I can notice that the smaller constituent could have the same shape as a big one. This is called *self-similarity*. The point is that with self-similarity, every small piece (leaf) of the full fern provides information on overall fern.

---

[10] See Section 1.10 (Uncertainty and Complex Systems).

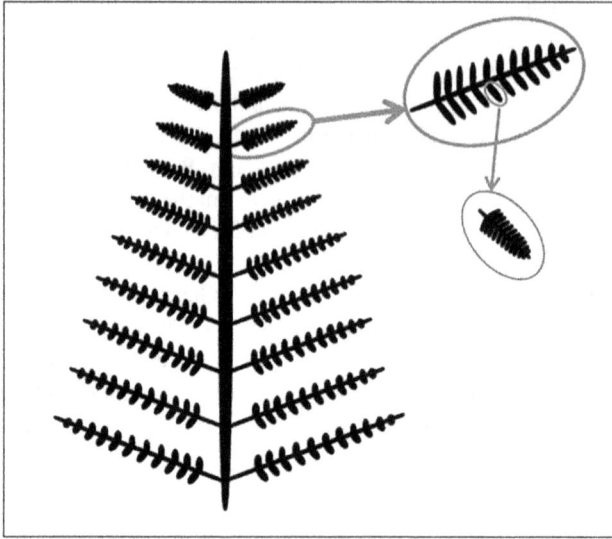

FIGURE 8.2 Fractals in Nature (fern leaves).

Benoit Mandelbrot (1924–2010), a Polish-born applied mathematician (lived in France and the United States), is the "innovating father" of fractals. He noticed that the Nature is full with fractals and started to study them. Today we agree that fractals are an intrinsic part of Nature.

When you see a fractal, you must say that it is beautiful!

There could be order in its fascinating image, but the formula that makes up this image does not belong to Euclidian geometry. The classical Euclidean geometry is quite different than the fractal geometry because fractal geometry deals with non-linear, non-integral systems. Even the fractal geometry can be used to check the non-linearity of the systems. In Euclidian geometry, the sum of angles in the triangle is always 180°. In fractal geometry, it can be even 270° (Figure 8.3).

As can be seen in Figure 8.3, the left picture is a planar triangle and the sum of all angles is 3 × 60° = 180°. On the right side is "rounded" triangle presented as planet Earth (a ball, highly non-linear shape) with Equator and two of its meridians. As it can be noticed, the angles between Equator and meridians and between meridians are 90° making the sum 270°.

So, Euclidean geometry is about shapes in the plane (lines, rhombs, ellipses, circles, etc.) and the fractal geometry is about algorithms. In this sense, "algorithms" mean creating a triangle on a ball. In general, the Euclidian geometry is linear and the fractal geometry is non-linear.

There are two basic characteristics of fractals. The first one is its self-similarity. There, each element building the fractal is similar to its "magnified" picture. A bigger fractal shape will be almost, or even exactly, similar (even the same) as building elements provide a repetitive pattern (see Figure 8.4).

Figure 8.4 shows a fractal known as Sierpinski triangle and it is built when each side of the triangle is divided by 2 and these points are connected to create four more

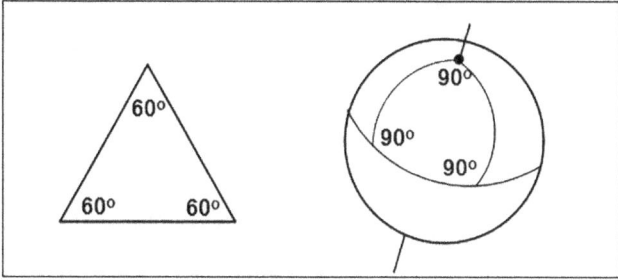

FIGURE 8.3   Euclidian (linear) geometry (*right*) and fractal (non-linear) geometry (*left*).

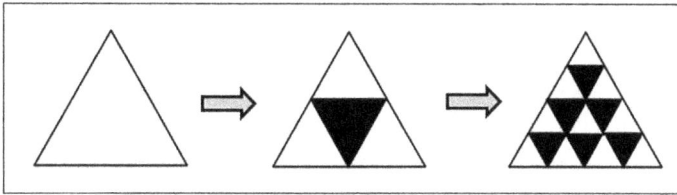

FIGURE 8.4   Sierpinski triangle and achieving self-similarity.

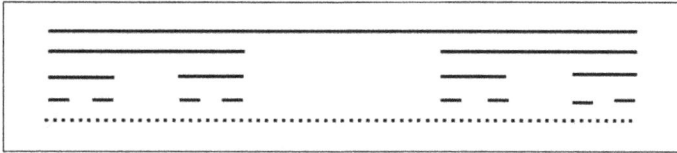

FIGURE 8.5   Cantor set.

smaller triangles. And each step is iterative: Continue dividing with 2 each side of all triangles and connect them creating smaller and smaller triangles.

The second one is that the fractals are quite different from the graphs of shapes that are considered in Euclidean geometry. Figure 8.3 can prove that there are two triangles, but they are significantly different.

There are many examples of fractals investigated by scientist, but from engineering point of view, the Cantor set is very much important. Cantor set (Figure 8.5) can be determined applying the similar algorithm from Sierpinski triangle to line with finite length: Divide the line into three parts and remove the medium segment. The same should be done with the remaining two segments and so on . . .

The Cantor set was used by Mandelbrot as a model for the occurrence of errors in a first electronic transmission lines for telegraph and for radio. At the beginning of the telegraph and radio communications, there were periods of transmission with errors and periods without errors. When these periods of errors were analyzed, it was determined that they contained periods without errors inside them. As the analysis of the transmissions were going on to smaller and smaller details, it was determined that such situations (as in the Cantor set) were excellent for modeling transmissions intermittency.

8.4   SENSITIVE DEPENDENCE ON INITIAL CONDITIONS

One day, Lorenz wished to look at short sequence of his equations which he ran previous day, so he switched on the computer. But instead of starting from the beginning, he chose one initial points and put the coordinates in the computer. The speed of computers at that time was tantalizing slow, so he left the room to do some other job. When he came back later, he saw a totally different graphic from the day before. It was the same situation, same data, same initial conditions, but picture on the screen was totally different???

He assumed that something is wrong with the computer, so he called a technician, but the technician said: Everything is OK. Lorenz then checked the program, but again it was OK. Anyway, he tried to run the equations few times and he always got the same graphic, but totally different than the graphic from previous day.

Trying deeply to investigate the cause for the difference, he noticed that the coordinates of initial conditions in the computer were stored by six decimals and during its previous tries, he put only first three decimals, neglecting the next three. Actually, if the initial condition in the first case was 4.786273, he put 4.786 in the second case. There was no other explanation than the one that these three missed decimals actually caused the difference in the graphics.

Although the values of these three decimals were so small (in our case it was 0.000273), the difference in the pictures was very big. He realized that the system which he has produced was extremely sensitive to small changes of initial conditions. He could not suppose that such a small difference (inaccuracy of fourth, fifth, and sixth decimal) which is an error of 1 over 10,000 would totally change the situation with result of the experiment.

Later this sensitive dependence on initial conditions was "transformed" by Philip Merilees who entitled one of the papers of Lorenz as "Does the flap of a butterfly's wings in Brazil set off a tornado in Texas?". After this conference,[11] the butterfly was synonym for Chaos.

These small "sensitivities to initial conditions" actually can be triggered by the presence of noise in our systems. So, whenever we think that we have adjusted a proper Working Point, the ever-present noise could interfere and system could change (if it is not in vicinity of stable fixed point). So, be careful: Although the system worked many times without any problem, you never know what will happen next.

The important question here is: How big is this sensitive dependence on initial condition in time?

It can be described by the following equation:

$$\delta(t) = \delta(0) \cdot e^{\lambda t}$$

---

[11] This happened in 1972 at the 139th Meeting of the American Association for the Advancement of Science.

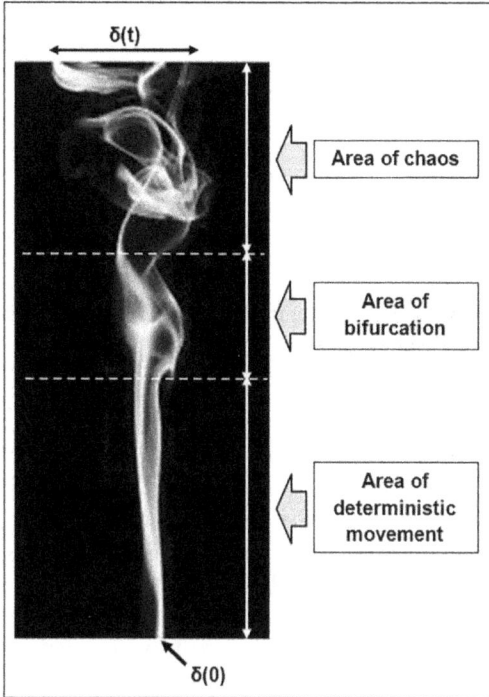

FIGURE 8.6 Different areas of non-linear dynamic motion of cigarette smoke (simple example of fluid dynamics).

Here, $\delta(t)$ is the distance between two close states of the system after particular time t, $\delta(0)$ is its initial condition (initial distance between these two states at the beginning which is in the range of 10^{-4}), e is Euler's number (2.17828 . . .), and λ is the so-called Lyapunov exponent.

As it can be seen in the formula, these two particles (very close to each other in the beginning) are exponentially distancing themselves. This could be also taken as axiom of Chaos: The two models of a dynamic system with three or more parameters, used in different simulation and starting with infinitesimally close initial conditions, will exponentially diverge from each other over time.

The simple example of this effect is given in Figure 8.6.

As it can be seen in Figure 8.6, the smoke of cigarette[12] starts from the same place where molecules of the smoke are very close to each other. This is initial condition for the system marked by $\delta(0)$. As the smoke molecules are traveling further up, following their own trajectories, they come to an area of possible bifurcations. Here, the molecules can change its trajectories, but it is not so evident. Going further up,

---

[12] Those who are looking for more details about mathematical expression of cigarette smoke may refer to "Mathematical modeling of cigarette smoke particles dynamics in high air change rate chambers like cars" by E.M. Saber and M. Bazargan (2010).

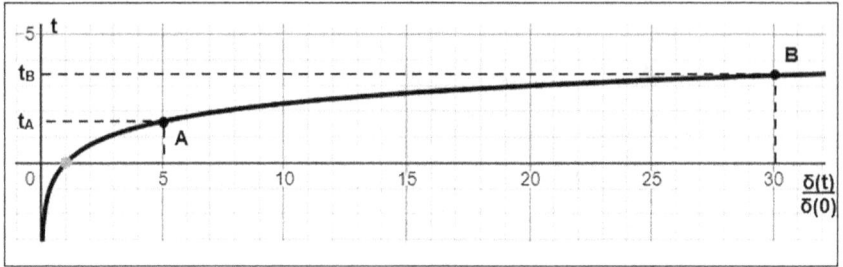

FIGURE 8.7 Graphic of the equation for interdependence of time and distance of molecules of cigarette smoke.

they are entering an area of chaos where distance between trajectories of these two molecules is quite big ($\delta(t)$) and, in addition, the movement is totally unpredictable.

The merit of increased distance (state of cigarette smoke) is Lyapunov exponent[13] and I could not find its value for the smoke of cigarette in the literature which was available to me. It is important to mention that there are plenty of Lyapunov exponents and each of them is different for different initial separations expressed by $\delta(0)$ between two molecules. It is not easy to calculate them, but in Figure 8.6 $\delta(t)$ is presented for the biggest one (corresponding to the biggest distance). This maximum Lyapunov exponent is important because it is used for the calculation of predictability of the chaotic system.

If λ is positive, it is Lyapunov exponent for chaotic system. In chaotic system, the trajectories of the states of the system run away from each other. If it is negative, it is a system with stable point (trajectories are taking direction toward the stable state of the system).

If we try to solve equation for sensitive dependence on initial condition to see what is the value of time t, we can obtain the following equation:

$$t = \frac{1}{\lambda} \cdot \ln \frac{\delta(t)}{\delta(0)}$$

Let's see what is the meaning of this equation . . .

The graph of the equation is given in Figure 8.7.

As it can be seen in Figure 8.7, I have marked two points on the graphic (**A** and **B**). Under the assumption that $\lambda = 1$ and using the equation for t above,

(a) the point **A** will have coordinates $\delta(t)/\delta(0) = 5$ and $t_A = \mathbf{1.609}$ (s); and
(b) the point **B** will have coordinates $\delta(t)/\delta(0) = 30$ and $t_B = \mathbf{3.555}$ (s).

If we compare the changes of the coordinates of **A** and **B**, it can be noticed that the ratio $\delta(t)/\delta(0)$ have increased six times (30/5) and t has increased only (approximately) twice (3.555/1.609).

---

[13] There is a huge amount of mathematics behind the Lyapunov exponent, but this book is not a place to present this mathematics.

If we go further[14] and calculate the value of t assuming that $\delta(t)/\delta(0) = \mathbf{5,000}$ (1,000 times more), then the value of t will be equal to 8.517 (the change will be approximately only 5.3 times).

And this is the main reason that any prediction regarding the future states of chaotic system will not be valid: For a short period of time (only 5.3 seconds), the distance of the molecules in the smoke of cigarettes will be 1,000 times bigger than the previous (initial or let's say "the starting") one. So, it can be seen that in chaotic systems, there is a huge non-linearity of uncertainty (it goes exponentially high) and this affects predictability. For the sake of truth, we can predict (theoretically, it is a deterministic system), but the uncertainty of this prediction will be huge.

There is no reason to go to engineering to understand the presence of uncertainty in the prediction of the behavior in chaotic systems. Take into consideration a weather forecast. It is a most common chaotic system and there (approximately) the weather forecast for only one day has uncertainty of only 10% (depending on the data available) and uncertainty for three days' weather forecast is 50%. So, do not consult the weather forecasts for five or seven days, they are highly uncertain and there is no use of them.

You may ask at this point: Where is the connection between the unpredictability and sensitive dependence on initial conditions with the chaotic systems?

Let me remind you what happened to Lorenz when he had experienced the totally different picture of his computer simulation neglecting the last three digits of the initial conditions. These three digits (extremely small value) resulted later in a totally different picture produced by simulation.

Having in mind that we cannot provide by measurements exactly known values of initial conditions (refer to Section 1.10!) to any process, in each measurement, we will have some uncertainty (errors) in the measured results regarding initial states of the system. This uncertainty (errors) will exponentially increase in chaotic systems making them totally unpredictable, even in a short portion of time.

Having chaos in your systems can be predicted (in which point it could start), but if it starts, it will be totally unpredictable in which state your system will go. That is the main reason for this book: To increase awareness between the Safety Professionals and engineering personnel what can go wrong with the non-linearity of their systems.

8.5   RESETTING THE CHAOTIC SYSTEM

As it has been explained in the previous chapters, the Chaos happens as highest category of effect in systems with three or more dimensions (variables) and under particular circumstances (initial conditions). Such a chaotic movement is flowing of a liquid (oil, petroleum, water, etc.) through the pipelines. For a particular speed of the liquid in the pipes, the capillary effect and the friction of the liquid with the pipe's walls will cause uncontrolled movement (turbulence) which will result in increased pressure on the walls and decreased velocity of the liquid inside the pipes. We do

---

[14] This is something which cannot be seen in Figure 7.7, but it can be calculated by the equation for t above.

not like this, because it can have damaging effect on the pipes and the flow will be turbulent.

Such a problem happens in the hydro dams where accumulated water is used to produce electricity. There, the accumulated water, under pressure, through the pipeline comes to the turbines connected to electric generators and turbine rotation produce electricity. If the speed of the water is too high, the chaotic movement in the pipeline will damage the surface of the pipes (whatever the material is used) and these irregularities of the surface of the pipes will further decrease the speed of the water. So, the water hitting the turbines with smaller speed will decrease the effectiveness of the generators which will result in the decrease of the produced electricity. If not repaired periodically and on time (which is a very much expensive job), the pipeline will be damaged so much that electricity cannot be produced.

In general, the Chaos, although innate in the Nature and in the engineering, is not good for systems. There is an eternal question: How we can stop Chaos if it happens? Could we just switch off the system and switch it on again? Could we reset it?

Unfortunately, it will not work . . .

The reason for having Chaos are the particular values of the variables and/or parameters and only for these values it will show up. So, the main point is to know when it could happen and to choose a Working Point (a stable one) which would be far away from the values of the variables and/or parameters which can trigger Chaos. In our case, with the flow of liquid in the pipes, we should choose a speed of transfer which would not produce Chaos or provide pipelines whose surface is extremely smooth and has low coefficient of friction for that particular liquid. These are constraints which could help.

If we switch off the system and switch it on again (reset it), even if the initial conditions are not the same, the Chaos will show up again, but the shape, the time, and its characteristics will differ. Do not forget: Different initial conditions will result in different shapes of the trajectories.

Could we avoid the Chaos in our systems . . . ?

Not really . . .

However, humans have found some solutions which could help them. Speaking about the weather, there are some meteorological Software programs for weather prediction where the forecast is based on particular algorithms which update themselves (something like rudimentary Machine Learning). They produce, by themselves, a model of weather which corresponds to the weather in that particular area. This model is based on the available observations regarding meteorological conditions made in that particular area.

So, although the weather is a chaotic phenomenon, these algorithms are capable to provide better forecast than the one which can be expected for this particular (chaotic) situation. Actually, when some of the current observations are not in accordance with the model, they will be neglected. This is something which is common in Statistics when some samples of the data (named as outliers) which do not correspond to the set of data used are simply discarded from the set (neglected in calculations).

However, the Chaos is (unstable) noise of non-linear dynamics and our Complex Systems are inherently non-linear due to use of Control Systems (which are mostly based on Feedback). There is no need to introduce big change in your system to produce Chaos. It is enough to design a system where some of the system's variables and/or parameters are close to the critical values and only simple and small change of environmental factors could trigger the Chaos.

Another thing is that trying to gather data about chaotic movement and use this data for statistically process will not help. Although the chaotic movement is deterministic, the statistic processing will treat it as random movement. It means that in the case of probability calculations of risks connected with Chaos, we should be extremely cautious: Statistics will not work.

So, whatever you do, be aware of the possibility to create a Chaos.

9 Non-linearities in Real Complex Systems

9.1 INTRODUCTION

Let's see how the non-linearities are present in our real Complex Systems. To be more precise and to emphasize again, they are present even in the simple systems, but in the Complex Systems, they are more critical due to their safety aspects.

The point worth emphasizing here is that the design of some non-linear system is simpler than the design of some linear systems. This is a little bit paradoxical statement having in mind the richness of non-linear phenomena, but in manufacturing industry, most of the products are non-linear and it is easy to produce them with the non-linear systems than with the linear systems.

In addition, the linear Control Systems would certainly require components with high quality to produce linear automation in the specified operation range, while the non-linear Control Systems may permit the use of less expensive components with non-linear characteristics, as soon as the operation is kept within tolerances.

The important criteria which apply to the non-linear Complex Systems can be summarized as follows:

1 *Stability* is something which is most critical. What is the point to have a Complex System which cannot provide stable operation? This stability is important not only for the model used for design, but also for the model installed at the customer's site (where the environmental conditions differ).
2 *Accuracy* is also important: What is the point to enter a command, if the system in use cannot execute it accurately? It is very critical for Complex Systems which provide tracking (aircraft, spacecrafts, ships, missiles, etc.).
3 *Speed of response* is also important. We would like to have immediate response of each command entered in the system. For some systems, especially in the medicine, the speed is a matter of death or life.
4 *Robustness (Resilience)* is the sensitivity to external influence factors (which maybe missed during the design). The system should be resilient to all these factors when performing the operation of interest.
5 *Cost* of a Complex System is determined mainly by the number and quality of the Control Systems used as Subsystems, which will put the desirable limitations (constraints) in the operations. These Subsystems and their components should be chosen with a sense of effectiveness and efficiency of their particular application.

DOI: 10.1201/9781003404811-11

9.2 THINGS TO BE REMEMBERED ABOUT NON-LINEAR SYSTEMS

Let's move forward and emphasize the most important things for real non-linear systems from engineering point of view:

(a) The linear systems are easier to solve using mathematical methods. The non-linear systems are not so easy to solve and sometimes it is even impossible to solve. They need very advanced mathematical methods, which sometimes give only approximate description of what is going on in the vicinity of the point of interest. That is the reason those non-linear systems are linearized in the vicinity of the point of interest and linear methods are used to assess the system.

(b) The non-linear systems do not need infinite time to jump into infinity. They can change its state very fast. For example, the explosion (infinity in regards of the volume) is non-linear phenomenon and the explosive can change its state in matter of seconds (finite time). So, for the operators: No time to react.

(c) The non-linear system could have many fixed points (stable, unstable, and hybrid), and as such it is a curse and a blessing: A curse because it is hard to analyze and obtain all fixed pints, so sometimes the non-linear system can be in a state which is totally unknown for us; a blessing because if there are few stable points, we can strive to design the complex System choosing one of these stable points as Working Point and others can be operational states when system will move, if it is faulty. This is called *Fail Safe* state. Under any circumstances, the unstable point may not be chosen to be a Working Point of the Complex System!

(d) The non-linear systems are prone to bifurcation. This is a process where the nature of fixed points changes: The stable one could transform itself into an unstable or hybrid one, and vice versa. This is a problem and to make efforts to solve it, we need good knowledge of the systems.

(e) The limit cycles in the non-linear system produce oscillations which do not depend on initial conditions. As such, they provide more flexibility in building the oscillators. The unwanted oscillations are known as vibrations, and as such they are a problem for the operation of the mechanical parts of the Complex Systems.

(f) The principle of superposition does not apply to non-linear systems, so the other solutions cannot be created to investigate the behavior of the system. That what you get, you work with that. In the scope of this superposition principle, it is also good to mention the frequency response of the linear and non-linear systems. If you bring to the input of the linear system two signals with different frequencies, the system on its output will also have two signals with different amplitudes, but with the same frequencies. If you put the same two signals with different frequencies at the input of non-linear system, at the output you will have these two signals and bunch of other signals with different amplitudes and different combinations of these two frequencies (intermodulation products and harmonics).

(g) The Complex System for some values of its variables and parameters can enter chaotic state (Chaos). As such, it is critical to pay attention to adjust the parameters not to allow such a situation. Also, there is need to adjust the parameters not to allow any variable to enter the Chaos in the range of its tolerances.

(h) The same initial conditions are not an issue for linear systems. Linear equation determines without any doubt the dynamics of linear systems and I can say, roughly, it is uniform. But for non-linear systems, two different initial conditions could produce two different dynamics (states).

(i) The Complex System where the non-linearity is present could enter different modes of behavior and even jump from one state to another autonomously, depending on the changes of the external influence factors. This could not happen with the linear systems.

It is good to mention here which engineering systems are excluded from the following considerations, as a consequence of being non-linear:

(a) Systems presented by partial differential equations
(b) Engineering systems with time delay in their operations
(c) Systems presented by differential equations with an additional stochastic (random) term inside
(d) Dynamical systems which contain discontinuities in the function or in the derivative

9.3   TIME RESPONSE OF THE COMPLEX SYSTEM

When the input signal $x(t)$ is introduced into a non-linear system with particular transfer function, we can notice that output is presented by two types of the same signal—the "Transient response" and "Steady-state response".

The Transient response is actually a temporal response; at a particular time, the changes inside will tend to become zero. So, as time passes, the Steady-state response is the one which characterize the output (Figure 9.1).

Figure 9.1 shows the response to the switching-on the power supply in our system in use. As it can be noticed, the output does not immediately change from 0 to the required voltage, but there is first a small overshooting and then after a particular time or after

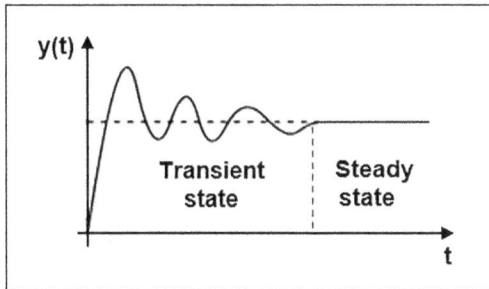

FIGURE 9.1   The responses to the power supply when switched on.

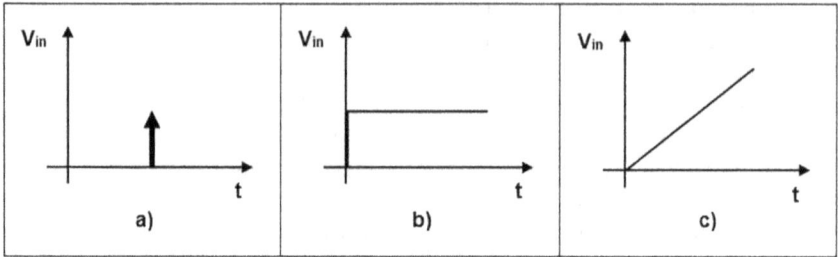

FIGURE 9.2   Three most used signals as inputs for analyzing non-linear system.

few dumped oscillations (Transient state), asymptotically, the voltage stabilizes on the required value (Steady-state). These oscillations do not necessarily exist. Actually, it is valid for underdamped system, where damping factor[1] is between 0 and 1.

This situation with transient states and steady states usually affect all commands and changes in the systems and in operations (processes, activities, etc.). It is important that each command need time to "travel" to a particular actuator and there is no momentary (immediate) response to every command, because the actuators also need time to execute each command. So, this should be taken into consideration during modeling and simulation during the design of the system.

In the beginning, the dynamic behavior of the system is primarily related to the Transient response and there the speed and its stability can be investigated. In the Steady-state, the accuracy of the voltage can be checked and, if not satisfactory, it would be a system error of the power supply.

To analyze these responses, in practice, the different test signals are used. Three of them are most important (Figure 9.2).

The first most used input for analyzing outputs of the Complex Systems is the Dirac impulse (Figure 9.2a). As it has already been mentioned in Section 4.4, this impulse is used to find the transfer function of the system (aka impulse response).

The second one is the step function (Figure 9.2b) and this is actually good for checking the delay imposed by the switching-on and switching-off the system. With this function as input, it is easy to determine time duration of the Transient state: When the increase of the output reaches 63% of the final value, then the Transient state starts.

The third one is the ramp function (Figure 9.2c) which is used to investigate the saturation of the output of the system and the level of distortion imposed by the signal which constantly increases.

9.4   THE MOST COMMON NON-LINEARITY IN INDUSTRY: THE FEEDBACK CONTROL

In general, almost every Control System (even the linear ones) introduces some non-linearity in the Complex Systems, but let's move to one very common and concrete example of non-linearity in Control Theory . . .

---

[1] More details about damping will be provided in Chapter 10.

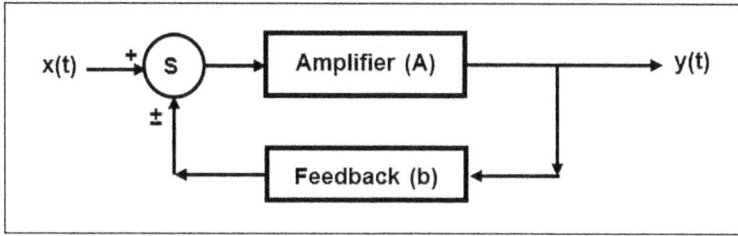

FIGURE 9.3 Feedback circuit used in electronics.

A good example of two-dimensional oscillations and limit cycles used as Control System, which were presented in Chapter 7, is the Feedback method. It is used in industry for control, but the example presented here is used in electronics for control or to produce oscillators (Figure 9.3). This is actually a closed-loop system elaborated from the point of control methods in Chapter 5, but seen from another context here. In this example, I will not use differential equations to describe the system, but there are two continuous variables which produce non-linearity of the system. So, this section includes a combination of explanations given in Chapter 5 and Section 7.7.

The simplified formula which depicts working of this system is as follows:

$$x(t) = \frac{A}{1 \pm b \cdot A} \cdot y(t)$$

Here, $y(t)$ is the output signal presented as total amplification of the input signal $x(t)$, S is the circuit to produce a sum of Input and Feedback signal on its output, A is the amplification of Amplifier, and b is the amplification of the Feedback circuit. The $\pm$ mark is defining the type of feedback (negative or positive).

As it can be noticed, the output signal is controlled by a combination (product) of amplification of Amplifier (A) and Feedback circuit amplification b. which means it is a system with two variables which is described by non-linear algebraic equation.

Depending on the value of the $\pm$ sign, there are three possible situations:

1 If we use + (it means that the phase of signal b is same as the phase of signal on the Output), the feedback is positive and it can be used to produce oscillations (closed trajectories or limit cycles).

2 If we use "−", it means that the signal b is an opposite phase from the signal on the Output, the feedback is negative, and it is used for control of the amplification of Amplifier.

3 Looking at the formula above, it is clear that $b*A$ may not be equal to 1. In such a case the nominator will be 0, which means the total amplification of the circuit will be infinite. In other words, the circuit will be destroyed (unstable limit cycle will bifurcate into spiral with radius which will progress toward infinity).

This Feedback circuit is a good example of stable "limit cycles" which happen exclusively in the non-linear dynamical systems. The limit cycle is a closed trajectory which is actually the state of system which is needed to keep it stable. Inside this closed trajectory the system is stable. It means that in our case from above, the amplification A is always limited, so it cannot destroy the amplifier (for such a purpose, we adjust the feedback to be negative). Outside of this trajectory (positive feedback), the system is unstable, it can produce the so-called *avalanche effect* during oscillations and it can damage the system.

I will not go into details of this two-dimensional system, simply because it is too complicated to make any point from safety perspective, but I would like to claim that in each continuous[2] system, three things may happen which are subject to changes or adjustment of parameters:

(a) In one-dimensional systems, bifurcations may happen.
(b) In two-dimensional systems, bifurcations and oscillations may happen.
(c) In three-and-more-dimensional systems, bifurcations, oscillations, and Chaos may happen.

As mentioned in Chapter 5, Feedback is the basis of almost all Control Systems used today in our lives.

9.5   MEMORYLESS NON-LINEARITIES AND NON-LINEARITIES WITH MEMORY

Roughly speaking, the non-linearities which are present in real Complex System can be divided into two categories based on their input/output characteristics: Memoryless non-linearities and Non-linearities with Memory. A basic difference between these two categories is that for non-linear memoryless systems, the output signal will be the same as that for the input signal, but it will be attenuated or amplified. For non-linear systems with memory, the output signal will be modified (distorted) and cannot be recognized.

9.5.1   MEMORYLESS NON-LINEARITIES

These are non-linearities where the output does not depend on the previous state of the signals which are brought to the input. So, the output will neglect the previous situation with input signals and it will produce a signal based only on the newly attached signal on the input.

Such well-known non-linearities are Dead Zone, Relay, Saturation, and Quantization.

9.5.1.1   Dead Zone

This is a situation when a system does not provide any signal on the output, although there are some signals on the input. Figure 9.4 presents two cases which could be found in industry.

---

[2] Please have in mind that the three things here could happen only in continuous systems described by non-linear continuous equations. The systems described with discontinuous systems presented by difference equations (mapping!) can experience Chaos even in one-dimension.

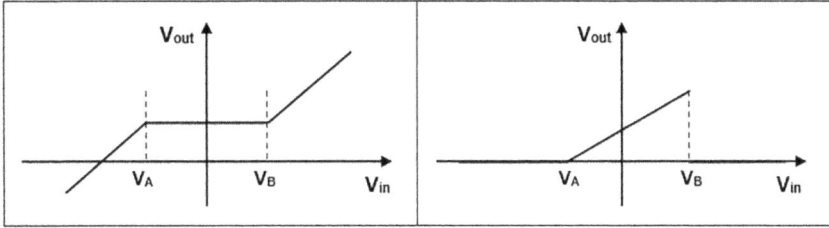

FIGURE 9.4 Dead zone non-linearities.

The first case (left side of Figure 9.4) happens with unstable systems. In the vicinity of zero (**0**), which should be actually stable fixed points (or using engineering dictionary, it is our system Working Point), if input voltage V_{in} changes in the limits between V_A and V_B, the output voltage V_{out} will not change. In this case, it will have some value determined by the adjusted Working Point. Outside the areas left of V_A and right of V_B, the output will leave the "save heaven" of the stable Working Point and the system will become unstable. If we like to keep our system operational, we need to limit the operation between voltages V_A and V_B. This is a non-linearity which can endanger our system, so we put constraint on the control of the system.

Such a case of Dead Zone happens also with the AC and DC motors. There is a particular value of the current and/or voltage which could provide movement to the motors. Below this value, the rotors, due to their mass and the present fiction, will not move if enough power is not applied. The similar phenomenon happens in the hydraulics with the valve-controlled actuators and in some of the hydraulic components.

The second case (right side of Figure 9.4) happens when the output V_{out} of our system should maintain the tolerances imposed to the input V_{in}. For the input between V_A and V_B, the output is linear. If the input is smaller than V_A or bigger than V_B, the output will be zero (**0**), which means that the system is Fails Safe. Another example of this case would be the operation of the wheels of the aircraft during landing and takeoff. During takeoff, they are down and after takeoff they move up in the fuselage of the aircraft. During landing, the opposite operation happens.

Such non-linearities can show up in the systems where there are sensors (which sense the value of input) and associated actuators (which adjust the output). The industrial robotics is a place where these non-linearities are very much present.

9.5.1.2 Relay

This is non-linearity which reminds us of switch, not only of relay. I would not explain what is ordinary switch, but the relay is an electronic component based on inductor and electromagnetism which is used to control other devices by switching-on and switching-off, depending on the algorithm applied by the Control System. The diagram of functioning is shown in Figure 9.5.

The relay is a very common component which is mostly used in power supply parts of the Complex Systems in industries, because it allows us to switch-on and switch-off high voltages and currents with small signals coming from the low-power

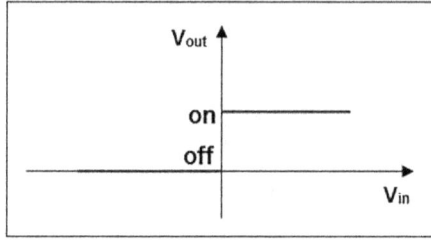

FIGURE 9.5   Relay (switch) non-linearity.

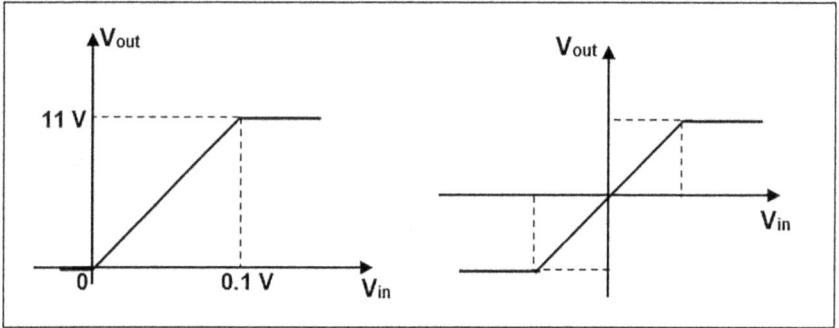

FIGURE 9.6   Saturation of the amplifier with inputs bigger than 0.1 V.

Control Systems. The same relay switching is applied to industrial robotics with "hands" when the "hands" have to be stopped in some particular positions. The controls of the spacecrafts (manned and unmanned) are provided by gas jets which are *on–off* devices.

However, this will not be ideal as it is presented in Figure 9.5. There will be a Transition State response and Steady-State response, as it was shown in Figure 9.1.

9.5.1.3 Saturation

Saturation is an example already discussed in Section 5.5.1 (open-loop control system) with the amplifier with amplification of 100 times and 12 V (volts) power supply. To not enter the saturation, the input signal should be lower than 0.1 V. If it is greater, the output signal will be "saturated" and full with distortions. If it is speech which is amplified, you will not recognize who is speaking. The diagram for two cases of saturation is shown in Figure 9.6.

As it can be noticed on the left side of Figure 9.6, all voltages on the input from 0 V to 0.1 V will change the output linearly from 0 V up to (approximately) 11 V. It is going approximately up to 11 V, although the power supply is 12 V. This happens because there are some other components which will "spend" some of the voltage of power supply. But for the increasing voltages on the input, the output voltage will stay always on 11 V, which means that the amplification of the amplifier will decrease and, in addition, the signal on the output will be distorted. This is all effect of non-linearity.

On the right side of Figure 9.6 is shown the saturation for sinusoidal signals used in analog and some digital communications. This is actually the saturation in reality.

The saturation is very much present in engineering and it is very common for Control Systems with imposed hard limits for actuators. For amplifiers, if there is no limit circuit on the input of the Complex System (AGC—Automatic Gain Control), the rise of input signal may create distortion of the output signal. It could also be caused by bad design or by error in the algorithm or procedure for operating the Complex System. The saturation is common in transistor amplifiers (reducing the gain and introducing distortions in the output signal) and in the servo motors (cannot increase the torques). For example, the hydraulic valves which are controlled with servo motors are saturated due to maximum flow rate of the pipes.

The disturbances due to saturations in unstable systems could produce unwanted oscillations and in stable systems it will cause delay in response to decreased gain of the system. The point is that the non-linearities caused by saturation will be noticed only for big signals on the inputs. Bigger inputs will cause bigger distortions. For small signals, the system could operate normally. A good thing is that saturation will affect only amplitude of the input signal, but the frequency and the phase will not change.

Aviation has experienced such situation with unwanted induced oscillations. The Swedish first prototype of their new military aircraft, JAS 39 Gripen, had crash during flight test in February 1989. The pilot has tested resilience of the aircraft to "pilot-induced oscillations" (part of the regular flight test). This is a situation when the pilot gives a series of opposite commands (up–down or left–right) in a very short period of time and the Software of that prototype could not cope with the changes due to saturations. Investigation showed that the design of the control laws of the aircraft control Software had deficiencies, because they do not assume that the Software can be saturated by the commands. The accident was also supported by the gusty winds as contributing factor. However, the pilot survived only with fractured elbow.

Although the investigation pointed to the Root Cause of accident, the changes in design, obviously, were not enough because four years later (August 1993), another JAS 39 Gripen crashed near Stockholm. The Root Cause was again "pilot-induced oscillations". Eventually, the problem was solved in 1995.

9.5.1.4 Quantization

Quantization is used all the time. It is a process very much used during digitalization of music, human speech, etc. It is connected with two types of the Subsystems known as ADC (Analog to Digital Convertor) and DAC (Digital to Analog Convertor). The ADC is transforming analog to digital signal and DAC is doing the opposite (digital to analog signal).

The input/output characteristics of the process executed by ADC is shown in Figure 9.7.

Quantization is transformation of one set with infinite number of elements (analog) into another set with finite number of elements (digital). As it can be noticed in Figure 9.7, all-possible voltages (on the input of device which will provide quantization) between V_A and V_B will be presented on the output as one voltage V_O. It means

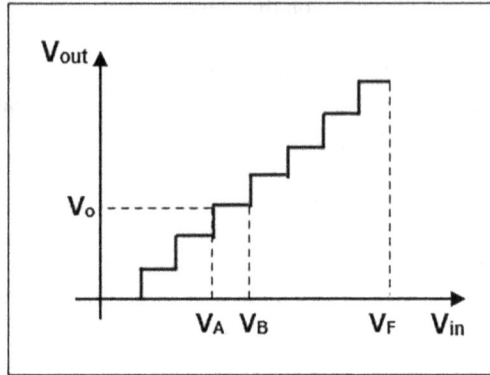

FIGURE 9.7 Quantization between the input and output signals.

that all (infinite) voltages from 0 to V_F on the input will be presented on the output only with eight voltages.

Today there is no system which does not use quantifications, because our engineering world is mostly digital and all the analog things need to be digitized to be used in computers (which are also digital).

9.5.2 NON-LINEARITIES WITH MEMORY

These are non-linearities where the output strongly depends on the previous state or even on the entire history of signals which were brought to the input. So, the output will remember the previous situation with input signals and it will produce a signal based on the combination of this history and the newly attached signal on the input.

The memory in the operation comes from the systems which have capabilities of storing the energy. This storage of energy is a big problem in Complex Systems, because it can provide instability and sustained vibrations.

Well-known non-linearities are Backlash in mechanical transmission systems used on gear wheels, Hysteresis is a magnetic and electronic system which contains capacitors and inductors.

9.5.2.1 Backlash

Due to imperfection of producing the gear wheels, there is always some gaps between them in transmission mechanisms. When the driving gear starts to rotate, there is a small gap in the next-in-the-row gear wheel. This gap is presented by the angle between the "teeth" of the gear wheels, so there is a small delay in moving the mechanism, which corresponds to the dead-zone. This delay could accumulate, if there are many gear wheels. It is presented in Figure 9.8.

In Figure 9.8, the gap in one direction rotation of the gear wheel is b and in the opposite direction it is $-b$.

As it can be noticed in Figure 9.8, depending on the direction of rotation of driver wheel gear, there are two possible gaps (b and $-b$).

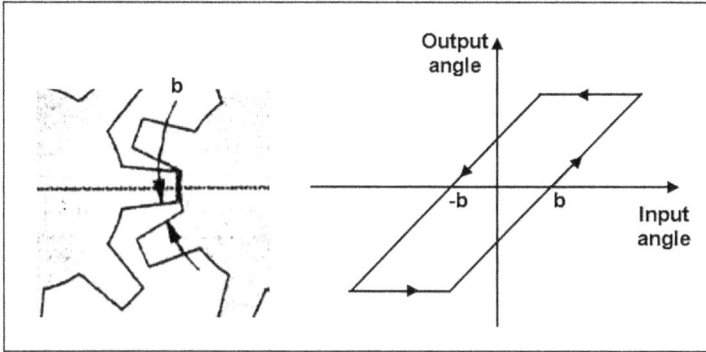

FIGURE 9.8   Backlash between two gear wheels in transmission mechanisms.

9.5.2.2 Hysteresis

There is one very known physical effect in magnetism which can be connected to non-linearity of changes of the magnetic material. It is good to mention it here, because magnetic effect is very much used in industry.

There are three types of materials with regard to their magnetic characteristics: ferromagnetic, paramagnetic, and diamagnetic.[3]

The ferromagnetic materials are attracted by magnets and they can be easily magnetized. Such materials are iron, nickel, cobalt, and their alloys. Paramagnetic materials are weakly attracted by magnets and such materials are aluminum and oxygen. Diamagnetic materials cannot be magnetized and they are weakly repelled by magnetic fields. Such materials are carbon and copper.

In the context of non-linear dynamics, ferromagnetic materials are most important for us. These materials can become magnets by themselves if they are found in strong magnetic field. This process is very much used in industry to produce magnets, but the process of magnetization is non-linear and the diagram of magnetization is produced by the so-called *Hysteresis curve* (Figure 9.9).

Figure 9.9 shows magnetization of ferromagnetic materials. On x-axis is presented strength of magnetic field (H) and on y-axis is presented level of magnetic induction (B) which is a measure of magnetization of ferromagnetic material in the process of magnetization.

We put the ferromagnetic material in the space where we can control magnetic field (position 0). Increasing the magnetic field, the magnetic induction inside the ferromagnetic material will increase and it will follow the non-linear curve. For particular value of H, the magnetic induction inside the material will reach point 1. The strength of the magnetic field at this point (h_c) is known as "saturated magnetic field". Increasing the strength of magnetic field H above this value has no effect on ferromagnetic material: It will not increase its magnetism and magnetic induction will be constant (saturated). Physically, I can say that h_c is the highest possible

---

[3] To be more scientifically correct, there are also antiferromagnetic materials (chromium and some types of glass) which have more complex interrelations with magnetic fields.

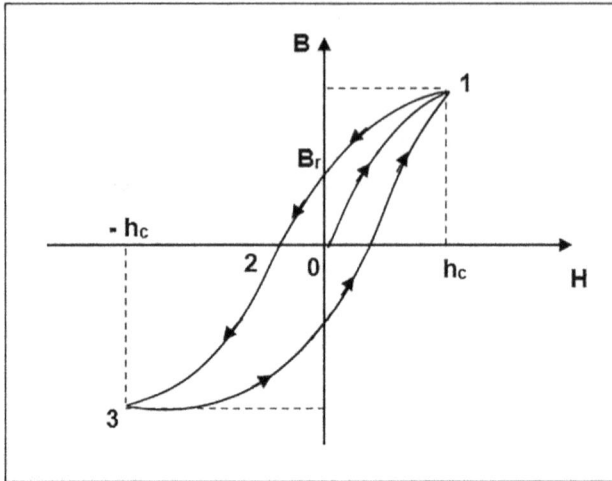

FIGURE 9.9 Hysteresis curve for the magnetization of ferromagnetic materials.

magnetism which can be achieved by a particular ferromagnetic material. From the point of view of non-linear dynamics, this is a stable point: If the magnetic field is moved away from the material, ferromagnetic material will stay magnetized by this value. If, in this position (1), we start to decease the strength of magnetic field, then changes will not stay on the same curve. Now, the changes of B will follow the upper curve in Figure 9.9. When H reaches the value 0, the ferromagnetic material will keep some magnetism inside itself and this magnetism (B_r) is known as "reminiscent magnetism".[4]

If we continue to change the magnetic field, but now by increasing it in opposite direction, ferromagnetic material loses its magnetism for a particular strength of magnetic field in opposite direction (point 2). This strength of opposite magnetic field is known as "coercive force".[5] This is a force imposed by outside magnetic field in opposite direction to cancel the "reminiscent magnetism" B_r.

If we continue to increase the magnetic field (H) in the opposite direction, we will reach point 3, which is actually reaching again h_c, but with opposite value (that is the minus in front of h_c). This is a situation where ferromagnetic material is magnetized again with the same magnetism as at point 1, but with opposite value (opposite poles). If at point 1, the ferromagnetic material was magnetized as *S–N* (South–North), then at point 3, it will be magnetized opposite, as *N–S* (North–South).

Decreasing the strength of magnetic field will produce the "mirror image" of the same curve as in the previous process. If the strength of magnetic field oscillates

---

[4] In some literature, you will find terms "retentivity", "remanence", or "residual magnetism". And there is nothing wrong in each of them . . .
[5] In some literature, you will find it also as "coercive magnetic field". I am not sure what is the correct term . . .

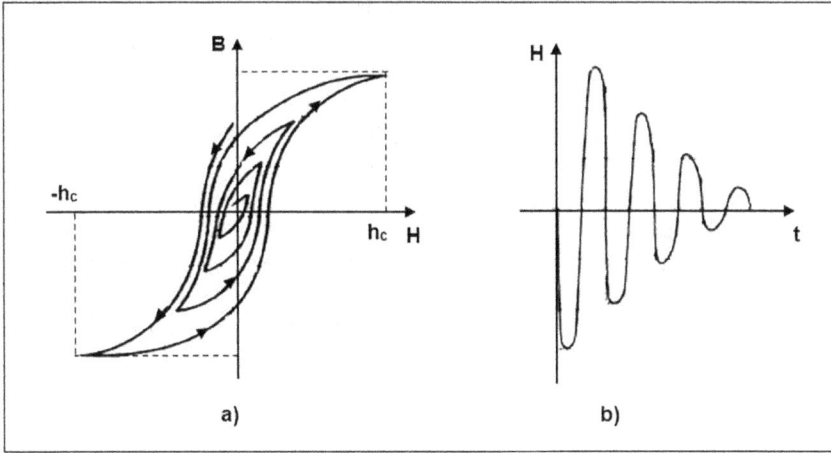

FIGURE 9.10 Hysteresis curve for demagnetization of ferromagnetic materials.

between h_c and $-h_c$, the value of magnetic induction will follow the curve (known as hysteresis curve) as given by the arrows in the diagram (Figure 9.9).

The process of demagnetization is given in Figure 9.10.

Figure 9.10a presents change of hysteresis curve for demagnetization when the strength of magnetic field is applied to the gradually decreasing oscillations, as presented in Figure 9.10b. This process is known as "Degaussing" and it is very much important in aviation. During flying the aircraft, there is a possibility for aircraft to be struck with lightning. In such a case, due to enormous current produced by lightning, the fuselage of the aircraft will be magnetized and if not demagnetized, it can affect the accuracy and precision of the navigation instruments (compass) inside. So, in each aircraft, there is device known as "Degaussing instrument" or "Growler". There is a procedure for use of this instrument on the ground and it can be done by engineer who is trained in the procedure. Simply, the improper use can magnetize some parts of the aircraft incidentally. In addition, there is so-called Magnetic Particle Inspection which is part of periodical preventive maintenance of the aircraft.

The hysteresis for demagnetization remembers the initial state of magnetized material, but the path which was used for magnetization is not the same as the path for demagnetization. Of course, it depends on the values of H which are used for magnetization and demagnetization.

9.5.2.3 Electronic Devices with Inductors and Capacitors

To be clear about this, I would also mention that in mechanical systems, there could be non-linearities with memory and these would be the systems with a mass with ability to store thermal or potential energy. To be clearer, these are systems supplied by linear or rotational movement where the energy is stored in inertia of the movement. However, the situation in electrical engineering is more critical because there are more such elements which can store energy.

In electronics, all devices where there are inductors and capacitors[6] as building elements are non-linear and with memory. In simple words, any device which is used for transmission of radio-frequency (RF) power (as text, audio, and video information) is non-linear and with memory. The reason for that memory are the energy-storing devices known as inductors and capacitors.

The memory is coming from the facts that for inductor, its current depends on the previous values of the voltage on its contacts and, for capacitor, its voltage is the result of the previous values of the current through it. It means that for the inductor, the value of the current will delay the voltage and the time of the delay will be proportional to the value of the inductance of the inductors. The same will be valid for capacitor, but there the voltage will depend on the values of the current through it and the time of delay will be proportional to the value of capacitance of the capacitor. This is for time domain, but for frequency domain, this delay can be seen as frequency-dependent gain and frequency-dependent phase shift of the signal. In general, there will be non-linear distortion of the signal.

This is more critical for the RF power amplifiers (PA), because these are devices used to amplify the signal for transmission through the antennae. The signal must not be amplified, but if we need to cover large area with RF signal, it should be done. These devices not only amplify the signals, but also amplify the so-called harmonics and intermodulation products of non-linearity.

These harmonics and intermodulation products[7] are distortions based on the frequencies which are combinations of the frequencies used to create the useful RF signal. Some of the intermodulation products belong to the same frequency band as the useful signal and they distort the information transmitted by that signal.

In simple words, the non-linearities in RF systems produce distortion of the input signal where the output is presented not with one, but with few signals, presented as combinations of the frequencies of the input signal. That is the reason that the non-linearity effects on the signal always need to be analyzed in the frequency domain than in the time domain.

The emphasis should be put more on transmissions of signals based on amplitude modulations (AM) than on those based on frequency modulations (FM) or phase modulations (PM). The reason is that for AM, the useful information is hidden in the amplitude of the signals.

For those who know that most of our communication is made by FM systems and computers and Internet are using PM modulations, please keep in mind that in aviation and space explorations, only AM is used. Due to high speed of the flying aircraft and space ships, the Doppler effects change the signals modulated by FM and PM, so only sustainable solutions for communication and data transfer are based on AM.

Actually, knowing that almost 90% of electronic systems today are digital, you can understand the effect of non-linearities in our lives . . .

---

[6] The three basic elements (components) in electric circuits are resistors, inductors, and capacitors. The resistors are not capable of storing energy inside. All these three elements alone are linear, but their combination in the circuits produce non-linear circuits.

[7] Harmonics are multiples of used frequencies and intermodulation products are linear combinations of the used frequencies. Any of them are unwanted effects of non-linearities in the electronic systems.

The only way to deal with non-linearity of the electronic systems is to provide linear circuits which can be achieved by the methods for linearization of the input/output characteristics of the system. This could be achieved, but the price will be decreased efficiency of the system or, in other words, it will increase the cost of everyday operations. So, some kind of compromise is inevitable.

Some of the techniques used for linearization are Feedback (closed-loop control system) and Feedforward Control System. The most used is Feedback Control System if the closed loop could provide sufficient incremental gain. It is achieved by up-conversion and down-conversion inside the loop, but it increases noise and loop delays, so it makes sense for the frequencies lower than 100 kHz.

Feedforward Control System is used mostly for higher frequencies and wideband amplifiers. There the distortion signal is sensed before it reaches the output, it is processed, and at the output, it is extracted again from the signal.

There are few other control methods to deal with non-linearities, but they are beyond the scope of this book.

9.6   TURBULENCE

Preparing this book, I read somewhere that the turbulence is (maybe) the last unsolved problem of Classical Physics. At the beginning of the 20th century, the Classical Physics was overtaken by Quantum Mechanics and Theory of Relativity. Most of the physicist moved their interest in these two areas and, really, there was not too much to be discovered in the Classical Physics anymore.

When we say "unsolved", it implies that there are many models that describe the appearance and maintenance of the turbulence and most of them depend on the situations where it appears. So, it cannot be modeled uniquely, but on a case-by-case basis.

However, the turbulence may be one of the most interesting phenomena of non-linear dynamics. Whenever there is some kind of movement of gas or liquid, the turbulence can show up, as such it affects a huge number of industries and in some of them, it can be pretty dangerous (aviation, railway, oil and petroleum, etc.). Research in the area of turbulence has shown that the turbulence can be (pretty successfully!) modeled as fractals which make it part of Theory of Chaos.

In general case, mathematically, the turbulence can be expressed by three-dimensional Navier–Stokes differential equations. It is easy for me to present these equations here, but there is no need for such a thing. The point is that the solutions of these equations behave very strange and they could be extremely unstable even with clear, smooth, and reasonable initial conditions.

So, I would not go into mathematical details explaining these equations, but I will focus mostly on some general properties of turbulence which any Safety Professional should know.

The low-speed flow of the water in the pipe is known as laminar flow. If we increase the speed[8] of the water, in some moment, as a result of value of the speed, viscosity of the water, and material of the pipe wall (which produce friction of the

---

[8] Please have in mind that the speed is a scalar and the velocity is a vector!

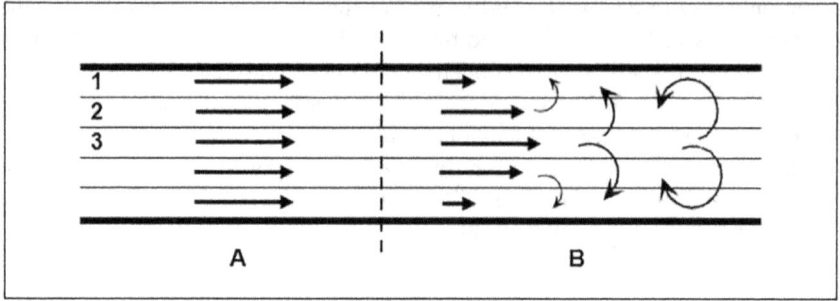

FIGURE 9.11   Turbulence of flowing water in pipe.

water with the wall), there will be change of the laminar movement into turbulent (chaotic) movement.

Let's explain the turbulence in more detail with an example of flowing water through pipe (Figure 9.11).

As it can be seen, I have (virtually) divided water through the pipe in five layers, but only layers 1–3 in Figure 9.11 are important. The lowest two layers can be considered as a mirror image of layers 1 and 2, if the layer 3 is the mirror. Please note that part A and part B are presented as difference in time, not as difference in space (different locations in the same pipe).

To explain it better, I will introduce two components of the velocity vector—longitudinal (parallel to the walls of pipe) and vertical (normal to the walls of pipe).

On the left site (part A), there is laminar flow of water which means that all layers have the same speed. Using the components of the velocity, the longitudinal component is same for all layers and vertical component is 0. Here, I can define a laminar flow as a flow where the vertical component of the velocity is 0.

The movement of the molecules of water in each layer is a straight line (only longitudinal component is present) caused by imbalance of the pressure from both the left side (high pressure) and right side (low pressure) in the pipe. How this pressure is achieved is not important. However, it can be by water pump installed on the left side, pushing water toward right side, or simply, by decreasing the height of the right side (inclining the pipe down from the right).

If the pressure imbalance increases, the speed of water inside the pipe will also increase. For some speed, which depends on the viscosity coefficient of the water and the material used to build the pipe (friction coefficient between water and pipe), there will be also an imbalance in the speeds of the layers (as shown in part B of Figure 9.11). This is known as "liquid shear".[9]

Look at the layers 1, 2, and 3 in part B of Figure 9.11.

It can be noticed that there are no same speeds of the flow in the layers.

Due to increased friction between the water and the wall of pipe, the vertical component of the flow will increase gradually and, at the same time, longitudinal component will decrease for each layer. It will produce different longitudinal velocities in

---

[9] There is a "wind shear" in aviation which is extremely dangerous during landings of the aircraft.

different layers. It means that the speed of layer 1 will not increase by the same value as the speed of layers 2 and 3. There will be an imbalance of speed in all layers, due to different vertical components of the velocity.

In other words, it means that the water in layer 2 will progress more than that in layer 1 in time. It will cause disbalance on the pressure inside the pipe in part B and the laminar flow will disappear by producing vortical (chaotic) movement of the water: The turbulence is born!

This vorticity will cause further decrease of the speed in part B and the longitudinal and vertical components in part B will behave randomly, changing their speed and directions non-stop. It all means that turbulence there will affect the efficiency of the transfer of the water through the pipe (decreased speed means less water will go through the pipe at the same time).

On the boundary between the water and the wall of the pipe, in the case of turbulence, the longitudinal component will disappear (its speed will be 0). It means there is no turbulence on the boundary in the pipe. There, the water will form a thin film on the surface of the pipe and the thickness of the film will depend on the viscosity and speed of the water and smoothness of the surface of the pipe wall.

The part of physics which deals with some kind of movement is called Kinematic and there are three kinematic characteristics of the turbulence:

1 *Irregularity:* There is a lot of randomness in turbulence, and as such the statistics play a very important role in explaining it. Turbulence starts when the movement of the gases or liquids reach a particular speed and a regularly repeating pattern in behavior cannot be discovered.
2 *Mixing:* With turbulence, there is a "mixing" of the particles of gases and liquids. It means that they change their locations with regard to each other (mixing tea with sugar and milk by teaspoon mixes the molecules).
3 *Vorticity fluctuations in three dimensions:* Turbulence in space (3D) results in high vorticity (a whirling motion of gases and liquids in 3D).

The turbulence is ubiquitous in our life. It was explained for fluids, but it can be found in other engineering areas (aircraft, cars, trains, technical devices, household appliances, etc.), biology (circulatory and respiratory systems of living organisms), and Nature (atmosphere, rivers, oceans, etc.). Even in cosmos, there will be turbulence in galaxies, stars, planetary interiors, etc.

Let's give here another explanation which is actually present in our lives: car and turbulence (Figure 9.12). What is mentioned for car applies also for submarines, boats, aircraft, and trains or any other technical system which moves through some gas or liquid.

This is something which is well-known by any man in the world: The turbulence behind the car when it is driven. This turbulence depends mostly on the speed and shape of the car. Higher speed and more rectangular shapes cause bigger "drag" caused by bigger turbulence behind the car. In reality, "drag" is the term used in aerodynamics providing explanation for the resistance of the aircraft to its movement during the flight. "Drag" is a resistance (burden) in the flight and it is one of the four main forces which affect the aircraft flight.

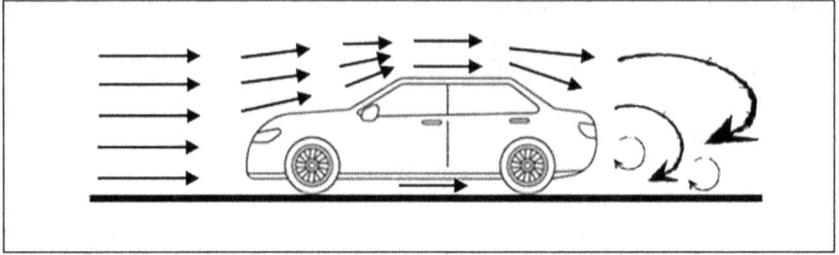

FIGURE 9.12   Turbulence behind moving car.

However, in the case of car, a bigger "drag" produces less speed and more fuel consumption. In these cases, the "drag" caused by turbulence can be decreased by providing more aerodynamic and "smooth" shape of the car.

There is another place where turbulence can show up in the car and this is combustion engine in the car. It occurs during the working of the exhaust valves in the engine. Every time when the valve is opened, it produces turbulence of gases which try to get out as soon as possible (due to the pressure inside the cylinders after burning the fuel and air mixture).

Eventually, I would give one simple example for solving the turbulence of a bullet in a flight. As anything else which move through air in a particular speed (car, train, aircraft, etc.), the bullet from the gun is prone to turbulence. But the engineers produced circular movement of the bullet through air by adding spiral grooves in a gun barrel. These spiral grooves are responsible for the bullet spinning. Unfortunately, it cannot be applied to the cars, trains, and aircraft . . .

10 Vibrations

10.1 INTRODUCTION

As mentioned in Chapter 7, in non-linear systems with more than two dimensions there are limit cycles. Sometimes they are wanted and intentionally produced (electronic and telecommunication systems) and sometimes they are unwanted. For the purpose of this book, I will call the wanted limit cycles oscillations and the unwanted limit cycles vibrations.

Unwanted limit cycles (vibrations) in non-linear systems could show up without any external influence, or in other words, they could be self-excited as evolution of a dynamics of the system (operation, process, activity, etc.). It means that it is an innate property of the non-linear system which could change its state toward unprovoked vibrations. An important thing to mention here is that the amplitude of these vibrations does not depend on the initial conditions (where the state of the system was previously). And to make things worst in regard to vibrations, they are not always easily affected by the parameter changes in the system. The point is that limit cycles represent cyclic motion which provide cyclic forces and these cyclic forces could be very damaging to mechanical materials.

When any part of the Complex System under consideration vibrates unintentionally, any part of that system could be displaced from its original static position equally in any direction. The reason for that is that the vibrations could "travel" and be transferred from one part to other parts. Here it is the interesting case: The average distance "travelled" by this motion is zero, but interchange of the energy is constant and, as adverse energy, it can cause damage if the system is not designed to survive these vibrations.

Roughly speaking, three elements which are part of each vibration are "Spring", "mass", and "damper" (Figure 10.1). We can say that they constitute a "vibrating system".

Of course, the "spring" is not a spring literally, but a part of the vibration system which stores and releases the potential energy. It could be anything which possesses particular elasticity.

Similarly, the "mass" is not a mass literally, but it could be anything which could move and provide inertia to its movement. This could be, actually, the part of our Complex System which could be subject of vibrations.

The "damper" is a part of vibratory system which contributes (intentionally or unintentionally) to the system to lose the energy and as such the vibration would vanish if not supported by some external source. The natural friction of the parts of the system during vibrations is an excellent example of "damper".

Keeping in mind these three elements of the vibrating system, we can describe it as spring-controlled, damper-controlled, and mass-controlled systems and it depends on which element controls the vibrations.

DOI: 10.1201/9781003404811-12

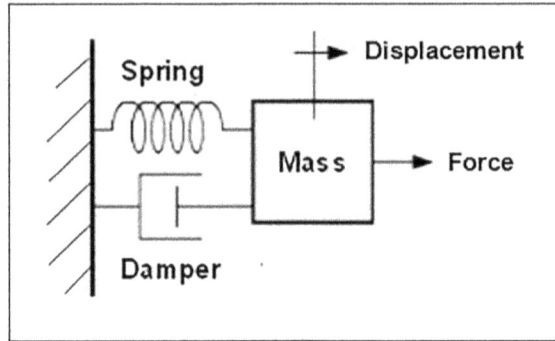

FIGURE 10.1   Three elements of the vibration system.

Here, I can state another problem with vibrations. The vibrations are supported by the system itself due to its non-linearity. If they are not caused by the change of the parameters or variables (intentional or unintentional) which will provide existence of limit cycles, the vibrations would vanish in time.

This non-linearity with vibrations causes another effect known as hysteresis. The magnetic hysteresis was already explained in Section 9.5.2.2 and it is noticeable during vanishing of the vibrations due to memory of the stored energy. This is similar, as explained in Figure 9.10, but the shape of the curve is elliptical. Hysteresis with vibrations is between the force applied and displacement of the system, which obviously is not linear.

In general, the damages caused by vibration in the Complex System in any manufacturing industry can be many:

- They would cause stress on the material and the imposed force levels may lead to faults of the parts. The faults could be sudden (immediate) or gradual (fatigue).
- Vibrations would cause wear of system parts and it will decrease the system performance. This will impose decrease of operational time of the system and increase a need for Preventive or Corrective Maintenance.
- In a production system with excessive uncontrolled vibrations, the product parts may be inaccurately produced, even beyond the tolerances, and as such they will be rejected as scrap.
- The extensive uncontrolled vibrations also cause an unstable installed production system (machine) to move away from its foundation. This is very much important in the cases where the system has small mass and the vibrations are triggered by (oscillatory) movement of its foundation (earthquakes and buildings, for example).
- The vibrations, as any oscillatory event inside the Complex System, could propagate in the form of waves, so it can be spread to other Subsystems or even outside and the waves could cause additional damage to other assets and to the environment.

The vibrations (oscillations) were discovered from the dawn of humanity and it was mostly through the strings of the bow and the tone which they have produced if touched briefly. So, for a long period of time (and today also), they were used for music instruments. After the industrial revolution in the 18th century, with the first machines, the bad influence of the vibrations was noticed and the "fight" against them will last forever. The point is that the vibrations in Complex Systems used in engineering and industry are ubiquitous, and as such they need due diligence in dealing with safety problems caused by them.

Among safety problems caused by vibrations, I would mention just few of them from Risky Industries—vibrations of the wings of the aircraft during turbulence, vibrations of the rockets during launch, vibrations of nuclear reactors during earthquakes, etc. These are areas where engineers extremely devoted attention and considerable testing to produce resilient Complex Systems.

But there is another problem with vibrations: They represent the repeating stress to the mechanical parts of the systems. And as such the wearing of the parts and the fatigue to the material could lead to catastrophic consequences. I would like to present one such example which happened on 17 August 2009 in the Siberian part of Russia.

Sayano-Shushenskaya dam and hydroelectric power station was built with a capacity of 6,400 MW of power. It used ten units for electricity production, each of them equipped with Francis's turbine. This turbine has a very good efficiency, up to 95%, but there is a flaw. Using it with 60–100% of the power, it is stable, but for achieving these values, it must pass the range of 40–60% power (a critical range), where there is instability. This instability is presented by vortex of the water inside the turbine which produces vibrations of all parts of the system which supports the electricity production. As such, operation in this range is forbidden for the Francis's turbine.

In the few months before the accident, the turbine of the Unit 2 from the power station has already shown increased vibrations during its functioning, so this unit was not working and it was put in reserve. A few days before the accident, there was a fire in neighboring power station, so the Sayano-Shushenskaya had to work in full power. Unit 2 was used only when there were picks in the power requests and, in four days, it was intensively put on and put off from operation many times. In all of these engagements, the Francis's turbine had to "pass" the critical range many times. These "passings" increased the strain due to vibrations at the basis of the turbine and the bolts (keeping it steady) broke down. The water from the pipe (which was 10 meters in diameter and 225 meters high) made a hole in the building and flooded the rest of the units. Seventy-five people died and the Sayano-Shushenskaya hydroelectrical power station was out of service for 4.5 years until it was fully repaired. I would not speak about the money spent for repairment.

Not only the machines, but also the humans are annoyed by extensive vibrations. The strong yelling and other strong noises (motors, engines, scratching the surfaces, etc.) are unpleasant situations for humans, very often causing significant discomfort. In the cars, buses, trains, and aircraft during flights, the vibrations are also very strong causes of discomfort when traveling.

10.2 RESONANCE

As said earlier, the vibrations are oscillatory movements caused by non-linearity of the structure or operation of our system and they are actually limit cycles. As any oscillatory movement, they are characterized by its amplitude, frequency, and phase, but from the viewpoint of bad things caused by vibrations, amplitude is the most important factor. If the amplitude of vibrations is bigger than the stiffness of our system in use, it is clear that the vibration will cause damage. Usually, the amplitude is not taken into account, but engineers use the so-called root mean square (RMS).

The RMS is used because it includes into itself the periodical shape of limit cycles. If we try to find the average amplitude of simple sinusoidal signal, it will be 0. That is the reason that RMS is calculated by the following formula:

$$A_{RMS} = \sqrt{\int_0^T \left[A \cdot \sin \left(\omega t + \varphi \right) \right]^2 dt}$$

Here, A is the amplitude, $\omega = 2 \cdot \pi \cdot f$ is the circular frequency, f is ordinary frequency, and φ is the phase.

The square root and the integral of the square of the oscillation take into account the periodic changes in the curve over time. As such, it provides information on how much energy is stored in the vibration which actually provides us with information on how powerful the vibration is.

The critical thing with any of these limit cycles is the resonance, which actually depends not so much on amplitude as on the frequency and phase. The resonance is a sum of the limit cycles which affects the system and this sum become very critical, if they are with the same frequency and the same phase. Sum of the limit cycles is possible always, even if there are signals with different frequencies and different phases, but if the frequency and phase of the signals are same, then the resonance will result in one signal which will continuously increase its amplitude until some of the component of the system fails or until the system is not destroyed.

The simple example of resonance is marching of the soldiers on the bridge. Having in mind that they march with the same "frequency and phase" (marching steps are in synchronization), the force caused by any of the soldiers stepping on the bridge during the march will add and after a short period of time can become bigger enough to destroy the bridge. The effect will be more evident if the frequency of marching is the same as natural frequency[1] of the bridge.[2] That is the reason why marching on the bridge is forbidden!

The "marching on the bridge" is perhaps the most critical case about the resonance with the limit cycles. But, in general, the unwanted limit cycles (vibrations)

---

[1] The natural frequency of the system is the inherent frequency of free oscillation of this particular system. This frequency is inherent and constant for each system and it depends on the physical dimensions of the system.

[2] In November 1940, in Washington, the Tacoma Narrows Bridge collapsed due to the vibrations caused by high winds. This is a well-known engineering design error where designers used new techniques for building a bridge. Unfortunately, they did not analyze the aerodynamical forces on the proposed design in the case of strong winds and hence the bridge fell down due to extensive swaying caused by the winds.

contribute very much to the wearing and fatigue of mechanical systems and they are eternally a problem there. The control in the system where the vibrations are present could be very hard to achieve. In general, each system (structure) has its own natural frequency and when the frequency of the imposed vibrations to that system (structure) matches its natural frequency,[3] it could lead to excessive deflections, faults, and destructions.

10.3 TYPES OF VIBRATIONS

Besides the definition as limit cycles, I can generally define the vibrations as unwanted periodic movements of the bodies or phenomena where there is constant interchange of the kinetic and potential energy. The interchange of potential into kinetic energy, and vice versa, happens when there is a change of the velocity of the "mass". The kinetic energy is that what causes the damage of the systems during the vibrations.

The kinetic energy is strongly connected with the movement of the objects and the vibrations, as limit cycles are dynamical periodical phenomena which cause displacement of the parts in the systems. As such, the quantity of this unwanted movement measured by the displacement caused by the level (force) of the vibrations is a merit for the fault of the parts and the failure of operations of the systems.

As explained previously, the vibrations could be caused by the change of some parameter or variable or by sudden mechanical hit or shock in the system. But it is worth mentioning that vibrations, although unwanted oscillations, could be intentionally produced to provide some benefit for the humans. Such examples are vibrations used to mix the things in chemistry, drugs in pharmacy, to allow smooth pouring of the concrete into the structure, etc.

The classification of the types of vibrations can be done through pairs of terms which could be used at the same time to explain any vibrations.

The first pair of classification would be free and forced vibrations. Free vibrations are these which, after imposing an initial force, continue to vibrate by themselves and they do not need external energy for that (their internal elasticity is a reason for that). Simple examples for this class are the musical string instruments. Of course, because they do not need external energy to be maintained, they will not last forever, but, depending on the imposed force and the friction with the environment, they will stop in time (the tone will vanish after a particular amount of time).

There are three types of free vibrations and they are shown in Figure 10.2. The displacement during vibrations is presented by two-head arrows.

The forced vibrations are those which are maintained by external energy imposed by periodical force with the same frequency. The external energy could come from the external force applied to the system or by externally induced periodic motion of the foundation that supports the vibrating system. Simple example for this case is the rotation of the motors and engines which cause forced vibrations. The forced vibrations can cause resonance, and as such they need to be controlled. In addition, there is also problem with forced vibrations in non-linear systems.

---

[3] Natural frequency of human intestinal tract is in the range of 4–8 Hz and any extensive oscillations in this range would cause trauma.

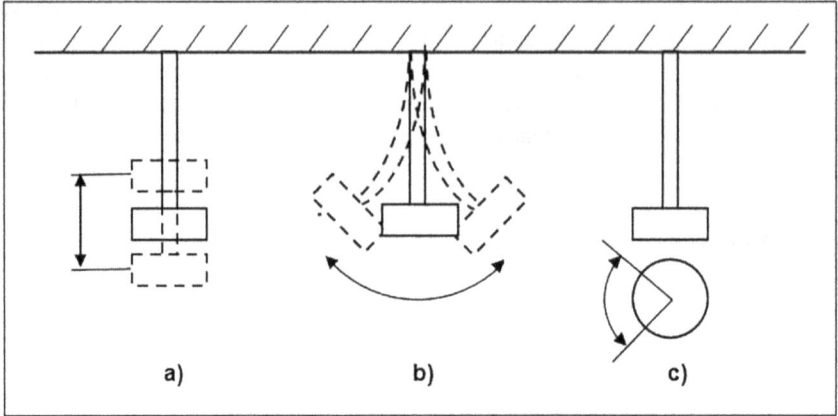

FIGURE 10.2   Longitudinal (a), transversal (b), and torsional (c) vibrations.

The force which will cause forced vibration with particular frequency, in the linear system, will provide response of the system in the same frequency. If there are few frequencies, response of the linear system will be only in these frequencies. There, the principle of superposition is valid. In non-linear system, the response of the system due to its non-linearity will provide many other frequencies and superposition principle is not valid there.

The second classification pair is damped and undamped vibrations.

During vibrations, if energy is lost due to friction between the vibrating parts or due to some other reason, the vibrations will disappear with time. These are damped vibrations. Our efforts, as engineers, would be to damp the vibrations (as much as it is possible) in our Complex Systems. If there is no reason to lose an energy, the vibration could last forever. However, such a thing could not happen in reality, but it is enough for these vibrations to last very long time, so that they could produce damage.

Going back to Chapter 4 where stability of Complex Systems was explained, let me remind you that I have explained the use of Laplace transform and there, the stability was investigated by s-plane where s was a complex variable expressed as $s = \sigma \pm j\omega$.

If you remember, ω (as imaginary conjugate) is the cause of periodical movement and σ (as real part) is actually a damping factor which provides boundary characteristic (limitation of amplitude of change) of the systems. If σ is less than 0 (negative), the solutions (roots) expressed by the term $e^{\sigma t}$ will asymptotically travel toward 0 and, accordingly, the system will be safe. It means that we need to artificially (during the design) provide damping factor σ less than 0 and possible vibrations will thus disappear.

Using Figure 10.1, I can provide the equation for damped vibrations;

$$m \cdot \frac{d^2 x}{dt^2} + \zeta \frac{dx}{dt} + k \cdot x = F(t)$$

Here x is the displacement, m is the mass, ζ (zeta) is the damping ratio (coefficient), k is the spring coefficient, and $F(t)$ is the force causing vibrations (for forced vibrations). If the vibrations are free, $F(t) = 0$.

Figure 10.3 presents undamped (left) and damped vibrations (right).

The so-called damping ratio or damping coefficient (ζ—zeta) is a dimensionless measure describing how the vibrations in a system will disappear after disturbance. This has already been mentioned in Section 4.6.3. Based on the values of roots (s) and value of ζ, there are three cases of damping:

1 If the roots are real ($s_1 = \sigma_1$ and $s_2 = \sigma_2$; $\zeta > 1$) it is called *overdamping*. In this case, damping is very much efficient and vibrations disappear with aperiodic oscillations almost immediately.
2 If the roots are a complex conjugate ($s_{1/2} = \sigma \pm j\omega$; $0 < \zeta < 1$)), the damping is called *underdamping*. This is a very common case when the vibrations disappear with periodic oscillations in time, but not immediately. Figure 10.3b presents underdamped vibrations which disappear with logarithmic decrement gradually.

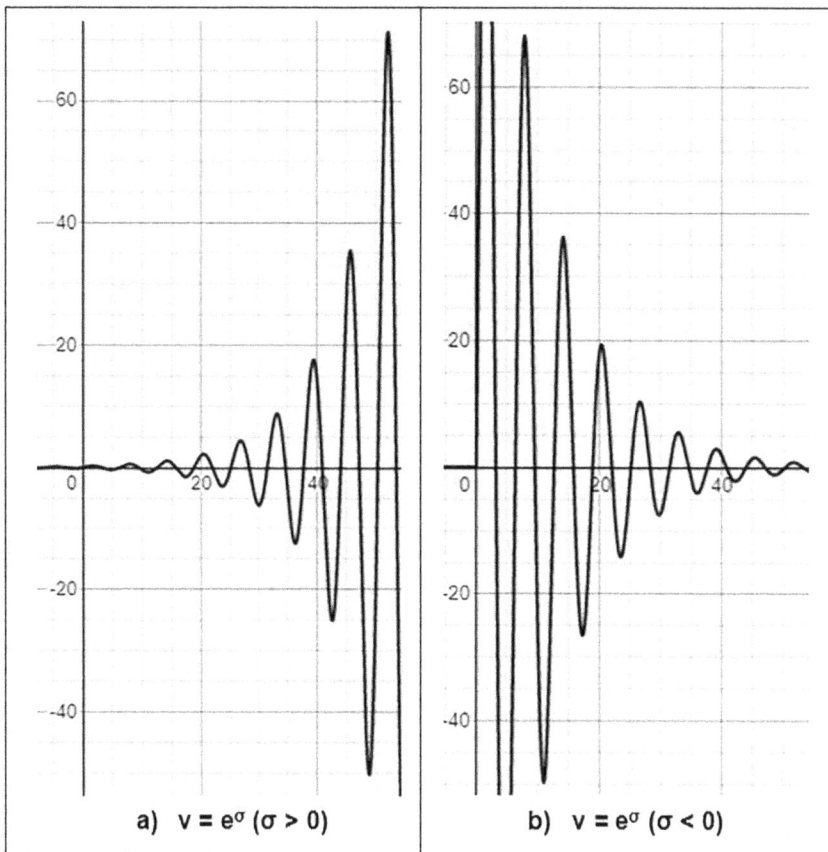

a) $v = e^{\sigma}$ ($\sigma > 0$)   b) $v = e^{\sigma}$ ($\sigma < 0$)

FIGURE 10.3   Undamped vibrations (a) and damped vibrations (b).[4]

---

[4] This figure is very much same as Figure 4.4, but transferred in the area of oscillations.

3 If the roots are equal (only one root $s_{1/2} = \sigma$; $\zeta = 1$), the damping is called *critical damping*. This could be the transition between "underdamping" and "overdamping" and it happens only with systems with single DoF. It is generally very useful because the system would move toward equilibrium in the shortest time without oscillations. It is very much used for car suspensions.

The third classification pair is connected by the mathematical presentation of the vibrations. If the vibrations could be expressed by linear equations, then they are linear vibrations. If they can be expressed by non-linear equations, then they are non-linear vibrations. In accordance with that what was explained in previous chapters in this book, ever type (linear and non-linear) follows the behavior of the corresponding linear or non-linear systems.

The fourth classification type is deterministic or random vibrations. If the characteristics of vibrations (location, frequency, force, etc.) are known at the moment and they can be predicted in the future, these are deterministic vibrations. If they cannot be predicted, these are random vibrations. The random vibrations are described by probability, if there is enough data for statistical use.

The fifth classification pair depends on the nature of vibrations and they can be periodic and irregular (random) vibrations. The periodic vibrations are with approximately constant frequency and they are permanent within the system. The irregular (random) vibration pops up at random intervals and can be with non-regular shape and frequency.

So, the vibrations could be any possible combination of these for classifications at the same time: they can be free, damped, non-linear and random or forced, undamped, linear and deterministic, etc.

There are also so-called transient vibrations which happen when the system is under some mechanical shock (hit by something) and these are vibrations which disappear soon after the shock. They could be treated as part of Section 9.3 with Transient state and Steady state (when the vibrations do not exist anymore).

10.4   NON-LINEAR MATHEMATICS BEHIND VIBRATIONS

It was mentioned at the beginning of this chapter that vibrations are unwanted limit cycles. Let's go back a little to clarify the mathematics behind the vibrations. In general, there are few types of limit cycles (oscillations with different frequencies!). But I will stay with the situation where there is only one limit cycle (Figure 10.4).

Figure 10.4a presents stable limit cycle which means that any point outside and inside the limit cycle will asymptotically move toward the limit cycle itself. It is a stable limit cycle and these are stable oscillations. If there are small changes in the frequency of the oscillations caused by any external or internal influence factor, the change will disappear very fast and the system will maintain the stable oscillations. If the limit cycle is intentionally produced, this is a dream of every designer: To create stable and resilient oscillator.

Figure 10.4b present a semi-stable (hybrid) limit cycle. Any point outside the limit cycle will asymptotically move toward the limit cycle, but any point inside the limit cycle will move away from the limit cycle toward the center (which means the center could be [not necessarily] the stable point). The vice versa can also happen: Outside points to move away and inside points to move toward the limit cycle. This case can be accepted by some designers as compromise, but it is not a good solution in general.

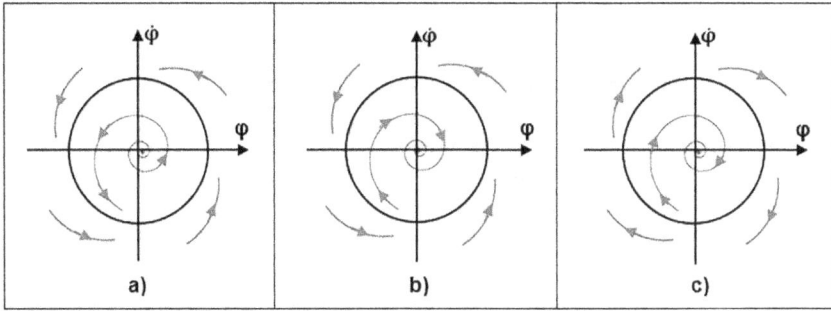

FIGURE 10.4   Different limit cycles in non-linear systems.

Figure 10.4c presents unstable limit cycle which means that any point, outside or inside the limit cycle, will move away from the limit cycle. So, any small disturbance of the frequency of the limit cycle will result in failure of the oscillator. If the oscillator is intentionally built, it would be a worst nightmare for any designer.

So, the vibrations are unwanted limit cycles which can show up and exist in the system, and as such they are problem. But this statement is flexible. If you have in your design of the Complex System unwanted limit cycle, then try to make it stable. If the state of your system uncontrollably changes into any unwanted limit cycle, the system will oscillate, but it will not necessarily be damaged. So, try to make it to withstand with the oscillations (resilient) or try keeping the amplitude small, so that the overall stress will not be damaging.

This is the main reason that the designers and operators of the Complex System must be familiar with all these things. The gathered knowledge about your system will help to handle all problems during the life cycle of the system with sustainable number of resources!

10.5   ANALYZING THE VIBRATIONS

The vibrations are damaging and as such we need to eliminate or mitigate them. There are two methods which can be used for that and usually the first one comes after the second one.

The first one is the mathematical method and it consists of a four-step procedure:

1 *Mathematical modeling:* This step is already considered, with all the good and bad things about it, in Section 2.3. I have read somewhere that the modeling helps with (approximately) 95% of the problems in the real-world vibrations. The point is to provide a model which will be used later to express the behavior of the system under vibrations as mathematical equations which can be used later for simulations. You need to be careful in defining the necessary number and distribution of masses, springs, and dampers. In addition, the input forces together with possible foundation motions must be defined. As said before, it is cost-effective, having in mind the availability of the Software applications (programs) and power of

today's computers, but it could produce only approximate solution which could be, in some cases, a catastrophic one. There are many cases where there are good models in theory, but they represent badly as systems in reality.

2 *Deriving the equations:* When the model is (satisfactorily) ready, the equations could be derived. These will be the dynamic equations which present the dynamics of the system and possible effects of vibrations to the well-being of the system. Here I must remind you of the term "degree of freedom" (DoF). This is the smallest number of coordinates which are used in equations to fully describe the system under consideration. But the DoF depends on the number of the masses in the Complex System and the coordinate system which is chosen to present the displacements. In the model, the DoFs should be defined to determine the modes which will contribute to the response to the implied forces or motions. If the modeled system has few DoFs, it means that there are few natural frequencies.

3 *Solving the equations:* The equations of dynamical behavior of the system must be solved to find the response of the system to the vibrations. This is also a mathematical process and different methods can be used (Laplace transform, iterative methods, linearization, matrices, etc.). The help of the computers with mathematical software is very useful.

4 *Interpretations of the results:* The solutions are used for particular simulations and these simulations will provide velocities and accelerations of different elements in the system. These results will be based on the resilience of the system and they could provide the requested information about the effect of the vibrations to the operation of the system. Of course, it must not be forgotten that it will be in accordance with the chosen model for the system!

The second method is building the system the best as you can and implement extensive testing for each vibration of the built system. This could provide excellent and realistic picture about the sustainability of the vibrations inside the system, but unfortunately it is a very much expensive method. The amount of the resources needed expressed as time, humans, equipment, money, and everything else is too big.

Speaking about the second method, there are three different measurement analysis which I would recommend to use:

1 *The Fast Fourier transform (FFT):* An FFT analysis is based on spectrum measurement of the vibrations and the measurement system calculates and presents on the screen the discrete Fourier transform[5] of the signal. This provides information about amplitudes and phases of the vibrations on different frequencies in measured signal. Having in mind that the measurement

---

[5] The fast Fourier transform is the algorithm used by computers to calculate the Fourier transform (which is similar to the Laplace transform, but is used mostly for periodical signals).

system is digital, the quality of measurements is directly proportional to the length of measurement and the sample rate used. The bigger the time and higher sampling rate, the better the results.

2 *Power spectral density (PSD):* The PSD is also based on the FFT, but FFT presents the amplitude of each frequency in the measured signal and PSD presents the power associated with each frequency of the measured signal. This helps in reducing the bad influence of the rate of sampling during measurement.

3 *Spectrogram:* A spectrogram is also used by computers and it is capable of calculating several FFTs. These FFTs are actually different individual segments of the time-domain signal. The point is to investigate how the frequencies of different vibrations in time change inside the signal.

There is economically forced synergy in implementing these two methods. First, considerable efforts in mathematical modeling and simulations are made and thereafter, with all adjustments, settings, and changes of the design, the system is built and tested.

However, this could be an iterative process and despite the modeling and simulations done with due diligence, the testing could show some things which need to be changed, adjusted, or set differently.

10.6   VIBRATION'S MEASUREMENT

It is a problem to understand something which cannot be measured. If there is no measurement, we cannot quantify it and cannot compare it with others. I have already explained mathematical theory behind vibrations, but they can rise unexpectedly and as such we need to measure them. As engineers, we are not always ready to deal with theory, so we usually start by measuring the phenomena and try to understand the root cause of the problem.

The methods for measurement analysis in the previous section could work if there is input which would present vibrations. There are two known ways to measure vibrations and these are achieved by different sensors:

1 *Accelerometer:* As the name says, this is a sensor which measure acceleration of a motion of the structure, and as such it is used to measure vibrations. There are piezoelectrical, piezoresistive, and capacitive accelerometers and names are given by the reason for producing electrical voltage/current. Figure 10.5 presents piezoelectric accelerometer. It is actually, a small electromechanical sensor that gives an electrical signal proportional to its acceleration. There is piezoelectrical crystal inside the sensor which produces oscillatory electrical voltage, as a result of the oscillatory movement caused by vibration. The vibrations are transferred to the Base which shake the piezoelectric crystal. Due to inertia, the Mass did not move and it produces stress on the crystal which produces electric voltage. The electrical contacts transfer the voltage to the amplifier and then to the measurement system;

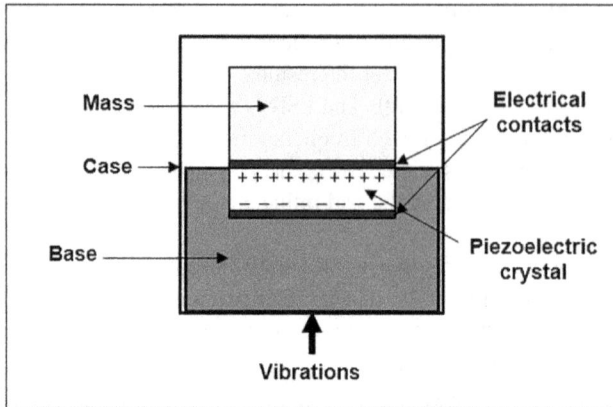

FIGURE 10.5 Accelerometer.

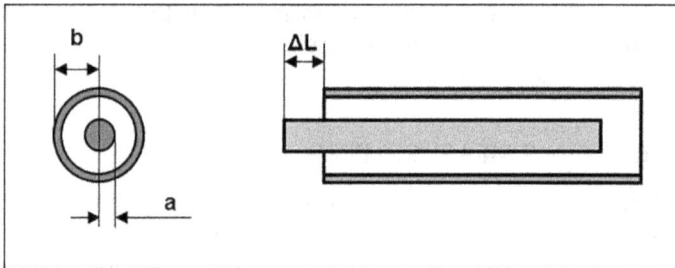

FIGURE 10.6 Cylindrical capacitive displacement transducer.

2 *Displacement transducer:*[6] It is a sensor which provides an electrical signal proportional to the displacement caused by movements of the structure. In the older literature, it can be found under the name "vibrometer".
There are many types—capacitive, inductive, lasers, etc. Figure 10.6 shows
a cylindrical capacitive displacement transducer. The change in capacitance
ΔC caused by displacement **ΔL** due to vibrations is given by the following
formula:

$$\Delta C = \frac{2 \cdot \pi \cdot \varepsilon_0 \cdot \Delta L}{\ln\left(\dfrac{b}{a}\right)}$$

---

[6] The transducer is a sensor which has as input one physical quantity and gives as output another physical quantity for the purpose of transmission and analysis. Capacitive displacement transducer changes
distance as the input signal into electrical (voltage) signal as the output.

The accelerometers are better at high frequencies. They are characterized with rugged and durable use, possess high stiffness, have wide dynamic range and fast rise time, and may be used in wide frequency range.

The displacement transducers are better at low frequencies. They provide unprecedented sensitivity, long-term stability, and Reliability, they are with robust and sturdy construction and with slim-line design.

However, both sensors are used to provide an input into one of the measurement systems used for this particular measurement.

10.7   SOLVING THE PROBLEMS WITH VIBRATIONS—DAMPING

Knowing that the vibrations are oscillations, it could seem that it will be easy to solve them. The periodic movement can be presented as sinusoidal movement and adding another sinusoidal movement with same frequency, same amplitude, and 180° shift in the phase, it will cancel the vibration. This is a beautiful theoretical solution applied to the systems, but in practice it will not work.

Implementing such a solution could help only if the knowledge of vibrations is available in the form of their frequency, amplitude, and phase. Having in mind that vibrations can also be dynamically changeable in time, the need for complete monitoring, in the time of the vibrations, will be needed. It means more complications in our system and more cost in the product.

So, the engineers are using other methods to deal with vibrations . . .

As it has been said in the previous section, there are also damped vibrations[7] and they naturally disappear in time. The artificial (induced by humans) damping is actually what we need to get rid from the vibrations. Damping actually presents dissipation of kinetic energy hidden in the vibrations and, by using safety dictionary, this dissipation is known as "dilution" where the risk is dissipated into something impotent.

In general, we need to make everything to make the vibrations disappear as soon as possible in our Complex Systems! Speaking in mathematical (non-linear) terms, the damped vibrations are actually limit cycles which we transform into vanishing spirals.

There are many methods which are capable to provide artificial damping of vibrations inside the system and they can be divided into two types—Electrical and Mechanical Dampers. Mechanical are more critical because electrical vibrations does not cause fatigue and in general, electrical devices are not prone to wearing in the levels as it happens with mechanical devices. In addition, the mechanical devices are inherently damped by its construction, so sometimes it helps. The metals in mechanical structures are very much prone to vibrations taking into account their elasticity and toughness. The dampers are usually materials capable of dissipating this vibration energy very fast, so they are very much effective, if associated with metal plates or panels.

---

[7] Transient states of the system (when change of the state happens) are damped vibrations and they are explained in Section 9.3. It means that vibrations show up always when a dynamic system changes its state.

So, these are some of the methods used to damp unwanted vibrations in the systems:

- *Dampers with viscous fluids:* These are shock absorbers used in the cars, trucks, and other land vehicles. There is a piston which moves inside a cylinder filled by some fluid (usually oil) with high viscosity. The damping comes from the friction of the fluid with the cylinder wall and depends on the velocity of the fluid movement inside the shock absorbers. The smaller velocity results with better damping and vice versa. There are also shock absorbers which use air, but they are not so efficient (velocity is bigger). This is a linear type of damping,[8]
- *Using covers with highly damping materials:* Rubber is such a material and if glued on the vibrating metal, it will damp the metal vibrations. Bitumen and high-damping foam are also popular. As example, please note that the floors and hoods inside the cars are covered by fat carpets which "kill" the engine vibrations transferred through the chassis, so the noise inside the passenger cabin is acceptable.
- *Coating on metal plates and/or panels:* There are coating materials (bitumastic, liquid rubber, elastomer, etc.) which can be applied to the surface of a metal panels and vibrations will be damped. This method is very popular by providing sandwich construction of thin metal shits with coatings inbetween them.
- *Changing thickness of used material:* This is a method when self-damping can be produced simply by changing the thickness of a vibrating panel. Usually, increased thickness would push the natural resonant frequencies of used panels toward higher frequencies and it will provide less vibration on lower frequencies.
- *Dry friction (Coulomb damping):* It is a method where used and unused surfaces (panels) with high stiffness in mechanical devices are made to rub together during vibration and friction between them provides linear damping. This type of damping can happen even when there is not enough lubrication between the two surfaces. In this case, usually energy is lost as a heat, so the surfaces could be hot.
- *Aerodynamic damping:* This happens during flight of the aircraft. Whenever the aircraft make some rotational flight maneuvers around its longitudinal (or roll) axis, the rotation of the aircraft around its center of gravity, the aircraft itself, through the wings, creates a restoring moment which is known as aerodynamic damping. This damping is a result of the difference between the flows (air pressure) above and below the wings, so the wings behave as viscous air dampers.
- *Structural (Hysteretic) damping:* This is damping[9] of rigid systems where the damping is achieved by the structural stiffness of the material used to

---

[8] Just for clarification, linear damping means the vibrations decrease linearly. It is mentioned that underdamped vibrations decrease logarithmically.
[9] Somewhere in the literature you can find a name Solid damping for this type of damping.

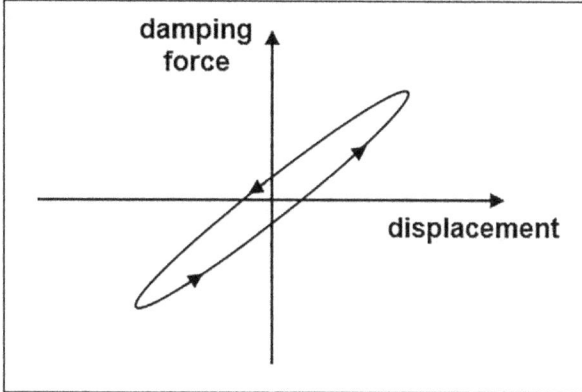

FIGURE 10.7   Damping hysteresis.

build a system. Any vibrations imposed on the system will not affect the system and its operation because in the structure of the system, there is enough internal friction between the materials used, so the system will be resilient. The volume of damping depends on the amplitude of imposed stress. In this case, the system, by itself, provides damping of the vibrations which happens following the elliptical hysteresis curve. Having hysteresis means that these are dampers with memory (Figure 10.7).

*   *Magnetic damping:* When the magnet moves in vicinity of the electric conductor, it will induce the so-called eddy currents inside the conductor. These eddy currents will create their own magnetic field which will oppose the magnetic field of the magnet and the magnet will decrease (damp) its movement. The dissipation of kinetic energy will produce heat and this has to be controlled. This effect is used to provide magnetic damping of the vibratory motion of the magnet. If the magnet is attached to something, the overall construction will be damped.
*   *Torsional damping:* This is a damping applied for torsion vibrations where the displacement is angular. It can be achieved mechanically by install-ing internal or external spring which will damp the torque caused by these types of vibrations.

So, to conclude, the dampers are important for several reasons. They support the operations and systems increasing the stiffness of the structures and, in the cases of external or internal vibrations, they reduce maximum displacement of the structural parts in the system. Most of the dampers use dissipation of adverse energy in the system and by their installation, the adverse energy entering a system is more evenly distributed, so it cannot endanger the operation or system. The point is that they are part of the Complex System and for them also, apply all requirements for monitoring and maintenance, as for other Subsystems.

Part III

Safety of Complex Systems

11 Achieving Safety of Complex Systems

11.1 INTRODUCTION

In Section 1.6, I have given a short explanation on how and why we have produced all these Complicated and Complex Systems and let's now see how they affect our safety.

As said many times before, the science and engineering contributed a lot to improve our well-being, comfort, safety, security, etc. It must be said that they provided (and still provide) many benefits, but, at the same time, all these Complexities have created new hazards. However, these new hazards provide lower risk, and as such these hazards are easier to control (making them beneficial).

But, let's be honest: The Complex Systems are inherently hazardous systems and that is the reason that we must focus our attention to defend them against faults and failures of their operations. Due to their Complexity, they have potential of catastrophic failure. Good thing is that in most cases, one single fault cannot trigger catastrophe, but could trigger other events which could build up an accident. It is impossible to eliminate or mitigate all hazards and risks, but that is the reason that monitoring all the time the Complex System's behavior and its operation is of utmost importance.

With Complex Systems, the Safety becomes a system property, not a component property!

The introduction of new hazards is not only a problem in the implementation of science and engineering in reality. The intention to provide more functionality of the systems (processes, operations, activities, etc.) in the industry has resulted in complex industrial systems. In the area of safety, we provide more safety measures within the systems and (again) we make them more complex.

This is something which I call a "Safety Paradox".

"Safety paradox" is explained by the answer of the simple question: Are more Complex Systems safer?

Yes!!! (We must confess that . . .)

The Complex Systems are intrinsically dangerous, and as such a lot of efforts are put into their production and operations to make them safer. The point is that all these efforts have resulted in increased Complexity of already Complex Systems. By making these systems more safe by being more complex, this "better safety" has produced other types of problems:

- We need more resources (knowledge, skills, experience, etc.) to design, produce, adjust, use, and maintain Complex Systems.

DOI: 10.1201/9781003404811-14

- If there is problem with these Complex Systems, we will need more knowledge, more skills, more resources, and more time to solve the problem(s).
- If the Complex System fails, the consequences of the failure will be such that we need to put more efforts (knowledge, skills, experience, resources, etc.) to eliminate and mitigate them.
- If the Complex Systems fails, after fixing the fault, we will need more time to make them operational again.

And all these things, by itself, create additional hazards!

OK, the questions which arise with these additional hazards are as follows:

What is the reason to increase Complexity of our systems in use when they provide additional hazards . . . ?

Do we really solve something . . . ?

Should we really feel ourselves safer . . . ?

The simple answer to these questions is: Yes!

We solve something and we feel safer. The reason for that is, although we produce more hazards, these hazards that we solve with increased Complexity are bigger regarding the frequency and consequences and those which are newly created are already known (in most of the cases), so we already have solutions how to deal with them. Maybe it looks funny, but we hardly (and successfully) defend the safety of the Complex Systems and these additional new hazards are "collateral damage" of these efforts.

The main point is that to provide safety of Complex System, we must be able to understand all of the hazards. Without proper (I would say, Excellent!) understanding of how these Complex Systems are designed, built, and how they operate, we cannot even start with Hazard Identification (which is the basis of Risk Assessment).

So, that is the reason that I have put a lot of efforts in the previous chapters to explain the reasons for making our systems Complex Systems.

11.2 DIFFERENT TYPES OF SAFETY

I can roughly divide safety in industry on two areas: Occupational Health and Safety (OHS) and Functional Safety.[1]

The OHS, in the modern times, is referred to as OHSE (Occupational Health, Safety and Environment). This is the area which takes care regarding working conditions everywhere—industry, banking, hospitals, schools, etc. These are the areas where the humans, assets, and the environment shall be protected at their working places. This is a very well-regulated area and dealing with safety requirements there is not a big issue if you follow the regulations. There are consistent regulations which determine the working conditions at the working places and there, for example. the fire protection is well-determined (number of fire extinguishers, hydrants, etc.) depending on the situation at the working place. The situation at the working place and its influence on the humans working depend on the volume of the hall (offices),

---

[1] As it has been explained at the beginning of this book, this is not a "Functional Safety" which is regulated with IEC 61508 series of standards!

materials (furniture, storage goods, machines, etc.) used inside, colors of the walls, purpose of the place (production, storage, testing, emergency rooms, etc.), number of humans inside, etc.

The Functional Safety is safety which is connected outside of the factory and it is coming from the functioning of the product (service) which is sold (delivered) to the customer. Whatever the products or services offered are, they must be safe for the customer (humans), for the assets and for the environment. This is the safety which applies mostly to the so-called Risky Industries. These are industries (nuclear, chemical, aviation, railway, shipping, pharmacy, medicine, etc.) where a particular adverse event can have catastrophic consequences for the humans, assets, and the environment.

The regulation in this area is not so well quantically determined and there is good reason for using only qualitative determination of the regulation. For OHS(E), the product (service) is produced (delivered) in a very controlled environment (factory, company, laboratory, etc.) where temperature, humidity, and other working conditions are same everywhere (for example, in Thailand, Norway, or the United Arab Emirates). Although, the outside temperatures and weather conditions at these three places are quite different, in every factory in any of these places, the temperature is controlled in the range of 22–25°C and there is protection from the rain, ice, and wind.

But requirements, for example, for the car sold and driven in any of these three countries differ considerably.

Producing a car which could satisfy all these different requirements could be very expensive and very much prone to contradictory requirements, so the automotive manufacturers adjust their cars to the requirements of the places where they are to be sold. For example:

- In Norway, the car must be ready for harsh winters and ice on the roads, the bottom of the car needs to be well-protected against rust, due to the salt thrown on the roads in winter, and the heating must be fast and considerably powerful.
- In Thailand, the car must be ready for high humidity in the air and at the roads, the rust protection should not be so strong, and the air-conditioning must be balanced to provide heating in winter and cooling in summer.
- In the United Arab Emirates, the car must be ready for high outside temperatures and huge amount of dust which comes from the desert areas, the rust protection below the car is not a problem at all (no salt on the roads and no humidity) and the air-conditioning should take care only for cooling because even during the winter, the temperature never falls below 20°C.

Another factor which differentiates the OHS(E) and Functional Safety is different social, educational, habitual, and religious status to the customers (humans) in these three countries. All these things contribute to the different styles of driving and behavior on the roads. In addition, the quality and conditions of the roads also differ very much, so each sold car in any of these countries put different challenges for the driver and for the car.

These are reasons that in Functional Safety the subjects are obliged to provide Hazard Identifications and particular Risk Assessment which is case-based. As a result of these activities, they must provide efforts to eliminate or mitigate the root causes or consequences connected with these hazards and risks.

The Complex Systems in this book are considered from the aspects of Functional Safety, because the consequences in this area could be very much catastrophic. Anyway, whatever is mentioned here regarding Safety of Complex System fully applies also to the Complex Systems in OHS(E) area.

11.3   DIFFERENT APPROACHES THROUGH HISTORY TO ACHIEVE SAFETY

Speaking about safety in the Risky Industries, I cannot neglect the different development "approaches" in time: Safety-I, Safety-II, and Safety-III.

Safety-I is the name of the safety which has been implemented since the beginning of the industry revolution until 2014 (approximately). In the beginning, Safety-I was defined as "absence of incidents and accidents" and later as the "absence of unacceptable risks".

This is a safety which had three main periods of development which I will explain with an example from aviation (Figure 11.1).

Looking Figure 11.1, it can be noticed that aircraft began to improve in technological meaning from the beginning of the 1920s. This is period where technology was "booming", the industry started to accept new technologies, As a result of that, the engineers thought that further automatization and control of the aircraft could help with the incidents and accidents. I can say that this is the time which was "cradle"

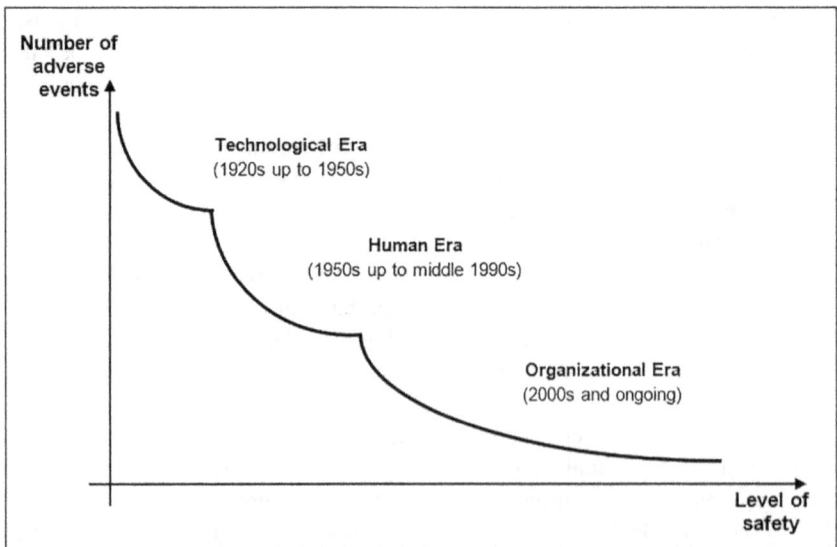

FIGURE 11.1   Safety evolution in aviation.

for the" Functional Safety" approach in manufacturing industry, today supported by IEC 61508. In addition, at the end of this period, as a result of increased Complexity of the equipment, the System Theory emerged. It resulted in the decrease of adverse events, but it was based only on higher Reliability of the equipment.

In the 1950s, more attention was put to data from investigations of the adverse events. At that time, there was a huge development of technology and aviation (as well as the industry) was benefiting from that. Subsequently, there were enough data about incidents and accidents to reach some conclusions. The aviation safety experts have realized that, with the considerable development of the aviation systems based on improved technology, more and more root causes for accidents were not connected with the systems faults, but with the human errors. So, there was a logical step to investigate the human involvement in aviation operations and reasons for human errors.

That was a period named as Human Era and, as it can be noticed in Figure 11.1, the additional decreasing of the adverse events was achieved. This is the time period where Human Factors were established as a scientific discipline and things were better.

Approximately in the mid-1990s, there was enough data to get some conclusions and the investigations showed that many of the human errors were actually triggered by deficiencies in organization and management in the companies. The need for more organized approach was proposed as one of the solutions. The regulatory requirements for aviation subjects to establish new organizational and systematic approach, to deal with safety, were enacted and they became obligatory worldwide.

All aviation subjects (airlines, MROs, manufacturers, ATC, CNS, AIS, etc.) were required to establish Safety Management System (SMS) which would guarantee the structured and systematic approach for improving safety of the aviation operations. This was actually a jump from old-fashioned Reactive Approach toward a Proactive Approach which (when enough data will be available) should move toward a Predictive Approach.

In the scope of the SMS, each aviation subject must establish processes through documented procedures to do Risk Management. The Risk Management consists of Hazard Identification in advance, particularly Risk Assessment for any of these hazards and appropriate Risk Elimination and/or Risk Mitigation for the causes and/or consequences through Preventive and Corrective Actions. It finishes with Risk Assurance which can be translated by "lessons learned" and "situation improved".

This Proactive Approach to safety is based on advance Risk Assessment by quantifying the hazards into risks by dedicating particular frequency (likelihood, probability) and severity to each hazard. There are some proposals to include also time of Fault Propagation, but this is more applicable for calculating Reliability. The Fault Propagation in Complex Systems could happen through a lot of interactions and interdependencies of Subsystems which usually share information, data, commands, etc. So, the fault of one Subsystem would, logically, also affect other Subsystems.

The main point with the Fault Propagation is that, there is chance to have a fault which will not affect functioning of the system in a disastrous way, but if we do not react properly in timely manner, this fault would cause other faults or failures which would affect the Complex System catastrophically. For example, if we notice that the temperature of the car engine is increasing during driving, it is a sign that something is bad with engine cooling. It could be a hole in the cooler, so the cooling liquid has leaked out, the cooling pump is faulty, or the fan is faulty, etc. However, we need to stop immediately and investigate the problem, because no cooling of the car engine will destroy it.

Honestly, I am not sure that it will bring some benefit in eliminating or mitigating bad (adverse) events. Time of Fault Propagation will matter only if there is no continuous monitoring, which happens in industries which are not classified as Risky. This may not happen in the Risky Industries. However, I am not against it, but you may decide by your own whether to include time of Fault Propagation in calculating the risks.

However, there is a method known as Failure Mode and Effect Analysis (FMEA) which was "innovated" by NASA in the 1960s, but today it is regulatory requirement for automotive industry. There, the Risk is calculated through Risk Priority Number (RPN), which is calculated by multiplications of the levels of Frequency, Severity, and Detectability. Detectability is quantification of easiness to detect the problem (fault, failure, flaw, deficiency, etc.) and as such it is more a matter of monitoring. However, it can also be connected to the Fault Propagation.

In the mid-2010s, a new approach to safety was introduced by the group of safety academics, where Erik Hollnagel[2] is most prominent. To make a difference with the Safety before 2014, the new approach was named Safety-II and the previous approach became Safety-I.

Safety, in this approach, is defined by scientists as "ability to succeed under different conditions" assuming that changes of conditions come from equipment itself, humans, assets, and the environment.

By the definitions of Safety-I and Safety-II, Safety-I is dedicated to "things which can go wrong" and Safety-II is dedicated to the "things which can go right". In other words, Safety-I deals with methods which try to rectify the wrong things and they are primarily dedicated to stop faults of equipment and failures of operations. It means that Safety-I is dedicated to stop the "things which can go wrong".

Safety-II deals with activities how to maintain the variations of normal operations to keep them within the tolerances with intentions to continue with these normal operations. So, Safety-II tries to improve the "things which are right", and in such efforts it is supported mostly by Resilience Engineering.

I cannot say anything against Safety-II and I think that Safety-II in cooperation with Safety-I can be an excellent improvement to the safety activities in industry. My problem with Safety-II is that we already have a management system in industry which is dedicated to make "things better" and it is called *Quality Management System (QMS)*. In my humble opinion, cooperation[3] between Safety-I and Safety-II can be very easily achieved by integration of SMS and QMS.[4]

Having this in mind, I support very much Safety-II, but it is something which is "innovated" by the guys who do not understand really what the Quality is!

But there is more: There is Safety-III!

To explain more about it, I will need a new section . . .

---

[2] Erik Hollnagel is Scientific Director at the Institute of Resilient Systems (Seoul, South Korea). In addition, he has professorships at few universities in Australia, Germany, and Sweden.
[3] Maybe "association" is better word in this sentence . . .
[4] For more details about integration of Safety-I and Safety-II, read my book *Quality-I is Safety-II: The Integration of Two Management Systems* published by CRC Press in 2017.

11.4 SAFETY-III . . .

Safety-III is a response to the new technologies which changes (improve) very fast. Nancy Leveson[5] is the promoter of this new approach and I must say, I like it. In Safety-III, the Safety is defined as "absence of unacceptable loses". There the Hazard is situation which can lead toward unexpected and unplanned loses.

The approach is based on resolving few factors which undermine our efforts to provide safety using new systems based on new technologies. As everyone has noticed, the new engineering systems become very complex, because they and the automation and control inside are not based only on mechanical and electrical parts, but there are also computers, microprocessors and Software. Now, these new representations of the new technologies are used to provide automation and control over processes used in the industry and over products used in our daily lives. In general, the digital electronics stretched the technology and it increased the overall computerization of our private and professional lives by creating the so-called, Information Systems.

However, Leveson, in his book *Engineering a Safer World* criticizes the Swiss Cheese model, although it is a model which actually explains and works with the barriers, which are extremely a good model for presenting constraints (Control Systems) in the Complex Systems. If you could remember, at the beginning of the book, the imposed constraints were held as reasons to design a Complex System.

These are some of my disagreements with Leveson:

(a) She said in her book that these are simple models, but the model would be created by the investigator who can make it simple or complex.
(b) Regarding her concern about "causality relationships" of the Swiss Cheese model, I can agree. But each barrier in this model must be equipped with few holes and the combination of these holes from different barriers is actually a measure for the interactions and interdependencies between them.
(c) Regarding the "subjectivity in selecting the events to include", I also agree. But it is a matter of the creator and the user of the model, not of the model.
(d) Same applies to the "subjectivity in identifying chaining conditions". It is a matter of the creator and the user of the model, not of the model.
(e) Regarding "exclusion of systematic factors" to stop them, there shall be an administrative or other controls (constraints, barriers, etc.) which can be showed in the model.

I think that there is no better methodology to deal with Complex System than Bowtie Methodology. Fault Tree Analysis (FTA) for pre-event analysis and Event Tree Analysis (ETA) for post-event analysis are excellent tools. Unfortunately, in the safety community, there is huge ignorance about all benefits and deficiencies of this methodology and, adding to the knowledge[6] needed to use it, most of the Safety Professionals ignore it.

---

[5] Nancy G. Leveson is a specialist in System and Software Safety and she is Professor of Aeronautics and Astronautics at Massachusetts Institute of Technology, USA.

[6] The lack of proper literature on Bowtie methodology was noticed by me 10 years ago and that was the reason for writing my second book *Bowtie Methodology: A Guide for Practitioners*, available at Amazon or other internet or real bookstores.

However, the point of Safety-III is that this approach in dealing with safety based on Safety-I and Safety-II cannot anymore support safety, simply because Software actually "imitates" humans and humans do not behave as mechanical or electronic parts. It means that redundancy (very much used in mechanical and electrical parts or Subsystems), as method for improving Reliability of systems, cannot be (anymore) the only thing to improve Safety.

Let me remind you: Reliability applies only to "physical" systems (equipment) and you cannot use it to express uncertainty of human behavior ("imaginary" system). Simply, by putting two same Software applications in one system, this is not redundancy and it cannot prevent the faults of the systems or failure of the operations.

There is another problem with the Software. The Software applications are prone to unexpected and unknown (never encountered before) bugs, so possible fault fixing of the Software can be done only by reinstallation of the same, but improved, Software. The faults or failures with the Software are not based on some type of wearing or breaking of the parts, but they are based on error in the code or by transmitting wrong information or wrong command. So, the system may work OK, but if the wrong command (or wrong data) is given, the failure of operation will happen, although there is no fault in the system.[7]

At the beginning of the book, I have mentioned NASA Mars Polar Lander disaster, which was caused by wrong human assumption and here I would like to mention another Mars disaster which happened on 19 October 2016.

The common ESA and Russian spacecraft to Mars, known as Schiaparelli, hit the Mars surface with assumed speed of 300 km/h due to unexpected rotation caused by parachute deployment. This rotation caused saturation of Schiaparelli Spin Measuring Instrument which provided error data in altitude estimation and the spacecraft crashed. Everything functioned exactly as planned in the spacecraft, but Spin Measuring Instrument's Software was confused by the received data showing negative altitude. All this resulted by command from Software to early release the parachute, although the speed of descent was very high and Schiaparelli was high above Mars surface.

Finding the bugs in the Software is not so easy as finding the mechanical or electrical faults. The laws of physics (mechanics and electrics) are strongly deterministic, but the Software knows very much to behave randomly.

Having in mind what has been said in Section 1.3, I can say that the Software is an "imaginary" Subsystem of particular "physical" system and as such it is responsible for interdependencies, interactions, and communication between the parts of the system. In other words, the Software will contribute very much to the Complexity in the systems in use, making them more complex.

Anyway, let's be honest: The Software is not so bad!

It really helps to improve automation and control of the Complex Systems. In addition, it is particularly very good for design purposes, because we can apply Stress Testing of the Software. It means, if we intentionally "brake" the Software

---

[7] In the cases when this wrong command is given intentionally, this became a case for Security Agencies!

with intention to find its operational limits, the damage will be zero: Simply the same Software will be reinstalled and everything will be same. Try to do Stress Testing with your mobile phone and you will understand what I mean . . .

All these things, using the words of safety, mean that we encounter different types of hazards based on the bugs in Software. The point is that these types of hazards were never met previously and as such we can call them Black Swans: They are totally unpredictable or better say, they are "unknown unknowns". So, our risk assessment methods, determined by approaches of Safety-I and Safety-II, cannot be so successful with these new types of hazards.

Another part of the mosaic is the role of the humans in connection with the Complex System. It is obvious that Human–Machine Interface (HMI) becomes a very important part in maintaining the Complex Systems. Complexity of the systems could be taken as another Human Factor (HF) which can affect human behavior. Complexity also triggers a need for a Team to control and maintain Complex System, simply because one person cannot have insight into all aspects of system's functioning.

Just one small example: Do you know how the NASA space missions (human or without humans) are controlled? With the Control Centers (Teams), employing hundreds of employees and each of them dedicated to this particular mission! The problem with the Teams is that they behave as a chain: The Team (the chain) is strong as its weakest member (ring) is strong.

So, having in mind everything I have said in this section, I can say that Safety-III is taking care of the technology and of its underlying science used to build the Complex System, associated with the social component consisting of organizational and human behavior. All these activities cover the so-called sociotechnical aspects of Complex Systems. It means that the Complex Systems must be considered as sociotechnical system. In general, the Complex Systems are presented as a mixture of sociological, human, cultural, regulatory, psychological, educational, and engineering aspects of working environment, which can contribute to adverse events. As such, all sociological aspects can be treated as Subsystems of the engineering Complex System.

11.4.1 STAMP (SYSTEM-THEORETIC ACCIDENT MODEL AND PROCESS)

STAMP is an expanded accident *causality* approach proposed by Leveson and it is a basis of Safety-III.

As it can be noticed, the word "causality" is in bold because this is actually a model which does not look for one simple Root Cause for explaining the accidents. This is a model based on System Theory and as such it explains the accidents as a miss or a failure of the particular safety constraint(s) imposed on the Complex Systems by the Control Systems. It means that accidents happen because some of the introduced and built (during the design) safety controls inside and outside the Complex System failed. In Safety-III, the Safety is defined as constraints imposed on the systems through particular controls provided by humans, Software, procedures, regulations, and/or Control Systems. If some of this control is missing or fails, the accident could happen.

The STAMP is assuming that the system is not static, but highly dynamic and it is based on three pillars—safety constraints, hierarchical control structures, and process models. Consequently, three major models were produced:

1 System-Theoretic Early Concept Analysis (STECA), to analyze the safety constraints imposed during the design
2 System-Theoretic Process Analysis (STPA), described in the next section in more details
3 Causal Analysis using System Theory (CAST) for post-event investigations when incident and accident happen.

With STAMP, the accidents are treated more as complex dynamic processes, due to the interdependencies and interactions between the Subsystems inside a Complex System. It means that most of the hazards shall be identified looking into these interdependencies and interactions and associated controls inside the system. It does not mean that faults of the system components shall be neglected, but actually "fault approach" is upgraded by the assessment of constraints imposed on the Complex System. By adding the Software, humans, and the environment in all these activities, STAMP would help in analyzing the safety of Complex Systems.

So, to clarify one more time, the STAMP is not about events, as the previous methods and methodologies look to find an event which would be considered as a Root Cause(s), but it is about missing or loss of constraints imposed on the system, Software, the environment, and humans. Actually, these are constraints imposed as "Control Systems" on all contributors to the operation of the Complex Systems. I put Control Systems in quotation marks because the real Control System applies to the equipment, but the Software, the environment, and humans are also contributors to the operation (processes), so they also need to be controlled.

There are roughly two types of constraints in the Complex System: Those imposed to provide smooth (normal) operation of the Complex System and those imposed to provide safety during operation of the system. Both are sometimes intertwined, so the clear difference between them cannot be found. If we go a little bit earlier in this book, I can say that there is nothing wrong with the modified Safety-II: The constraints, to provide normal operation, are a matter of Quality and the constraints to provide safety are a matter of Safety. Anyway, in this book, I will focus mostly on the safety constraints. However, you must have in mind that good Quality is a predecessor of good Safety.

The character of the constraints can be physical and operational.

The physical constraints are based on the physical (engineering) laws used to make the system work and they can be done by the Control Systems. For example, these can be devices for current limitations, temperature (thermostats) or pressure controllers, voltage protectors, inertial devices, etc.

The operational constraints are also very much important, because they can be provided by Software algorithms, procedures, and regulations. The operation is very often monitored and controlled by humans (operators), and as such the documented procedures provide written instructions which explain what and how to monitor, control, and maintain and they explain,[8] at the same time, what is not to be done. In addition, the operational constraints can be imposed also by the management decisions within the company, by governmental directives and state laws, and by the regulations from Regulatory Bodies in each industry.

To emphasize the importance of following the procedures, I will give one excellent, but sad, example: The Alaska Airlines flight 261 crash in September 2017 near the coast of California in Pacific Ocean.

The root cause of the crash was breaking of the jackscrew on the tail horizontal stabilizer of McDonald Douglas MD80 aircraft, due to improper use of procedure for applying grease to it. The procedure to do it last 4 hours, but the maintenance team of Alaska Airlines did it in 1 hour. It happened not because they found some new more efficient way to do it, but because they actually did not understand the instructions in the procedure. This misunderstanding was due to poor training and they, intentionally, skipped some of the steps.

As explained in Section 1.8 and Figure 1.2, the hierarchy in the Complex System is one of the important characteristics. But there the hierarchy was explained only for engineering Complex Systems. Providing safety must also involve the interdependencies and interactions between the Complex Systems and the environment and humans. The hierarchy should actually include the Government, Regulation Bodies, and company management. These subjects are pretty much involved in the operation of the Complex System due to their job for enacting specific regulations and their decision-making processes. So, speaking about hierarchy for STAMP, the point is that it has wider perspective.

Having in mind that the basis of the STAMP are constraints imposed by the Control Systems, it is clear that different levels of hierarchy are controlled by different controls and, usually, the lower layers are controlled by higher levels. The control is provided by the commands enforced from the higher levels in the hierarchy and accidents could happen due to lack of the proper commands from the higher level, the late receiving of the command, misinterpretation of the commands from the lower levels, and wrong execution of the commands by the lower levels.

Having these in mind, it is of utmost importance to provide excellent communication with feedback inside the same hierarchy levels and between different hierarchy levels. In simple words, between different levels, there is need for a command to be communicated clearly to the lower level and there is need for a feedback from the lower level to the upper level that the command was executed as stated. There are many accidents which have happened because command was issued, but the operator just assumed that it was executed correctly. In reality, the command was not executed at all or it was executed just partially, and further activities just amplified the accident.

A simple example of this is one of the biggest accidents in the Baltic Sea which happened on 28 September 1994. On that date, the MS Estonia, a cruise ferry, sunk

---

[8] Actually, in my humble experience, there is a huge deficiency of procedures in the industry! Most of them do pay attention to what must be done, but they do not state what must not be done. Even during the training, how to use procedures is not mentioned.

41 km from Finnish island Uto claiming 852 victims. The reason for sinking was the locks on the ferry's bow door which failed and the door, which vehicles use to board the ferry, separated from the rest of the vessel, pulling the ramp behind it ajar.

The water poured into the deck, causing slowly sinking of the ship. The point is that the captain and the pilot of the ferry on the command deck did notice that there is light on the command deck, but this light was showing that the power is applied to the door's motor (which need to close the door). However, they could not see the front of the ferry and they did not notice that the door is not closed (although the proper command was given) and they started the trip with the open door.

The third pillar is the process model.

It is very much important that the designers and the users of the Complex Systems are looking at the same process model. But there is a problem. Usually, the designers are informed what their design must satisfy, but their understanding of how it can be done does not always comply with the user's understanding.

As said before, the humans are those who will take care of the Complex System after installation and for them the appropriate understanding of HMI is very important. Very often, the designers and the users have different impressions about offered solutions. Having in mind that the safety is achieved during the design, the burden of the designers is bigger. So, they do need to pay attention to the system when it will be in use and they must provide Poka-Yoke[9] for the system designed by them, to prevent possible operator errors.

The process models are coming from the System and Control Theory. For each system to be analyzed, there is need to build a particular model which will be used for analysis. If we would like the Control System to provide particular constraints (to control the system), then it must have a model of the process provided by the system in use. Only in this case, it can provide successful control.

And this is the problem with the modeling: Two analysts who would like to analyze the same system, they could produce different models of this same system. The model is actually a description of the system as understood by the analyst. So, even the slightest misunderstandings of the system with any of the analysts would result in different models. The point is that the understanding of the process model is not the same among the designers and users of the system.

So, the most important thing with the models is that, whatever be the model, it must contain the same information of the system: The required interactions among the Controllable influence factors, the information about possible effect of changes of uncontrollable influence factors, possible future states of the system based on the present state, and the directions where the dynamics of the system could bring the state of the system in the future, if not controlled.

In the scope of post-activities of the STAMP, the Preventive and Corrective Actions are based on creating improved or new safety constraints which will improve to overall control the Complex System. Considering the Complex System is dynamic, these controls must be adaptable to time and in space to be effective in the future. This adaptability could be part of the Resilience engineering.

---

[9] Poka-Yoke is an error-proofing method mentioned in Section 3.5.4.

Having this in mind, the STAMP looks on the faults of the Subsystems or components inside the Complex System, but it has a wider area of analysis which also covers the failures of the Software, environment, human, and organizational behavior.

11.4.2 STPA (SYSTEM-THEORETIC PROCESS ANALYSIS)

Whatever be the system which needs to be assessed for safety and whatever be the industry, the first step is always Hazard Identification. In line with what was explained for Safety-III, Leveson proposed new Hazard Identification method called *STPA (System-Thoretic Process Analysis)*.

She is not satisfied how other (old-fashioned) methods and methodologies work with Complex Systems, especially in the area of design errors, failures which come from the interactions and interdependencies of the subsystems and components, Software bugs and flaws, employee's and manager's decisions, and social, organizational and governmental factors contributing to the accidents.

The main difference from old-fashioned methods regarding the analysis of hazards is that the STPA does not consider only the faults of a component or Subsystem or failure of operators, but it also pays attention to the existing hierarchical relations in the Complex Systems.

STPA has only two steps:

1 Identify *only* the possible hazards which are coming from inadequate control of the Complex System:

Looking at the first step, it can be noticed that Risk Assessment starts with some sort of "rudimentary" Hazards Identification. It means that there is a realistic assumption that the Complex System was designed thoroughly and with due attention to all requirements. What is checked for hazards are only faults or/and failures of imposed controls inside. So, whatever is the brain-storming session, the focus is put on the Control Systems and their incapability to provide required control to keep the operation normal and safe.

Checking for hazards after the design is finished, it can produce good picture of the associated Control Systems used in design. If something is wrong and not appropriate, the STPA should find it and particular actions (changing of the design) could be implemented in these early stages, before testing of the prototypes start.

The point is that, STPA shall be implemented before the system is designed and as such it should be part of the design. It means that, as the first draft-design of the system is done, before the testing, the STPA should be executed.

In general, the most often hazards used in STPA which could happen due to not maintaining the constraints provided by the Control System are coming from few Unsafe Control Actions (UTA):

(a) The incorrect (wrong) command is given to the system in use.
(b) The correct command is given, but it is not followed by the system (Hardware fault);
(c) The required control command is not given (Software or human failure).

(d) The proper command is given, but the timing was wrong (too early or too late).

(e) The execution of the command finishes too soon or its execution lasts too long (timing problems or Hardware fault).

2 Determine how any of the hazards identified in step 1 could develop further (possible causal scenarios shall be considered in this step):

This is a step in STPA where there is need to identify all Causal Factors for any of the hazards identified in the previous step. There is need to select an Unsafe Control Action (identified hazard) and identify what might cause it to happen. After that you should try to develop accident scenarios and, for each of them, identify possible controls and mitigations. In addition, when this job is done, you need to identify under which circumstances or situations the control actions may not be followed or executed properly.

Presented like these two steps above, it seems that STPA missing is one big part of Risk Assessment and this is the quantification of hazards into risks. This is the part where, for each hazard, you need to determine its likelihood (frequency, probability) and the possible consequences. So, the STPA covers only controls and constraints in the Complex Systems and regarding the risks, it is a little fuzzy.

So, although it is composed strongly for the use in the Complex Systems, the imposed or changed controls, as a result of STPA, must be based on possibility of elimination of hazards and mitigation of risks. If you have not determined them, the question is how you would know that you have to mitigate them.

This could look as a deficiency, but not necessarily. This approach is very good with new technologies and newly innovated systems, where there is no data enough to produce likelihood (frequency, probability), and as such probabilistic methods do not provide any value.

It seems that this approach is not without flaws and it is not holistic. In my humble opinion, this approach complicates the overall process of Risk Assessment. The structure of activities does provide complicated view of interdependencies and interactions, which allow humans to provide errors and to miss something. Maybe it can be used for post-event analysis (crash investigations), but not for preliminary Hazard Identification.

However, I would prefer to use standard Hazard Identification process and include inside the constraints imposed to keep interdependencies and interactions within the tolerance limits. Whatever you think about it, you must confess that the loss of any control function (constraint) shall be treated as Hazard. Do not forget, it is by definition a situation with capacity to produce harm to humans, assets, and the environment.

Maybe this is the place to analyze the focus on constraints with STAMP. The basic question here is: How you can control a *fault* of Subsystem and/or component inside the Complex System?

The constraints imposed by the Control Systems are mostly for keeping the normal operation inside the tolerances, because it will provide a good product (service). The only control imposed to prevent the faults inside the Complex Systems is based on monitoring and appropriate redundancy and this is the Reliability issue with safety consequences. It is clear that the fault of Subsystem and/or component will endanger the interdependencies and interactions, but the fault will be a Root Cause for the failure of the operation, not the constraints.

So, whatever focus you put on constraints imposed by the Control Systems, the faults may not be neglected, because we control constraints, but we cannot control faults. For such an approach, the Hazard Identification process shall start with possible faults and later put the focus on possible deficiencies and flaws of constraints during the design. Eventually, there is need to investigate the interdependencies and interaction of the Complex System with humans and with the environment.

Whatever you think about Safety-III, first comes the fault(s) and everything else comes later!

In Safety Cases, I would always prefer to use Bowtie Methodology (FTA and ETA combined) as Risk Assessment methodology for Complex Systems. There, all interdependencies and interactions could be clearly defined, together with the possible faults. Also, the non-linearity can be modeled, because the feedback loops can be included in the FTA.

The point is that the STAMP as approach and the STPA as a method, compared with other Risk Assessment methods, just change the subject of risk analysis which is now focused mostly on constraints imposed by the Control System, humans, and the environment. I cannot see some advanced benefit. But, using them in combination with faults and the Bowtie Methodology, it could be a winning combination.

12 Root Cause

12.1 INTRODUCTION

All that have been presented until now in this book deal with Safety Assessment before the Complex System is installed (design) and after it becomes operational. This is something which is requested by the Regulation in every Risky Industries where each system before being operational must be subject of Safety (Risk) Assessment for the hazards and risks for its particular installation and operation. This type of Safety (Risk) Assessment must take care of all environmental conditions at sites as well as the cultural, religious, educational, and social aspects of the employees who will operate, monitor, and maintain the Complex System.

But all systems from time to time fail and there is need to investigate the reasons and modes of their faults and/or their operational failures. This is something requested also by the Regulation in Risky Industries, simply because the consequences of these faults and failures could be catastrophic. The main target of each investigation is to find what has gone wrong and what would make difference in the development of these faults and failures. In other words, the investigation must point to Root Cause of fault or failure with the intention to provide Preventive and Corrective Actions so that these events never happen again! This process is called "Lessons Learning".

12.2 DEFINITIONS OF ROOT CAUSE

You can find many definitions of Root Cause in safety literature and, in the last few years, there has been a lot of controversy about it. The point is that, actually, these different definitions are reasons for controversies.

Different Risky Industries provide different definitions, but in some of them, the definition of Root Cause is missing. In aviation, for example, although the regulation imposed by ICAO (International Civil Aviation Organization) explicitly requires implementation of SMS (where the Risk Management is one of the pillars), it does not provide clear definition for Root Cause in two of its most important safety documents: Annex 19 (Safety Management) and DOC 9859 (Safety Management Manual). Maybe this is a good place to mention that the Root Cause has no exclusivity only in Safety areas, but it is fully applicable also to Quality areas in manufacturing industry. There, the Root Cause is a reason for providing products or services with poor quality.

In nuclear industry, there is IAEA (International Atomic Energy Agency) and they have published document "Root Cause Analysis Following an Event at a Nuclear Installation" (IAEA-TECDOC-1756) where several different Root Cause Analysis (RCA) tools, methodologies, and techniques are presented for effective post-event investigation and analysis. Going through the document, it can be noticed that all these different tools, techniques, and methods provide different definitions for Root Cause.

DOI: 10.1201/9781003404811-15

In another IAEA document, entitled as "Root cause analysis for fire events at nuclear power plants" (IAEA-TECDOC-1112), the Root Cause is defined (only for fires, I assume) as deficiency in the surveillance program in the nuclear plant. In this document, IAEA presents ASSET (Assessment of Safety Significant Event Teams) methodology[1] for fire safety.

In manufacturing industry, there is definition of Root Cause as a fundamental breakdown of system or failure of operation (process, activity, etc.) which could be stopped and/or resolved, if the Root Cause is determined on time.

In US Department of Energy, there is a document entitled "Root Cause Analysis Guidance Document" (DOE-NE-STD-1004–92), where there is definition for Root Cause as a cause which is capable to be corrected, and as such it would prevent recurrence of incidents and accidents or similar safety events.

By this definition, the Root Cause does not apply only to these events, but it has generic implications to a group of possible other events. As such, Root Cause is the most fundamental aspect of the cause that can logically be identified and corrected. This "group of possible other events" is a series of events (at the same time or one leading to another) which can be identified and all together, combined, should be pursued until the fundamental and correctable cause has been identified.

The importance of this definition is that it is very much similar to the definition proposed by Leveson in her book *Engineering a Safer World*. Leveson discusses the belief that an event or a chain of events can be considered as a Root Cause, due to intrinsic subjectivity in determining this event (or chain of events) as a Root Cause. According to Leveson, the Root Cause is based on Unsafe Control Actions (UTA), which are actually a set of sufficient and necessary conditions for adverse events to happen.

I do not disagree with "subjectivity", but in most of the cases, the Root Cause could be defined as initiating triggering event and it can be objectively determined. For example, the faults of critical systems are usually not triggered by something subjective, but they happen due to:

- lack of our knowledge about present uncertainties;
- the inability to measure particular characteristics;
- the physical laws caused by inaccuracies of measurement systems;
- use of wrong materials or manufacturing methods;
- unproper settings or providing operation beyond tolerances;
- undetermined or undermined stress and fatigue of the material;
- unknown non-linearity; and/or
- to missing constraints (controls).

This triggering event (Root Cause) could initiate actions from Humans which could resolve the problem or they can make it worst. The point is that no actions will be initiated, if this initiating triggering event does not happen. Speaking about that, at the beginning of my "safety career", I have read that every such fault (event), which is

---

[1] IAEA coordinates the ASSET service for its members as an international mechanism to analyze and provide specific and generic lessons for the enhancement of the level of fire safety in nuclear power plants.

unknown to the operator, will increase the probability of making another error in the following actions undertaken by the operator. Actually, the probability of implementing bad action for unknown event in such a situation is bigger 30–80 times (depending on the situation) than the probability to experience the first triggering event. This is actually going in favor of the Domino Model[2] of having adverse events: The adverse event starts with first abnormal event and, like a Domino effect, it develops in a series of events directing toward incidents and accidents. This could support the theory about Fault Propagation.

Explanation for this increasing probability to react wrongly is very clear. Most of the operations in industry are covered by procedures, so if something bad happen, there is particular procedure to be followed. But if you are not aware in advance about that what could happen, then there is no procedure. It cannot be covered by the procedure, simply because you did not know that such an event could exist.

So, if something unknown happen and there is no procedure, the operator must improvise. Of course, this improvisation could be right or wrong. Knowing nothing about what just happened (Black Swan Event!) and having no procedure will cause the operator to be under stress. Thinking reasonably under stress is not easy and this is the main reason why the probabilities for wrong decisions, in such cases, will be significantly increased.

I would define the Root Cause as an initiating (triggering) event which (I can say) cause an accident, or at least, it is the one which triggers the series of future events which would (eventually) lead toward incidents and accidents to happen.

The point is that, sometimes, there is a need to have few more associated events which will contribute to the accident, but the initiating (triggering) event would be the one which when happens will trigger an accident to happen. In the scope of this definition, these other events would be "contributing events": Necessary for accident to happen, but not sufficient enough for accident to happen.

In 1972, Willie Hammer, in his book *Handbook of System and Product Safety* (published by Prentice-Hall in 1972), initially discussed concepts of "initiators", "contributors", and "primary hazards" in the context of hazard analysis, which is in line with my proposal of initiating (triggering) event (initiator by Hammer) associated with "contributing events" ("contributors" by Hammer).

However, even Hammer noted that finding which event (in the series of events) is or has been directly responsible for an accident is not as simple as it seems. This triggering[3] event, which can be treated as a Root Cause, could be clearly determined, if the answers of the questions what, when, where, how, and why (regarding the accident) are known.

The triggering event, not necessary, can be something bad (fault, error, etc.). Very often, "the road to the hell is paved by good intentions", so it can be almost everything which would provoke some kind of action from the system: Nature, systems, humans, the environment, organizational, or sociological factor, etc.

---

[2] Herbert Heinrich was American safety engineer who published the Domino model in 1931. This model is one of the Sequence of Events (SoE) models.

[3] I prefer the word "triggering" over "initiating" for this definition.

12.3 EXAMPLES TO SUPPORT "MY" DEFINITION OF ROOT CAUSE

However, there is a gap here, if you are not cautious. The problem is that you need to be reasonable how far away in the past you will go to determine this triggering event. There are many situations when pilot has got heart attack during the flight. And in some of them, the copilot handled situation and safely landed the aircraft. But on 24 August 2019, Emily Collet (acrobatic pilot) had heart attack and the aircraft crashed near Stonor, Oxfordshire (UK). She had flown the Pitts S-2A Special acrobatic plane with her student and, when this happened, the student was not able to safely land the plane. Both died in this accident.

There is another similar example from aviation: The Tarom 371 flight which crashed near Bucurest (Romania) in March 1995. The Airbus A310 was a pretty new aircraft which was in good condition except one small problem: During the take-off, when the change of trust of engines shall change from Takeoff power to Climb power (slightly decrease of the engines power), the throttle of engine number 1 (left engine) was going back to Idle. It was noticed early by the pilots and it was reported, but the mechanics could fix it only temporarily. However, there was a procedure where, if this happened, one of the pilots has to put back the throttle from Idle to Climb. It has happened 12 times before this crash and it was resolved by the crew without any problems.

However, on this flight it happened again, but the captain was incapacitated from something which could not be determined by investigation after the accident (maybe heart attack?). So, he could not do it. Unfortunately, the first officer, who was flying the aircraft, could not do anything and the aircraft went into rolling and hit the ground. All passengers and flight crew died . . .

So, the reason for crash was an unbalanced power between engines 1 and 2 which was the cause of rolling, producing an uncontrolled flight into terrain. But if the captain was not incapacitated, he would have fixed the problem very easily. The point is that the triggering event and the Root Cause was captain's incapacitation. No other contributing factors, except the throttle problem and cloudy weather (maybe?) could be assigned to this crash.

In addition, there are many incidents when bird(s) enter the jet engine(s) of commercial aircraft. What is the Root Cause in this case?

The US Airways flight 1549 which landed in Hudson River (New York city) was one of such cases where the Root Cause was clear and there were no contributing factors. Of course, we will not go back to understand why the bird entered the jet engine. Sometimes, the bad things happen to good people.

For all such situations (however different they were), the bird entering the jet engine is the Root Cause (triggering event) of the following incident or accident. There are no constraints or controls in such a case.

Maybe it is good time to mention here that the problems with the bird strike in the aircraft engine could be solved only by "toughness" of the Resilience Engineering. I have read somewhere that Rolls-Royce (aircraft engine's manufacturer) in its search for robust engine have tested their products by throwing dead birds in the aerodynamic tunnel when engine was operating with full power.

The point with all these examples is that, for sure, we can assign the heart attack of the pilot as the Root Cause for the accident. But many of the Safety Professionals would go further by asking why she/he had a heart attack, So, the blame would be put on hie/her diet (too much cholesterol in her/his blood), high blood pressure, less mobility and no sport, etc.

However, you may go far away in the past and you will always find some previous events which could explain the next event, but the question is: Is this reasonable? Should we go back down in time to the creating of a world by the God (as it is explained in the Bible)?

Of course not!

Whatever we assume that has triggered this heart attack, it is not relevant, because this was a triggering event for this particular accident. At the same time, there were many people who have had a heart attack in the world. Some of them died and some of them survived, but this particular one was a Root Cause for this accident. So, it is reasonable to assume that the reasons for this particular attack are not important for this accident. We cannot post rules about food of the pilots and their private lives, hoping that this will not happen. This is highly unreasonable!

I am sure that many of the safety professionals will disagree with this definition (Leveson would be one of them), but let us be honest: I agree that there are companies in any industries which have organizational culture which actually "feed" (or at least "make easy") the adverse events to happen, but in any particular case, there is an *initiating (triggering) event* which, if not happened, would not cause the adverse event to happen.

From another side, there are also many companies in any industries which have organizational culture that actually "feed" (or at least "make easy") the adverse events to happen, but the accidents never happened to them. There are many such real cases and my definition is very much practical for them!

Of course, the definition of "initiating (triggering) event" implicitly includes in itself that this event is usually followed by other events which would guide toward incident or accident. So, although there is an "initiating (triggering) event", there are still chances to stop bad things to happen if the operators react in a timely manner with proper Corrective Action. This statement could widen the Root Cause definition by trying to explain "what actually happened".

Another thing which supports the definition of the Root Cause as "initiating (triggering) event" is the fact that catastrophes (accidents) happen by a series of many connected events. There is simply no one thing (event, fault, failure, etc.) which could cause catastrophe. All these are actually going into favor of Swiss Cheese Model for barriers, because simply using barriers (imposed controls) for the series of events triggered by the "initiating (triggering) event' would stop the accident.

12.4   PREVENTIVE MAINTENANCE AND ROOT CAUSE

A very good help to finding a Root Cause in advance, with regard to the possible faults in the Complex System, is the Preventive Maintenance. During this periodical activity, many of the mechanical parts are changed in advance (not because they are faulty) and electronic circuits (Control Systems, sensors, and especially, monitoring circuits) are checked for proper functioning. This is important for the Control

Systems because during normal operation, very often, we have no information how they function correctly. The newest Complex Systems has the ability to also control the functioning of their Control Systems, which makes a job more convenient.

As example for Preventive Maintenance, I would mention regular car maintenance which happens approximately every 10,000 km. We bring our cars to the workshop and mechanics there change the oil, filters for oil, air, and fuel. Additionally, they check other things (light bulbs, brakes, pressure in tires, etc.). It does not mean that something is wrong, but it means that something can be wrong due to limited capacity of filters and dirtiness in the oil. If not changed, it can destroy the engine or something else. Of course, it does not mean that the accident cannot happen 5 minutes after the Workshop, but at least we are confident that, before the accident, everything was OK.

In addition, there are different types of notifications regarding the state of our system (just remind yourself of Section 1.8.5). In the case of Warning (yellow or orange notification), there is need to check what is going on, because Finding the Root Cause for Warning can help very much in preventing real faults of the systems.

12.5   FIXING ROOT CAUSE AFTER FAULT/FAILURE

If there is Alarm (red notification), it means that the system is faulty and as such it cannot provide normal operation any more. There is need in such a situation Complex System to be switched off automatically, and if it is not, the manual switching off by the operator shall happen immediately. Sometimes, faults can be benign, without big damage in the system, but sometimes, they can propagate and bigger damage could happen. That is the reason for switching off.

It is important to find a Root Cause for the fault and it is a job for the Maintenance Team. They are educated and trained for fault-finding missions and they (usually) can use help from the Complex System itself. Many Complex Systems are equipped with Built-in Test Equipment (BITE) which can help with fault finding. How the BITE information will be used depends on the knowledge, skills, and experience of the Maintenance Team.

It is worth mentioning that BITE cannot always provide the reason for fault, but it can provide information what is faulty. That is the reason that Maintenance Team shall establish which Subsystem fails first and find the reason for that. Sometimes, there can be more faults at the same time and it is challenging to find a Root Cause for each of them.

You should be careful with the BITE reading in such a case because the same information can be triggered by two or more faults. Such a situation is known as "Dragon-King". It means that there are two faults which belong to two different Subsystems, but they trigger the same information in BITE.

12.6   DETERMINING THE ROOT CAUSE IN
POST-EVENT INVESTIGATION

As said before, the investigation to find the Root Cause is strongly connected with investigations after the adverse event happened, but some of the methods used for that could also help with Safety (Risk) Assessment of systems (operations, activities, processes, etc.) before adverse events happen.

This is very much applicable during the design, when the design testing showed that things do not work as planned. As such, there is discipline known as Root Cause Analysis (RCA) which is actually a problem-solving activity implemented as during-design activity or as post-event investigation. It is usually supported by many methods, techniques, and methodologies. The RCA process is used to gather knowledge and to achieve an understanding of the possible (or happened) adverse event or design flaws, its causes, and, the most important: To stop it to happen again (or to fix the design)!

The RCA is based on a requirement that the multifunctional Team (including designers), which is involved in it, must put efforts, based on the objective evidence, to understand the fundamental or underlying cause(s) how the event (could) happen(ed) and to look toward the systematic solution(s) to the immediate and (possible) future problems, by preventing their reoccurrence.

The RCA is a methodology dedicated not only to identification of the Root Cause, but also to determine the contributory factors which can be treated as missing constraints in engineering and/or in management of the company. It could also trigger measures of risk reduction strategies, associated with producing action plans based on Preventive and Corrective Actions.

For those purposes of the RCA, there is need for information how the things developed starting from triggering event. The information that are needed are as follows:

- What was the situation with the system before, during, and after the adverse event?
- How were the employees involved and what actions were undertaken by them?
- What were the environmental factors on the site and what was their influence on the adverse event happening?
- Any other information relevant to the situation.

All these things should be documented because in the cases of victims or disaster, the public and Regulatory Bodies must be presented with evidences. In addition, the documentation may be needed for the future.

In the scope of the Complex Systems, determining the Root Cause through RCA could be a tantalizing job and there is also need for multidiscipline Team to deal with it. Some of the things explained in this book could provide help in these activities.

Let me use one single example regarding determination of the Root Cause. It is the accident of USAF B2 (stealth bomber) Spirit of Kansas which happened at Anderson Air Force Base in Guam on 23 February 2008. Luckily, the crew survived (ejected themselves), but the aircraft was destroyed. This was one of the most modern USAF aircraft (worth US$2 billion) and finding the Root Cause for the accident was very important. Before the takeoff, the crew noticed on the display the warning

AIRDATA CAL, which is actually a call for calibration of the sensors. They called maintenance technician, he came and executed calibration of the sensors and warning disappeared. They started with takeoff, but the Master Caution warning was on and it disappeared after few seconds. The pilots decided to continue with the takeoff, but moments after the aircraft leaned on the left side, the left wing touched the ground, the pilots had to eject themselves, and the aircraft crashed.

Investigations showed that call for calibration (AIRDATA CAL) was shown due to problems with Pitot Tubes (sensors for speed). Previous night there was storm in Guam and aircraft has been on the runway without any protection. The moisture from the rain entered one of the Pitot Tubes and it caused disbalance compared with other sensors (there are 24 such sensors in the B2). In the same weather condition, all sensors must have approximately same indications and it is achieved by calibration of all sensors. The calibration of the sensors was done, but the "faulty" sensor still kept the moisture inside.

The problem escalated before the takeoff when the pilots switched on the Pitot Heaters which need to protect Pitot Tubes from freezing during flight. The heat evaporated the moisture from the "faulty" sensor and the disbalance showed up again. That was the reason for Master Caution signals to be ON again.

On-board computer tried to fix the problem, but the choice which it made was wrong: From 24 sensors as input into 4 Control Systems, the computer disabled 2 good Control Systems and chose 2 with bad data from "faulty" sensor. This is an aircraft which is structurally different from other aircraft, because it is actually "flying wing" without standard fuselage and standard rudder, horizontal stabilizers, and elevators. As such, it cannot be flown without computer. The wrong choice of Control Systems by on-board computer caused wrong reading of aircraft pitch and aircraft speed and it caused rolling of the aircraft causing the left wing to hit the ground. The aircraft was disintegrated and burnt into flames.

The point is that maintenance engineers in Barksdale Air Force Base, Louisiana, have noticed that in the cases of huge humidity, the Pitot Tubes could show wrong readings and they provided informal procedure to switch on Pitot Heaters for not more than 40 seconds and the moisture will evaporate. If this is done, there will be no need for additional calibration and problem will be solved. But no one make efforts to formalize this procedure and put it into Flight and Maintenance Manual, so just guys in the Barksdale Air Force Base were familiar with it.

Question is: What is the Root Cause for this accident?

Obviously, following my definition for Root Cause, in this case it will be the fact that the aircraft was left all night on the runway in time when there was a huge tropical storm with big rains. For many of you, the Root Cause would be the moisture in the Pitot Tube or even the non-spreading of the information about the unformal procedure, but these could be only contributory factors. By following my definition, it all started with the rain. If there was no rain, although the aircraft was parked on the runway, the accident would not happen. So, rain was the initiating event. All others contributory factors (aircraft at the runway, not spreading the informal procedure, wrong choice of the computer, decision of the pilots not to abort takeoff, etc.) were just factors which contributed to the accident. Without them, maybe the accident would not have happened, but the rain was the initiating event. Other factors just supported the path toward the accident.

12.7   WHAT IF THE ROOT CAUSE IS NOT SO EASY TO FIND?

Important thing to remember is that we are not doing post-event investigations to provide a blame to someone, but to stop these things to happen again (Lessons Learned concept). Practicing Safety for the sake of blaming is very wrong.

In the cases where there are Complex Systems, there are also complex operations. Their involvement in equipment (humans, assets, environment, Software, etc.) is big and during accidents usually the equipment is destroyed catastrophically, so no BITE is working and no objective data can be gathered.

In aviation, the "black boxes" are containing information regarding the aircraft systems performance (FDR—Flight Data Recorder) and communication in the cockpit (CVR—Cockpit Voice Recorder) between the pilots. So, these two devices can help, if not destroyed. But in other industries, things are more complicated. There are no "black boxes" in chemical, pharmaceutical, and other industries.

However, even in the cases where Root Cause cannot be clearly determined, the "contributive events" can be determined and in such a case, the Investigation Report must analyze them and propose measures for elimination or mitigation.

It is a little but similar as in medicine. There, if the doctors cannot determine clearly the disease, they do not give up. They try to "fight' the symptoms in the patient, hoping that the disease will be determined before it is too late.

12.8   THE HUMANS AND THE COMPLEX SYSTEMS

Maybe this is good time to say few words regarding the humans and Complex Systems, because we may not neglect their role in designing, testing, operating, and maintaining the Complex Systems. Neglecting the human role as designers of Complex Systems and focusing only on operations, the humans are actually involved in safety of the Complex Systems as "double actors".

The first role is connected with the normal operations. The operators are operating system trying to keep its normal operation continuously and they actually "defend" the Complex System from adverse things to happen.

In this role, they are included into functioning of the Complex System mostly as "watchers": Monitoring all the time operation and product (service) and affecting, from time to time, the Complex System by some small adjustments (if needed). The human operators are actually "adaptable monitoring and control Subsystem" of the Complex System, because they have power and duty to keep the system (as last line of defense) into normal operation by providing necessary changes (adaptations, settings, etc.) of the system based on current situation and needs.

Today's reality with Complex Systems in industry is that the operators share control of systems with Control Systems, which is a part of automation. In such reality, the humans found themselves into positions of higher-level decision-making and making some adjustments in the form of commands to these Control Systems. These activities generate some new types of human errors which are characterized with more errors of omission and smaller number of errors of commission.[4]

---

[4] Error of omission is error which happen due to omission to provide some command, adjustment, activity, data, recording, etc. Error of commission is error when what is provided (data, command, adjustment, activity, recording, etc.) is incorrect (wrong).

Some of these adjustments which very often happen on moment-by-moment basis can be explained as follows:

- Changing the configuration of the system in order to reduce wearing of vulnerable parts;
- Adjusting Complex System to changeable nature of the operation or due to a new product;
- Providing maintenance modes for fixing the faults and recovery of the normal operation after Corrective Measures;
- Monitoring indications for early detection of changed system performance in order to allow faster processes for production or increasing the resiliency;
- Etc.

But at the same time, their second role could also be as "actors-producers" of the failures of operation. I said "failures of operation", assuming that the contribution of trained and competent operators in the Complex System's faults should not happen. It means that their knowledge and understanding of the operation is excellent, so under any circumstances, they will never put wrong command or wrong data into the system to cause its fault or damaging its components.

Obviously, all this is about human performance within the operations conducted by the Complex Systems. The point is that it could affect any operation in Risky Industries. I do not know how it is going in other Risky Industries, but I know that ICAO, as highest Regulatory Body in the world in aviation, has recognized that. In 2021, they have published ICAO Doc 10151 with the name *Manual on Human Performance*[5] (HP) for Regulators. This is a document which "takes a system's perspective on human performance, and it brings to focus the human contribution to the global aviation system".

In this document, ICAO presents five HP principles based on investigations how the performance of humans is influenced by different factors:

1 *People's performance is shaped by their capabilities and limitations:* I would add here also the "knowledge", because it is "mastermind" of "capabilities and limitations".
2 *People interpret situations differently and perform in ways that make sense to them:* This is something already explained in more details in Section 2.11. Look at again for "narrative fallacy" and "platonicity".
3 *People adapt to meet the demands of a complex and dynamic work environment:* I would not add anything here . . .
4 *People assess risks and make trade-offs:* Yes, but it is based on their knowledge, understanding, skills, attitude, etc., which is easy to say than to perform.
5 *People's performance is influenced by working with other people, technology, and the environment:* This is covered with something called Human

---

[5] ICAO defines human performance (HP) as a way how people perform their tasks. It actually represents the human contribution to the system performance.

Factors (HF) which is actually scientific discipline investigating different influences (sleeping deprivation, fatigue, drugs, stress, etc.) which could trigger human errors and decrease human performance.

Statistics says that in aviation, 70–80% of all accidents are human errors. The reasons for these errors are actually humans who does not produce faults of equipment, but they react improperly to the faults which happen due to some other influence (system fault, for example). It means that something happens to the system and the operators are not prepared how to react or they react with delay. However, whenever you look for human failures in connection with Complex Systems, always the decisions made are with well-intentions, but unfortunately, they are followed by disappointing consequences.

The reason for that, especially in the cases of Complex Systems, should be explained by the fact that there is considerable uncertainty of the influence factors with these systems. So, the decision-making process, in many rare situations, is based on particular probability and it is similar to gambling.

Unfortunately, the outcome of the gambling can be valued only after the event. Usually, if the gambling results with an adverse event, the critics is huge about chosen solution and if nothing happens, nobody will be praised.

Associated with these uncertainties regarding Complex System's functioning are also the "latent failures" which could be connected not only with the system, but with the managerial and organizational structure of the company. These are actually hidden error-capable situations, which could pop up unexpectedly and the surprise caused by them would affect response to them. The biggest problem with these "latent failures" is the fact that they change constantly and some of them disappear and some of them show up instantaneously.

It is obviously that all humans employed in the company for operating the Complex System will not have same knowledge, experience, and skills, so there is need for particular human balance in each shift during the production process. At any time of system operation, there is need to have presence of highly competent person regarding the system in use and this person is responsible for all decision-making process during operation. Due to fluctuating of the workforce, each company in the Risky Industries must strive to have availability of 50% more competent persons than needed for normal operations.

It is the best if the companies strive all time to improve the competence of the employees and this should not be treated as economic burden. This need for better competence is another thing which goes against humans in companies.

Previously (last century), the average time to transform scientific discovery into a technical product was 30 years. Today new technologies become available on the market in 2–3 years. In addition, after the product is placed on the market, it can become obsolete in 5–6 years. All this race to place a new product on the market faster decrease the capability of companies to carefully test designs and products on time and help them to understand the product behaviors and possible risks.

These fast changes of the new technologies also put the employees and users of the new products in a bad position regarding keeping pace with the new technologies. As the systems become more and more complex, the need for additional knowledge in the form of higher university degree or additional specialized training is inevitable, especially for the employees. If you think it is expensive to send your employees to trainings, seminars, and conferences, then try to calculate the price of production scrap based on operator's ignorance in your company.

The particular attention should be put on situations where there is need to change the type of product or service, which would trigger changing of the settings of the Complex System used in production. Reason for the change could be different, but the point is that humans are, in general, reluctant to any changes. As such, there is need to provide good training regarding the change.

The training shall explain:

- Why the change is needed?
- What and how will be changed?
- How the change will affect daily (weekly, monthly, etc.) operations?
- Which departments (units, etc.) will be affected by the change?
- Who will be responsible for the change?
- What will be the backup if change fails?
- Who will provide support after the change is implemented?
- What are the new hazards and risks triggered by the change?
- How the monitoring and control will be affected by the change?
- What will be Preventive and Corrective Actions and Maintenance Procedures after the change?
- Anything else which will help employees to accept the change and motivate them to provide necessary support during activities needed for the change.

General training shall be provided for all employees, but additional (specialized) training could be provided for different departments depending on the Nature of the change.

These are the problems which operators encounter during their work, but maybe the most important thing which should help to operate Complex System is the Human–Machine Interface (HMI).

Whatever the Complex System is, the HMI shall be simple and shall be designed in such a way that it will enable the humans to operate it, without any Complexity or uncertainty. The monitoring part shall provide just necessary indication with capability to change the screen window, if operator needs more details.

The control part shall be designed in such a way that there are not many commands on each screen, but they need to be grouped together with the particular Subsystem monitoring indicators. In such a way, applying the command, the operator will immediately see its effect on the indicator. The command for emergency switch off the system shall be available to any screen!

Let me give you one example from aviation: The American pilot (Captain) Warren Van der Burgh, in cooperation with the Training Department of American Airlines, conducted analysis of modern accident and incidents. The data showed that 68% of them were caused by automation mismanagement by the pilots in the cockpit. It showed that the aviation training has made pilots too much dependent on the automation created inside the cockpit. Blind confidence in the auto-pilot and other automation systems inside the aircraft spoils the pilot's capability to fly manually in situation when some of the systems fail. It means that the pilot's training shall be based not only to use the Complex Systems for "fly-by-wire", but they must also be trained to switch off the systems and fly manually.

The increased confidence into automated systems in the aircraft decrease the pilot's confidence to fly the aircraft without those systems. As a result, the pilots may not be able to regain the control of the aircraft spoiled by the faulty automation system. This is something which affects many Risky Industries. The confidence that the automation embedded into Complex System will take care of everything and not having capability to handle the system when the automation is faulty are the recipe for disaster.

As I have said at the beginning of this book, this is a book which needs to provide knowledge of the Safety Professionals to have better understanding what is going on with their Complex Systems. This knowledge could help, but it cannot replace the correct understanding. Putting equality between the knowledge and understanding will not work. Somewhere on LinkedIn I had read a post that old Mayans had a knowledge how to calculate the movements of the planets, but they did not have understanding why they moved in the sky.

So, my message to the Safety Professionals: Try to understand every aspect of functioning of the Complex System and this could happen only if there is a good knowledge about the science and the technology hidden in the design of this same Complex System.

But have in mind that this is not your goal!

Your goal is to use this understanding to anticipate what wrong can go with the system and be prepared in advance for these situations. There is tiny line between what is normal operation and what will be abnormal operation. Actually, the Safety Professionals must be competent to determine when the normal operation is close to pass this line and, if the line is passed, they need to know how to react accordingly!

13 Design of the Complex System

13.1 INTRODUCTION

Whatever is the idea of building any kind of engineering system, the goal of the design process is to produce effective, efficient, predictable, stable, cost-effective, and reliable systems that meet prespecified requirements in different working (internal and external) conditions. For this purpose, the designers are provided with help by a bunch of industrial standards and regulations.

I do believe (and I hope that everyone would agree) that the Safety starts with design process. Very few things which are mistaken during the design process could be rectified later. Even if these rectifications are conducted later, the costs of the overall process for rectification will be pretty much expensive.

So, the main point during development of a new system is to focus on the design process, because this is the place where the basis of the Safety is established.[1] To be clear: The Safety does not finish here! It is a process which starts with the design, but continue with installation, commissioning, operation, maintenance, and decommissioning of any system.

Here, I can state that overall life cycle of any Complex System could be divided into three parts. These parts belong to the Reliability Assessment of the Complex System and they are presented by the Reliability Bathtub (already presented in Figure 3.5 and explained in Section 3.5.1).

13.2 THE DESIGN . . .

The design starts with requirements!

There are bunch of the requirements which any system shall satisfy and all of them shall be taken into account during the design. The problem is that some of these requirements are too expensive to be satisfied by the Hardware design and some of them are unreliable. So, a particular level of compromise must be present.

The point is that during designs of new systems, especially those where new technologies and methods are used, there is lack of data and lack of clear experience how some of the requirements should be satisfied. In such cases, the designers must assume the functioning of the ready system and these assumptions would differ from the assumptions of the operators who will operate the system. Using the words from modeling in Chapter 2, the designers and operators will have quite different understandings how the "model" (system) should operate.

---

[1] This is even emphasized in ISO 26262 standard, where considerable attention is given to design of the Software and Hardware.

DOI: 10.1201/9781003404811-16

I would like to point here to one example, a tragic one, which resulted with crashes of the aircraft due to differences in assumptions of the designers and operators. This example is about two crashes of Boeing 737 MAX 8 in Indonesia (Lion Air Flight 610, October 2018) and Ethiopia (Ethiopian Airlines Flight 302, March 2019).

The aircraft was equipped by anti-stall system, the so-called Maneuvering Characteristics Augmentation System (MCAS) which, due to changed aerodynamics triggered by the position of the wings and changed weight structure of the aircraft, should help pilots with aircraft stability during the flight. Unfortunately, the MCAS provided wrong commands in these two flights and the incapability of the pilots to switch off the MCAS on these two aircraft caused crashes. The number of the victims was 346.

Boeing has designed MCAS to be activated only if three things happen simultaneously:

1 There is an excessive angle of attack on the wings (nose of the aircraft up).
2 The auto-pilot is off.
3 The flaps are retracted.

This is extremely rare situation and Boeing's designers has assumed that in situation of malfunctioning, the pilots will understand what is going on and they will switch off immediately the MCAS (approximate time to do it is 3 seconds).

The first time (Lion Air flight), when MCAS incidentally switch on due to wrong readings of altitude and velocity sensors, the pilots tried for 10 minutes to "fight" the commands given by MCAS. The captain (Pilot in Command—PIC) was using manual trimmer to cancel the MCAS command for nose down. He tried it 34 times, but he did not notice that each time, the MCAS was responsible for pushing the aircraft down. All these efforts were without success and the aircraft crashed at Java Sea.

After this crash Boeing provided updated instructions worldwide what to do with MCAS, but obviously it was not enough. Four months later, another aircraft crashed—the Ethiopian Airlines flight 302. The Root Cause of the crash was (again) wrong activation of MCAS. After that the aircraft was grounded for almost 2 years.

The main point here was that Boeing assumptions that the pilots will notice particular problem with MCAS and solve it in 3 seconds was completely wrong. The pilots of both aircraft struggled to understand what is going on with the aircraft for more than 10 minutes. In addition, the MCAS, as a system, was not even mentioned in the Manual of Procedures and as such there was no procedure to follow.

The designing of the Complex System is also hierarchical job: There are many levels, (horizontal and vertical) and it can be done sequentially or simultaneously by different teams. In the most of the cases, for the design is used "top-down" approach, which means that we are moving during the design from desired functionality of the Complex System toward means which need to be used to provide this functionality. Every design process starts with satisfying the system requirements through modeling and simulation and finishes with extensive testing in the laboratory and in the operating environment of the designed and prefabricated Subsystems and the Complex System in total.

Not always, the output of testing process is successful certification. If the required certification cannot be achieved, there is need for change of the design and overall process is repeated. Full process is, actually, a repeatable loop where even the starting specifications and functionalities could be changed. It makes the design iterative process: There are repetitive cycles of designs and testing.

The certification is made through the processes of Validation and/or Verification. If the system functionality must satisfy something which is new and not covered by standard(s), then the applied process is called Validation. If there is(are) standard(s) for the system functionality, then the applied process is called Verification. Very often, these two processes are mixed and applied together.

Regarding the design of the Complex System, it was mentioned before that there is a Hardware which is supported by Software. The Hardware behave like a "muscle" and Software like a "brain". This is, actually, an integration of both, which create Complexity through devices and programs (algorithms).

The Hardware is responsible for resilience and modular designs and the Software takes care for complex computational and decision-making responsibilities regarding the control of normal operation and Safety of the system. During the design, different Teams deal with the Hardware design and Software development and their job is not only to produce Hardware and Software with particular quality, but also to provide their synergy in the normal operation. The Hardware and Software would be designed by different design teams, but these teams must communicate and cooperate on every hierarchy level and any phase of the design process.

This importance of the Hardware and the Software design has been noticed by the Regulatory and Standardization Bodies in Risky Industries, so they have been addressed by few standards. Very often, these standards are more like guidance materials, than regulations, but the Regulatory Bodies usually use the documentation from the design process to certify the Software and Hardware of these Complex Systems.

In aviation such standards are DO-178C (Software Considerations in Airborne Systems and Equipment Certification) and DO-254 (Design Assurance Guidance for Airborne Electronic Hardware). In automotive industry, such a standard is ISO 26262 (Road Vehicles—Functional Safety) which has 13 parts and more than 600 pages in total. This standard covers the electronic and/or electric systems (Hardware and Software) within the road vehicles.

I would provide few sections later, where I will consider all these things for the design of Complex System.

13.3   UNCERTAINTY AND ITS INFLUENCE ON THE DESIGN OF COMPLEX SYSTEM

The uncertainty has been mentioned and roughly explained in Section 1.10 and it is one of the biggest "partners" of the Complex Systems.

Having in mind that uncertainty is a "prescription for error", I would like to point here how these uncertainties in the Complex System are propagated to get (approximately) the uncertainty of the whole system. Having in mind that important thing is the measurement of variables and parameters (to provide operation within their tolerances), these calculations are borrowed from Metrology.[2]

As we understand, the output of the Complex System is a function of many variables and parameters and as such it can be presented by the equation:

$$y = f(x, a, b, c, \ldots)$$

Here, $y = y(t)$ is the output, $x = x(t)$ is the input, and $a = a(t)$, $b = b(t)$, $c = c(t)$, etc. are the variables and parameters (internal and external influence factors) which can affect the operation of our Complex System. All of them are functions of time (t), but due to simplicity, I intentionally missed the time in the formula.

The point is that any of these variables, parameters, and input can introduce uncertainty in our system, which will propagate from the input to the output. It means that at the output (y), this uncertainty will be present. To make this output uncertainty small, we need to keep all other uncertainties small. The tolerances, as explained in Section 3.2, are the limits (constraints) where the uncertainty (for each variable, parameter and input) must be allocated. There, μ (mean) is the measure for "accuracy" and σ^3 (standard deviation) is the measure for "precision".[4] Or better say, σ (standard deviation) is used to express uncertainty. Having this in mind, the propagation of uncertainties could be expressed by this equation:

$$\sigma_y^2 = \sigma_x^2 \left(\frac{\partial y}{\partial x} \right)^2 + \sigma_a^2 \left(\frac{\partial y}{\partial a} \right)^2 + \sigma_b^2 \left(\frac{\partial y}{\partial b} \right)^2 + \sigma_c^2 \left(\frac{\partial y}{\partial c} \right)^2 + \ldots$$

Here, σ_y^2 is the square of standard deviation (uncertainty!) of output y, σ_x^2, σ_a^2, σ_b^2, σ_c^2, etc. are squares of standard deviations (uncertainties) of input x, variables, and parameters a, b, c, etc., respectively, and $\left(\frac{\partial y}{\partial x} \right)^2$, $\left(\frac{\partial y}{\partial a} \right)^2$, $\left(\frac{\partial y}{\partial b} \right)^2$, $\left(\frac{\partial y}{\partial c} \right)^2$, etc. are the squared portion of the influence (contributions) of changes of each variable, parameter, and input (x, a, b, c, etc.) to the output y.

As it can be noticed, the uncertainties (which are actually standard deviations or expressions of the potential errors in the Complex Systems) are added quadratically (Figure 13.1). The dots in the angles of Figure 13.1 are actually the markings for 90° angles.

As it can be noticed in Figure 13.1, the overall sum of uncertainties is not obtained by adding linearly, but quadratically (Pythagoras theorem!), and as such it is not so big as it would be if addition was linear.

---

[2] Metrology is a science for measurement. There is BIPM (Bureau International des Poids at Measures), the international organization for measures and measurements, in Paris (France) which provides internationally accepted rules and regulations for measures and measurements. The most important document there (JCGM 100:2008) can be downloaded from www.bipm.org/en/committees/jc/jcgm/publications.

[3] Let me remind you again: The same letter σ has different meanings in Metrology (standard deviation) and in Control Theory (Laplace transform, damping factor).

[4] Do not forget: The Metrology (science for measurements) does not recognize the terms "accuracy" and "precision"! Only the "uncertainty" is recognized there! These terms are used in this book only for "teaching purposes" (narrative fallacy!).

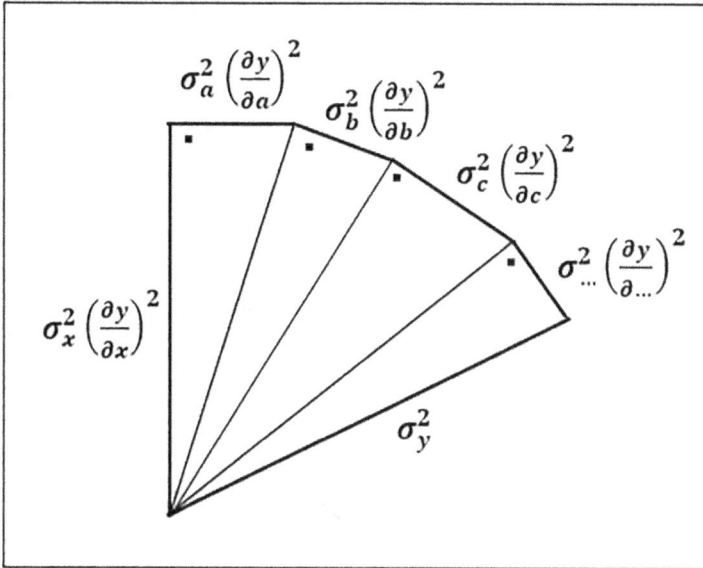

FIGURE 13.1   The quadratic sum of standard deviations (uncertainties).

From the presentation above, it can be noticed that contribution of the input x uncertainty would be the biggest, but this is valid only in this case, because I made such a choice in the figure. However, the biggest uncertainty in the Complex System is the one which will contribute the most in the overall uncertainty in the output y of the system.

In addition, it is worth to go further in explanations of the equation from above. As it can be noticed the squares of partial derivations present the influence of the changes of variables, parameters, and input to the output of the Complex System. They provide some Complexity of the equation, but have in mind that, if there is no influence, i.e., the change of these factors, it will be 0 and, as such it does not have any effect on the output. If these parts of the equation are 0, they will disappear from equation.

So, the theoretical approach for finding uncertainty during the design of any Complex System would be as follows:

1 Decompose the system on its Subsystems and find all uncertainties of any variable, parameter and input with significant contribution to the system operation. This could be done through Reductionism.
2 The one which is biggest, shall be decreased by any means, in the phase when the Subsystems are designed and before they are assembled (through modeling and simulations) into Complex System.
3 The point is that these uncertainties may change (decrease or increase) when the whole Complex System is assembled due to interactions and interdependencies between the Subsystems. So, these uncertainties of the Subsystems

should be checked again when the System is assembled. It means that it should be an iterative and a balanced process.

4 It does not mean that other variables, parameters and inputs should not be decreased, but they will not contribute significantly to the decreasing of output uncertainty.

It would be a theoretical approach and, as such, it is good to be done. This is also the stage where the tolerances should be determined.

When the prototype of the system is produced, there is a need for testing which could provide more information about uncertainties and tolerances, so this is a phase where final adjustments will be done. If requested adjustments are not easy to be made, maybe the small changes of the design should be considered.

13.4 DESIGNING A SYSTEM SCIENTIFICALLY

The first thing which can be stated here is that everything which was mentioned in the previous few chapters about stability and dynamics of non-linear systems could be integrated in one mathematical explanation. We can present (roughly) any Complex System mathematically as a system with (at least) two differential equations:

$$\frac{dx}{dt} = \dot{x} = f(x, u, w)$$

$$u = g(x, y, w, r)$$

Here, x is the state of the Complex System under consideration, $\dot{x}$ is the dynamics (change) of that state, u is the control input into the system[5], w is the unknown parameters and possible disturbance which needs to be controlled by the Control System, r is the reference stored in the Control System, and y is the output of the Complex System.

The upper equation applies to the Complex System itself. It shows that the change in dynamics of the Complex System $(\dot{x})$ under consideration depends on the following:

* Present or previous state of the system (x);
* The input from the Control System (u); and
* The disturbance caused by influence factors or by unknown parameters (w).

The lower equation applies to the implemented Control System as a Subsystem to the Complex System under consideration. As it can be noticed, its output (u) (the input into Complex System) depends on the output (y) of the Complex System (which is controlled value for the Control System), the reference (r) for the operation of the Complex System, and the disturbance (w) which is sensed by its sensors (Figure 13.2).

Do not get confused because u is not a derivative in the equation above. It is not dynamics (change) of the Control System. It is, actually, a command which change the state of the Complex System to reject the disturbance and to make the Complex System resilient. The Control System should be "programmed" to particular "control

---

[5] Each Control System shall be presented by one equation.

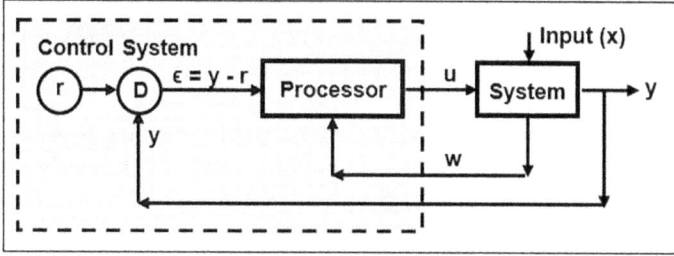

FIGURE 13.2   Example of simple Complex System built by Control System and system in use.

law" which, based on the sensor's inputs in the Control System, will provide appropriate command to keep the performance of the Complex System within the tolerances.

If there are many disturbances (and usually there are), there will be few Control Systems, which will control the behavior of the Complex System affected by each of these disturbances.

The real challenge in designing of the Complex System are actually unknown disturbances (w) and unknown parameters. Having in mind that they are the so-called "unknown unknowns", they will cause errors ($\epsilon = y - r$) in operations. The good (resilient system) design of the Complex System will happen, if the Control System take care for the errors without knowing what is the disturbance w. Knowing w is good, but it is not always achievable . . .

However, the errors in design cannot be presented always by probability of happening, because they sometimes cannot show up during design and testing activities. Unfortunately, many times the bad design showed up when the product (service) was delivered and it cost money and, sometimes, lives. It is not strange, these design errors to be presented as operator errors during use of the system. This happened with Boeing 737 MAX 8 aircraft, when Boeing tried to blame the pilots for the accidents.

These are the design methods that could provide stability in some vicinity of the stable fixed point(s). However, we would like to provide stable system over a large region of its operation for several fixed points. This can be achieved, if we try to linearize the system around each fixed point and design a Control System for each critical fixed point.

This design methodology can be described by the following steps:

1 Linearize the given nonlinear system about several fixed points, which could be controlled by few parameters and/or variables.[6]
2 Design a family of linear Control Systems for each fixed point to locally stabilize the system around each of the points. Each family of Control Systems should be based on "control law" which will provide particular commands, if necessary.
3 Construct a gain-scheduled Control System. The gain-scheduling is a method of providing area of stability by Control System for one stable point

---

[6] Linearization was a method for analysis of non-linear systems explained in Section 7.8.

which includes at least one neighboring stable point (controlled by another Control System). Designing Control System which could take care for gain-scheduling will make this system Master and other individual controllers will be Slaves.

4 Verify and validate the operation of the whole Complex System in correspondence with gain-scheduled Control System and its synchronization with other individual Control Systems by simulating a nonlinear closed-loop model (for the whole Complex System).

5 Think how and try to optimize the overall design of the Complex System.

13.5   TAGUCHI DESIGN (DESIGN FOR ROBUSTNESS)

One of the pioneers[7] of Quality is Genuchi Taguchi (very much respected by me!). The concept which he has established in Quality areas could be very much used in Safety areas also. He dealt with something which he called *Robust Engineering*, but today it is known as Resilience Engineering (although the partisans of RE are not ready to give him credit for that).

Among many of its contributions to the Quality is the Taguchi Design, a concept for designing a robust product which will be "tough" enough to deal with many of disturbances during use of the product and "elastic" enough to deal with some of the disturbances which are not known.

There are three phases in the Taguchi design: Concept (system) design, Parameter design, and Tolerance design.

The beginning phase of design of the new product, by suggestion of Taguchi, is called the Concept design. It is a phase when the designer looks for the best concept which can satisfy the requirements for the purpose of the use of the product (system).

It is the phase when we need to consider already known concepts, but there is a need to be innovative and creative also. Already known models are not always good for the Complex Systems, so the innovation here could make difference. This is the phase where the structure and configuration of the system and Subsystems (together with building elements and materials) could be considered and decided. Also, this is the phase where the internal interdependencies and interactions between Subsystems (inside the Complex System) and external interference between them and the environment are considered.

When the basic concept during Concept design is defined, the second phase starts. Parameter design is a phase where we the required "toughness" and "elasticity" of the Complex System should be achieved. The intention of this phase is to provide choices of working point and parameters which would shape the operation of the Complex System. So, this is the phase when we choose which parameter(s) we will use to control the operation and which parameters we will keep fixed not to cause failure of operation. It could be achieved by dealing with possible effects of internal and external disturbances, which are responsible for the variability of parameters.

---

[7] The pioneers of the Quality are known as Quality Gurus. The W.E. Deming is the first one. Others are Walter Shewhart, Genuchi Taguchi, Joseph M. Juran, Kaoru Ishikawa, Philip Crosby, etc.

As I have mentioned before, there are interdependencies and interactions between Subsystems inside the Complex System and there are interactions of Subsystems with environment. These interactions in the literature could be found under the name "noise". This noise is responsible for variability of the parameters inside the Subsystems, so there is need to design parameters to minimize this variability.

Now, let's go back to Chapter 1 and Figure 1.1. There, you can notice two types of influential factors: Controllable and Uncontrollable. Controllable factors are those who are used to control the operation (variables) to allow the system to do the job which is designed for. Uncontrollable factors (parameters) are those (internal and external) which we keep fixed and if they change beyond the tolerances, the Complex System and/or operation will fail. So, all these factors in the design process are considered by particular values and I will use this presentation here.

Also, it is good to remind yourself what is presented in Chapter 7. If you remember, there, in every diagram presented there is the parameter x presented on x-axis and change of this parameter x (presented as its velocity $\dot{x}$) on the y-axis.

The similar diagram is presented in Figure 13.3 (as it can be noticed, it is a non-linear function) and I will use the figure to explain two possible situations.

The first situation is when we choose the parameter x to control the operation ($\dot{x}$) (Controllable influence factor—variable) and the second one where parameter x is uncontrolled parameter (internal or external influence = Uncontrollable influence factor – parameter) and its change will cause the failure of operation $\dot{x}$.

In the first situation, the change of the variable x is used to control the operation ($\dot{x}$).

It can be noticed in Figure 13.3 that, for the same change of the parameter x ($x_2 - x_1 = x_4 - x_3$), the changes of $\dot{x}$ are different. As it can be noticed, for the same operation, the variability p_1 is bigger than p_2, so obviously we will define the Working

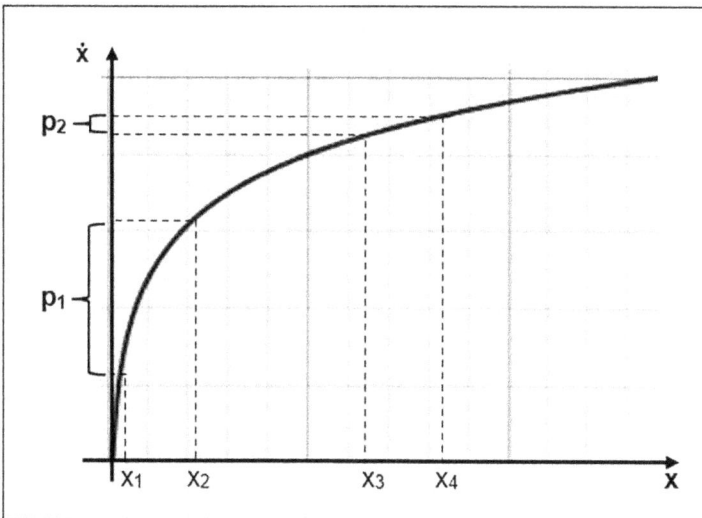

FIGURE 13.3 Speed and/or volume of change of operation ($\dot{x}$) triggered by the change of parameter x.

Point of the parameter x to have operation depending on the design concept: Do we need big or small change of the operation $\dot{x}$ controlled by parameter x?

If we need big changes, the working point of x will be in the area $(x_2 - x_1)$ defining operation in the area p_1, or in other words, Complex System will work in area when the parameter x will cause big changes of operation $\dot{x}$.

If we need small changes, the working point of x will be in the area $(x_4 - x_3)$ defining operation in the area p_2 or in other words, the Complex System will work in the area when the parameter x will cause big changes of $\dot{x}$. This is actually basics for robustness (Resilience Engineering): Big changes of the parameter to produce small changes in operation.

In the second situation, applicable for Uncontrolled influence factors, the working point will be chosen in the area where x is in $(x_4 - x_3)$, because it will cause small change of operation $\dot{x}$.

Defining of variables and parameters and setting the Working Point of the Complex System (operation) is done through experiments and particular testing in the environment where the Complex System will operate, so most possible environmental interactions with variables and parameters will be included in testing.

The third phase of Taguchi Design is Tolerance definitions which will define the tolerances. Keeping the parameters and variables within the tolerances will make a difference between the safe and unsafe operations of the Complex System. This is also a phase when designer chose which components and materials will be used for the Subsystems, how we can control them, and what will be the requested quality to be achieved. This is also the phase when the particular balance between the operation, the Safety, and the price should be determined.

13.6   DESIGN FOR SAFETY

Design for Safety is a second type of engineering design of Complex Systems and it actually complements the Complex System design. The point is that during design process, the engineering designers try to provide system which will be used to satisfy some working (operational) requirements for some job to be done. Their focus is on the produced system to be effective and to be efficient (to spent less resources) during its operation. The second requirement for the system is to be safe for use.

This second requirement actually undertakes the first place in the Risky Industries: What is the point to have effective and efficient system, if it is not safe?

In the first phase of the design process, the designers pay attention to the fault of components or Subsystems which will build the Complex System. In this phase, the knowledge of the Reliability is important because it will help with the faults of components or Subsystems. This is part of the Hardware design when components and materials which will be used are determined.

Also, this is a phase where the Working Point of the system is chosen and the tolerances for it are determined. The point is that, this is the phase where all controls to keep the Working Point within the tolerances are designed and they will provide required constraints in the operation of the Complex System. In addition, the operational, administrative, and regulatory constraints for operators (humans) and for the environment are established. These few types of constraints could be a part of the compromise made as decision of the manufacturing company, because it will also have financial impact on the Complex System. Usually, the better the component and the material, the higher the price.

So, if we need to achieve Safety of the Complex System, we need to think about it during design process. It means whenever the draft-design is ready on paper, the team of designers (involved in draft-design) and the team of knowledgeable and experienced Safety professionals (involved in operation in similar or same systems) must sit down and assess the Safety of the draft-design.

It is very important to have also "knowledgeable and experienced operators", simply because whatever the designers have tried during the design process, their perspective of the system's operation is not the same as perspective of the operators who already worked on similar systems in such an environment. Whatever information was given to the designers, always something is misunderstood, the attitude to information is different, or some parts of the information are missing.

The meeting(s) between these two teams should be in the form of brainstorm session, where designers will mostly listen and "knowledgeable and experienced operators" will mostly talk. This is also the place where involvement of the humans and the environment in operation of the Complex System must be highly considered. The wise approach is that in these meetings determine all Controllable and Uncontrollable influence factors and check what is done to keep them under control in different conditions.

Of course, that there will be strongly need for compromise in this area, because system must be reasonably cheap, as it must be reasonably safe! This factor of compromise could affect the design even later, when the system needs to be offered on the market or to the customers. The ALARP[8] (As Low As Reasonably Practicable) is good approach in this area. Do not forget: The absolute Safety (zero risk) is not possible and for everything else, you should be very clever, brave, and a little bit (it is wise to be) conservative.

The point is that after this session(s), the design will change in some areas and this is normal. The designed system can fail, but the design must be such that the system Fails Safe. In other words, if the system experiences some problem, it must move in a stable state without providing any damage or loss. This is the main characteristic of the good design!

Another thing which is very important here is that any type of design (not only Taguchi or Design for Safety) uses the fact that the Complex System can be controlled and optimized just by few control variables.

Of course, whatever the design is, it needs to be tested. So, the compliance of the variables and parameters with intended operation of the Complex System is determined by modeling and simulation. But the design cannot be released for production without *thorough testing in the real conditions (environment!) and by proper analysis of the result of these experiments.*

Taguchi, by himself, has provided also a method for design of the experiments. But, although is pretty effective and efficient, it is also pretty much complex method.

---

[8] ALARP is a method of achieving compromise (for the system under consideration) between the risks controlled and time and money invested in design, operation, and maintenance of the system. Of course, the risks with highest severity may not be subject of compromise!

That is the reason that it is not so popular between the engineering designers today. However, it is only chosen design of experiments method which considerably fulfills the Resilience Engineering requirements for Complex Systems today.

As you can notice above, there are bold letters to finish the sentence about experimental validation and verification of the design. This is not random mistake, but it is done intentionally to emphasize the conducting the testing experiments under real environmental conditions.

To better explain this, I will present here one very interesting case of accident involving Embraer aircraft . . .

On 5 April 1991, the Embraer EMB 120 Brasilia aircraft (flight 2311), owned by Atlantic Southwest Airlines (ASA), crashed during lending in Brunswick (Georgia, USA). The investigation showed that the crash was caused by the failure of a worn part in the Hamilton Standard 14RF-9 propeller unit (quill) controlling the angle of the blade of the left propeller. Although, there was another Safety measure for such a situation, it was not deployed and the aircraft went out of control, because the blades of the left engine, improperly presented a flat face to the wind, creating unsustainable drag.

The point is that NTSB investigators did an experiment in the laboratory of Embraer company in Brazil, which showed that the additional Safety measure (needs to stop pitching of the blades in dangerous position when there is a problem) worked excellent. Anyway, one of the investigators insisted the same experiment should be done during the flight.

The Embraer managers agreed and during the test flight, this Safety measure did not work (the pitching of the blades did not happen during the flight). Simply, the testing in the laboratory was deprived from all vibrations and oscillations which are burden and stress for the aircraft during the flight and, obviously, the design which worked excellent in the laboratory did not worked in the real conditions (environment).

13.6.1 Design of Hardware and Software for Road Vehicles

The ISO has produced a family of 12 standards under the name ISO 26262:2018 Road Vehicles—Functional Safety.[9]

This standard provides a reference for life cycle of vehicle production which can increase Safety of vehicle operations. It includes rules for Hardware design and Software development and it covers electrical and electronic systems, providing Functional Safety for vehicles. The standard considers the vehicle life cycle, which is defined as design, production, operation, service and decommissioning of vehicles (cars, motorcycles, buses, trucks, etc.). ISO 26262 assumes that the company already has a development process for design and the standard provides additional constraints to your process to satisfy particular Safety aspects.

The V-Model (Waterfall Model), used in design based on ISO 26262, is a model with 9 steps where 4 + 4 steps[10] are executed in parallel (Figure 13.4). Actually, dur-

---

[9] This standard is actually adaptation of IEC 61508 standard for road vehicles. Although it is mentioned in this section, it is actually a standard for Hardware. Only Part 6 applies to Software.
[10] In the literature you can find different variations of this model, but I think there is nothing wrong with that. Every Software Developer may choose his own model. This model is also known as Validation and Verification model.

ing dealing with the steps from left, the preparation of the steps from right is ongoing (gray arrows).

In addition, this standard provides a risk-based approach, where the criteria of risk assessments are in the form of ASILs (Automotive Safety Integrity Levels). These ASILs apply also to Software. It is important to mention that these ASILs are borrowed from Safety assessment covered by IEC 61508. Also, you can use FMEDA (Failure Mode, Effect and Diagnostics Analysis) to determine real values of SILs. By using FMEDA, the failure rates may be derived for each important failure category for Hardware and Software.

In this particular case, a very interesting thing with the ASILs is that the risk is presented as a combination of Probability, Severity, and Controllability. The Controllability is estimation how the risk can be controlled by the development of Software and its synergy with the Hardware.

The values for Frequency, Severity, and Controllability for ISO 26262 standard are given in Tables 13.1 and 13.2.

Determination of particular ASIL level means that for that level, you should apply some appropriate techniques to eliminate or mitigate the risk in that area. In Table 13.3, one possible example of determined ASIL levels based on their average probability to happen by using FMEDA is given.

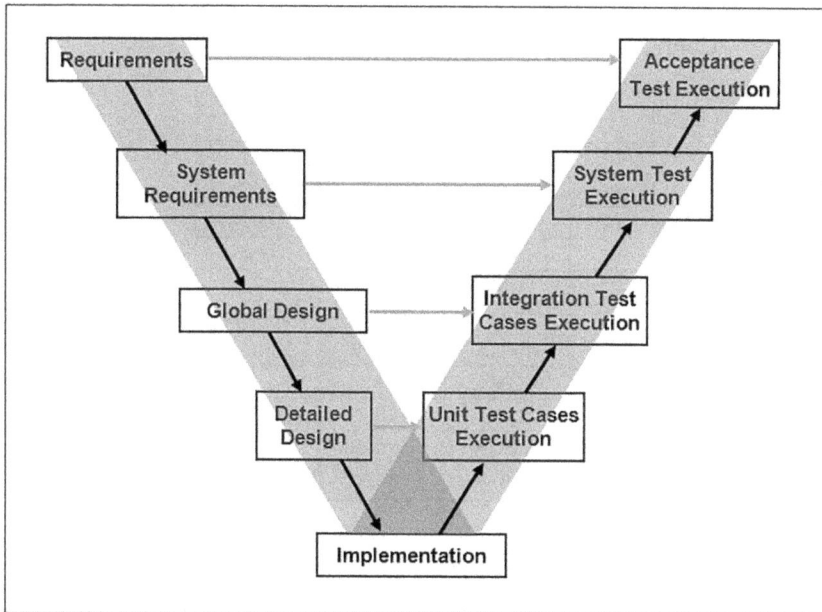

FIGURE 13.4 V-model for system design.[11]

---

[11] In literature available on the Internet. you can find slightly different V-model. I, personally, appreciated this one, but there is nothing wrong with others too.

TABLE 13.1

Exposure, Severity, and Controllability Criteria for ASILs

Probability				
E0	**E1**	**E2**	**E3**	**E4**
Incredible	Very low probability	Low probability	Medium probability	High probability

Severity			
S0	**S1**	**S2**	**S3**
No injuries	Light and moderate injuries	Severe and light threatening injuries (survival probable)	Life-threatening injuries (survival uncertain), fatal injuries

Controllability			
C0	**C1**	**C2**	**C3**
Controllable in general	Simply controllable	Normally controllable	Difficult to control Uncontrollable

TABLE 13.2

ASILs Depending on Probability, Severity, and Controllability

		C1	**C2**	**C3**
S1	E1	QM*	QM	QM
	E2	QM	QM	QM
	E3	QM	QM	ASIL A
	E4	QM	ASIL A	ASIL B
S2	E1	QM	QM	QM
	E2	QM	QM	ASIL A
	E3	QM	ASIL A	ASIL B
	E4	ASIL A	ASIL B	ASIL C
S3	E1	QM	QM	ASIL A
	E2	QM	ASIL A	ASIL B
	E3	ASIL A	ASIL B	ASIL C
	E4	ASIL B	ASIL C	ASIL D

* QM stands for faults which can be prevented or handled by quality management activities.

All 12 parts[12] of the standard ISO 26262 are more than 600 pages together and I would not provide more details here.

[12] Part 13 was in process to be published at the time when this book was written.

TABLE 13.3

Example for Probability Levels Based on FMEDA

Probability Level	Average Probability
E0	$\geq 10^{-8}$ to $< 10^{-4}$
E1	$\geq 10^{-4}$ to $< 10^{-3}$
E2	$\geq 10^{-3}$ to $< 10^{-2}$
E3	$\geq 10^{-2}$ to $< 10^{-1}$
E4	< 1

13.6.2 Design for Aviation Hardware

My professional background comes mostly from aviation area, so I have always used aviation as representative of Risky Industries. In my tries to make my points I often use some examples from this area. I do believe that it is similar in other Risky Industries, because they are also highly regulated areas.

Since modern aircraft are extremely Complex Systems and comprise a variety of Safety-critical applications, Safety and security must come first. Safety deals with non-intentionally produced adverse events and security deals with intentionally produced adverse events. For Safety, there are regulations with requirements which needs to be satisfied and for Security there are legal laws which humans needs to be abide.

From the Safety point of aviation business, the regulations cover almost everything which could provide better Safety. There are organizations which deal with designing the Hardware and Software for Complex Systems used in aviation. Two such organizations are RTCA (Radio Technical Commission for Aeronautics) and EUROCAE (European Organization of Civil Aviation Engineers). The first one is in the United States and the second one is in Europe, but both of them cooperate and the regulatory documents which they produce are same in 95% of their content. Of course, both of these organizations protect the manufacturers of aviation systems in their geographical areas.

There is document named DO-254 (produced by RTCA and coordinated with EUROCAE) which has title Design Assurance Guidance for Airborne Electronic Hardware. As the title says, this document provides rules how to design aircraft Hardware systems without endangering Safety of the operations. At the beginning of DO-254, there is classification of electronic hardware items for the aircraft into simple or complex categories and later it provides objectives and processes to ensure the systematic design of these classes.

The DO-254 is actually more a guideline for design and assurance of airborne electronic Hardware than regulation, but it is accepted by aviation industry as regulation. It applies to electronic (Hardware) items like LRU (Line Replaceable Units), PCB (Printed Circuit Boards) assemblies, PLC (Programmable Logic Controllers), application-specific integrated circuits, integrated technology components, etc.

The DO-254 requirements provide a DAL (Design Assurance Level)[13] framework to designate different Safety levels to different parts of an aircraft Complex Systems,

---

[13] Somewhere in the literature you can find it as Development Assurance Levels.

based on their safety criticality. Then it explains the series of processes and expected deliverables to ensure that these elements are safe during operation.

These DALs are graduated by the letter meaning that DAL A would be the most critical system and DAL E would be less critical system (Table 13.4). The criticality of the system is assessed by examining the aircraft and other interacting systems.

Required processes for Hardware design which, in accordance with DO-254, would provide required level of Safety are as follows:

1 Planning
2 Requirements capturing
3 Conceptual design
4 Detailed design
5 Implementation
6 Product transition, acceptance testing, series production
7 Validation and verification
8 Process assurance
9 Configuration management
10 Certification liaison

To prove that aforementioned processes have been applied through the design process, the design team needs to prepare the following documents:

• Plan for Hardware Aspects of Certification
• Hardware Process Assurance Plan
• Hardware Configuration Management Plan

TABLE 13.4
The DALs for Aviation Hardware Based on DO-254[14]

Design Assurance Level (DAL)	Consequences	Target System Failure Rate	Examples of Critical Systems
Catastrophic (Level A)	Crashes, deaths	$<1 \times 10^{-9}$ chance of failure/flight/hour	Flight controls
Hazardous (Level B)	Possible crashes, deaths	$<1 \times 10^{-7}$ chance of failure/flight/hour	Braking system
Major (Level C)	Possible stress, injuries	$<1 \times 10^{-5}$ chance of failure/flight/hour	Backup system
Minor (Level D)	Inconvenience	None	Ground navigation system, in-flight entertainment
No Effect (Level E)	No Effect	None	In-flight entertainment

[14] Processes and documentation in this section are presented in Table 11.3 with permission from Intland Software GmbH (Stuttgart, Germany) and the processes are undertaken from their paper "The Basics of DO-178C&DO 254".

- Hardware Design Standards
- Hardware Verification and Validation Standards
- Hardware Archive Standards
- Hardware Requirements
- Conceptual Design Data
- Detailed Design Data
- Top-Level, Assembly and Installation Control Drawings
- Hardware/Software Interface Data
- Hardware Traceability Data
- Hardware Review and Analysis Procedures
- Hardware Review and Analysis Results
- Hardware Test Procedures
- Hardware Test Results
- Hardware Accomplishment Summary
- Hardware Process Assurance Records
- Hardware Configuration Management Records, Problem Reports

The point is that, during the design process, there is a need to follow some practical rules which could help with the Safety of the designed product. These practical rules could be stated as follows:

- During the design of the Complex System, you have to use definitions and terminology which is unambiguous to you, your team, Supervisory board, Regulator and Customer.
- It is important to assure and document the traceability of development process from the beginning to the finish of the design (start to production). This will help with the design, but mostly it will help with the costs, if something needs to be changed later.
- During the design, the requirements for the Complex System should be unambiguously clarified between the design team, production team, regulator and Customer.
- The design process and all documentation must be in order, so that later everyone can find what one needs and understand what is this about.
- During the design and production process, the right tooling and measurement instruments and methods must be used and the information about it must be available to the Customer. If there is need for special tools or special instrumentation for maintenance purposes, it must be also submitted to the Customer, as part of the deliverables.

Usually, obligation of the manufacturer does not finish with the installation and commissioning of the Complex System on the site. There is one-year or two-years warranty period where bad design could show up. The good design will show up also during these 2 years, but the good design also will be presented during the operational part of the life cycle of the product: The product will last longer without too much maintenance.

13.7   STECA – SYSTEM-THEORETIC EARLY CONCEPT ANALYSIS

STECA was mentioned in Section 11.4.1 (STAMP [Systems—Theoretic Accident Model and Process]) as the first activity conducted for Safety Assessment during the design. Actually, the STECA approach is based on causality model presented by STAMP. However, dealing with it, I understood that it is very similar to Taguchi Design. The only difference is that Taguchi Design is more focused on the robustness and STECA is mostly focused on the failures of interdependencies and interactions between the Subsystems, humans and environment.

However, similar to Taguchi Design, this is a true method where approach is to think of Safety before and during the design of the complex System.

This is a method which uses systematic tools for possible scenarios for Hazard Identification and Risk Assessment and, as such it is based on activities defining the system requirements, technological, structural and material choices for modeling and analyzing the system under consideration. Also, the process shall take care for undocumented assumptions presented as Black Swans ("unknown unknowns").

The method treats the incidents and accidents as events caused by wrong interdependencies and interactions among the Subsystems inside the Complex System and between the Complex System and the operators and the environment. These events in the Complex System happen mostly due to errors in design, Software errors or human errors, although the faults of Hardware are not excluded.

As such, it is holistic, because it deals with all Complex System. It is a model proposed by Leveson and it is based on three elements:

1 Defining the Concept of Operations (CONOPS)
2 Producing a model which would fit the requirements for design the Complex System
3 Executing Complex System's analysis based on accepted model

13.7.1   CONOPS

This is the first element in the design process, where all requirements regarding the design of the Complex System shall be defined and accepted as Concept of Operation. The requirements could come from the intended operation, available technologies and materials, standards and regulations.

This element deals mostly with the concept how the requirements will be satisfied by the Complex System for this particular operation. There is nothing strange if there is need for some compromise to be done in this place, although it can be done also during dealing with other two elements.

The CONOPS shall be complete when appropriate Hazard Identification and Risk Assessment is done. It must take care for all aspects of the operation with clear definitions of operational and Safety responsibilities and it must guarantee particular coordination and consistency of proposed operational and Safety measures.

13.7.2 PRODUCING A MODEL

Based on the considerations from the first element and produced CONOPS, the designer shall produce a model of the Complex System which will be a subject for simulations, testing and first draft-design (prototype).

This is the place where the structure of the complex System will be defined, as well as the number of the Subsystems which will conduct and control the operation, having in mind the realizability of the CONOPS.

Of course, that building of the model will start with Subsystems which would execute the particular operation (process, activity, etc.). When these Subsystems will be defined and optimized, there is need to move into next step, which is determining the controls, which needs to provide constraints of the operation and Safety of the system. The third step will be choosing the methods for control in the form of particular Control Systems.

All these three steps are actually the part of the design where the system model is considered part by part.

The next step will be integration of all parts into one model and trying to optimize it. In this, the fourth step, the sustainability of the model is checked theoretically by treating it holistically and checking the possible merging of some of the Subsystems or components. This is also the place where all hierarchies, interactions, and inter-dependencies between the Subsystems shall be defined end determine how they will be realized.

If there are some problems in this step, the model can be easy changed and adapted to new reality.

The open mind is very much important here, because the design is creative job and, as such, there are many models which could fit the determined CONOPS. The most important thing is to provide good and safe operation (process, activity, service, etc.), and after that make the system effective, efficient and sustainable.

The knowledge of the similar CONOPS and new theories and technologies could have very big part in this element.

13.7.3 MODEL ANALYSIS

The designed model shall be a subject for thorough analysis which can be executed through Hazard Identification, Risk Assessment, simulations of the operations and anticipated environmental conditions. The Hazard Identification shall be done for individual Subsystems and later, for the Complex System, as emergent property created by it.

The model analysis shall take care for few important things:

(a) Are there any unspecified assumptions?
(b) Is there some information which is neglected, incomplete or inconsistent?
(c) What were the results of the Hazard Identification and Risk Assessment?
(d) What compromises shall be done in the CONOPS and the model and how they will affect the determined Hazards and Risks?

(e) How the Software, environment and human operators will affect the operation of the Complex System? and

(f) What will be the practical realization of the Complex System?

The model could have no use, if it is not tested through analysis, simulations and building the first draft-design (prototype). The draft-design shall be a subject of thorough testing in the laboratory and, later, in the real operational environment. However, before and after the draft-design is tested, there could be some adjustments or changes in the Complex System's structure or hierarchy.

Here, the final CONOPS and final building and operational structure will be produced as preparation for manufacturing of the Complex System.

14 Software Development[1]

14.1 INTRODUCTION

I have mentioned in Section 1.3 that the Hardware is "physical" part of the system and Software is an "imaginary" part of the system. The Hardware is something tangible with physical characteristics and any of these characteristics could be measured. The Software cannot be touched and, as such, not any of its characteristics can be measured and determined with certainty as it can be done for the Hardware. The Hardware can be influenced by the environment (temperature, humidity, pressure, etc.) and most of these elements will decrease the performance of the Hardware.

The Software is not susceptible for these elements, but it highly depends on the human performance. Whatever you think about computers equipped with the Software, they are stupid machines:[2] The humans are those who put the data and commands in the computer based on their decision-making capabilities. So, the Software is part of the system, but it is also more prone for human errors and as such, more critical for functioning of the Complex System.

There is another thing which makes the Software very critical for Complex System's operations. It has been shown that most of the Software-related faults and failures in Complex Systems are caused by the flaws in the requirements specifications or they were not clearly determined. The point is: If there is a wrong requirement about specification, it cannot be detected, because it will pass every model, simulation and experiment for checking the Complex System.

There are many methods and methodologies which can be used to provide Risk Assessment of Software. I would not go into details and I will not propose any of them. Just I would mention most used (as per literature on Internet). The most used methods are Software FMEA, PHA, Software FTA, and they (applied together) can be found under one name as HAZAPS (HAZard Assessment in Programmable Systems) methodology.

However, I do not think and do not support approach that the Risk Assessment of the Software should be done by ordinary Safety Professionals. I do believe that it can be done by the Safety Professionals which have very strong background in Software Development and Software Testing. If there are more of them, as members of the multifunctional team for Hazard Identification and Risk Assessment, it will be better!

---

[1] Regarding this chapter title, I have been advised that software cannot be "designed", but can be "developed" and I have accepted it!

[2] For the time being, this is a valid statement, but nobody knows what Machine Learning (ML) and Artificial Intelligence (AI) will bring to industry and to our lives in the future.

DOI: 10.1201/9781003404811-17

14.2 FAILURES OF SOFTWARE

In general, the problems caused by Software failures differs from the problems caused by Hardware faults. The main point of this difference is the Complexity of the relations between the failures and consequences when the Software fails. Another thing is that, if there is fault with the Hardware, the Complex System will stop. One Hardware fault in one Subsystem may induce damage in the Hardware in other Subsystems. But if there is error with the Software, the Complex System may stop or may behave inappropriately. In most cases, it will not result in Hardware fault (but it could happen).

Important thing to understand about the Software, is that it can contribute to accidents on two ways. The first one is if there is omission to do something which is required. The second one is if there is commission or, in other words, to do something that should not be done. The second one could have three possibilities:

1 Do it at the wrong time (problem of programming).
2 Do it in the wrong sequence (problem of programming).
3 Do it by using wrong data (problem of operator or problem with integrity of data).

You can find in literature on Internet that Reliability of the Complex System included in Safety-critical processes[3] should be in the range from 10^{-5} to 10^{-9}. The value of 10^{-9} is actually NASA standard not to have fault in 10-hour flight. FAA is using the same value per flight or per hour. UK requirements for Nuclear Safety, actually request values of 10^{-7} (in 5,000 hours) not to have fault in nuclear reactors during their operations. This "unreliability" of the Software is very much used for cyberattacks, so particular protection of these systems is contributing further to their Complexity.

Another thing is that the Software is highly non-linear in its development and its application. The basics of programing is based on logical circuits (electronically produced) and they are discrete circuits, so by its nature, they are non-linear. To operate these circuits, they use Boolean Algebra and it is also non-linear. In this book, the non-linearity takes big part and let's explain more about Software non-linearity.

In addition, in "physical" systems the effect usually directly follows cause, but in Software development and its applications in operations, the cause and the effect often seem to be not so directly connected.

The point is that Software can contribute very much in the values of Reliability mentioned above, but it is not easy to achieve them if there is no particular dedicated attention to Software development. The simple example (already mentioned in this book) of this are the Boeing Software problems with their newest version of Boeing 737 MAX 8 aircraft which is the freshest reminder about Software importance in proper functioning of Complex Systems.

The core of the problem with Software Reliability is more historical. The Software started to be used in the 1980s for industrial automation and there is no historical data to calculate accurately (whatever this word means in this context) the Software

---

[3] The values mentioned in this section are taken from the article "Software Safety: Why, What, and How" by Levenson available on the Internet. The article has been published in *Computing Surveys*, Vol. 18, No. 2, June 1986.

Reliability. Let me remind you: The Hardware-based automation was used from the beginning of the human evolution, but mostly it "exploded" after industrial revolution in the 19th century.

Next thing about the Software is that it works in synergy with the Hardware and response of the Software on Hardware malfunction, due to fault, adverse environmental conditions, or wrong data, cannot be predicted even by testing. Imagine that there is a need for command to be issued by the Software in a complex process and the command is issued, but due to fault of the hardware, command is not executed. How the Software will behave if there is feedback about the inexecution of the command, we cannot predict, if it is not solved by the Software algorithm for such a situation.

Also, the Hardware can be redundant and the Software cannot be made to be redundant. However, in this area (comparing Hardware and Software) I must state that Software can be tested by "crash test" and Hardware testing in such a manner is too expensive. During "crash test" the Hardware will be destroyed, but the Software (if "destroyed") can be just uploaded again from the backup copy.

There is one good and expensive example which affected the Japan Aerospace Exploration agency (JAXAO in February 2016 when they have lost their satellite Hitomi.[4] Due to some problems with Attitude Control System, it showed spinning of satellite, although it was OK. They tried to stabilize the satellite and they succeed in that, but when they loaded new thruster control parameters (which were not tested), the satellite broke up as result of uncontrolled spinning which overpass the material limits.

Actually, in the Software, the change of one single character in programming or in data used as input, can cause total outage of the application. Wrong bit in the register, counter or in the memory, can cause Software to behave not as planned and expected. There are many other types of changes inside Software that seem they should be simple, but that can have a big impact on operation. For example:

(a) Updating a minor bug by available patch;
(b) Not updating it although there's a security patch available;
(c) Replacing a shared Graphic User Interface (GUI) component;
(d) Making-up used model for Software development;
(e) Etc.

Associated discipline in providing engineering practices, especially in Software development, is extremely important. These are some recommendations[5] which needs to be addressed:

- Make one change at a time on already developed application.
- Keep everything versioned together in a proper source control system.

---

[4] The satellite's original name during its development was New X-ray Telescope, but at the time of launch it was renamed as ASTRO-H. After the launch, after its solar panels were deployed, it was renamed again as Hitomi.

[5] These recommendations are taken from www.curiousduck.io/non-linearity/, established and maintained by Elizabeth Hendrickson.

- Treat configuration as a code.
- Automate all regression tests.
- Explore to discover unintended consequences.
- Deliver the product incrementally and frequently.

To be honest, achieving such discipline is not easy. It requires both: The clear understanding and full support of leaders in the company, as well as a compromise in the root level of programming. The leaders are in extremely important obligation to foster a culture that values a good Software development and to compromise (when necessary) which is important part of the company's life.

There shall be built-in control of functioning of the Software through particular monitoring routines and there should be a possibility for recovery routines embedded inside. When malfunction of the Software is detected, the Software cannot provide a requested operation or it cannot be executed within a prescribed time limit, the Software shall stop and recover. Recovery techniques are usually two types: Backward recovery and Forward recovery.

The Backward recovery are based on returning the system to a prior state and then continue with an alternative step. The Forward recovery would stop the program and try to fix the problem (if it is internal state of computer or faulty state of the process). If the problem can be fixed, the Software will continue program execution. If not, it will stop into "Fail Safe" mode and it will inform the operator.

It is wise if there are routines which will put the Software into particular "safe state"[6] any time when problems are detected. This state shall be accessed from any step of the process execution, when the problem is detected. Simple example for such a "safe state" is blinking yellow light at the street traffic lights when they are not working properly.

Eventually, it is the reason to check any Software through building a simulation. You cannot model all the complexities of the system in design, but maybe the designed simulation can help developers to develop an understanding how to reduce non-linearity. Of course, this will affect all good and bad things about modeling and simulations explained in Chapter 2 in this book.

Today, as part of any Complex System, there shall be an Emergency Button launching Emergency Routine which (if pressed) shall be executed by the Software. Previous Emergency Buttons were designed to cut the power of the system, but today there shall provide safe stop of the process where everything which is done in previous steps, will be preserved and it can be recovered without any additional damage to the system or product (service).

So, having this in mind, I can say that the Software development and its testing is more expensive,[7] more uncertain and more time consuming. It needs more resources and developer attention during its development, because it is not only code writing.

---

[6] The "safe state" is not the same as RESET of the system. RESET will bring the Software to the beginning of its execution and "safe state" will continue from the step where the problem was detected when the problem is fixed.

[7] Speaking about money, a study conducted by the US National Institute of Standards and Technology (NIST) states that cost of failures caused by Software malfunctions in the United States is (on average) about $59.5 billion each year.

It must take into considerations the system's Complexity which the Software use for control over processes to manage the constraints (imposed by this same system's Complexity).

The Risk Assessment of the Software started firstly in the Risky Industries which were subject of considerable automation. As I have said previously, the "unreliability"[8] of humans was solved by adding automation to the equipment and, with the development of computers, the Software became part of that automation.

The Nuclear Industry was dedicated to improving Quality of the automation systems there and, at that time, there was no Risk Assessment of the Software. In other Risky Industry, it was similar, but the aviation changed everything. Approximately, at the beginning of this century ICAO brought initiative to deal with Risk Assessment of all operations there and it was done through regulation about requirement to implement Safety Management System (SMS) in every aviation subject. The Regulatory Bodies worked on that and the deficiencies due to possible malfunctions of Software were recognized as one of the possible hazards.

14.3   SOFTWARE IN AVIATION

The Risk Assessment, as part of SMS, in aviation started from operations. Different aviation subjects, especially those who provided services (airlines, airports, ANSP, etc.) started with implementation of SMS. However, most of these subjects were not included into Software development, except maybe, by writing requirements what the Software must do.

Aviation Safety is based and practiced by managing risks in aviation. This includes preventing the aviation accidents and incidents through research, educating aviation personnel, passengers and the general public. In addition, it is dealing also with the design of aircraft and all supporting aviation infrastructure on the ground and at the sky.

14.3.1   ESARR 6

In Europe, in area of Air Navigation Services and Air Traffic Control (ATC), there were six documents prepared by EUROCONTROL[9] known as ESARRs (European Safety Regulatory Requirements). The ESARR 6 (Software in ATM[10] Functional Systems) is the one dealing with the Software. As it is written in its Abstract:

> ESARR 6 deals with the implementation of software Safety assurance systems, which ensure that the risks associated with the use of software in Safety-related ground-based ATM functional systems, are reduced to a tolerable level.

The ESARR 6 was maybe the first document which regulatory addressed the Software in industry and it deals only with ground-based ATM systems. It is small document, in total 20 pages, but it is very important. As it can be noticed, only

---

[8] Let me remind you, the Reliability is strictly connected with the equipment.
[9] EUROCONTROL is a European Organization for Safety in Air Navigation Services.
[10] ATM stands for Air Traffic Management.

"Safety-related" systems should be considered by this regulation. It means that Hazard Identification and Risk Assessment of aviation operations should decide which system is Safety-related.

ESARR 6 also introduces a "functional system" as "a combination of systems, procedures and human resources organized to perform a function within the context of ATM", so it is related not only to engineering systems, but the total approach is holistic by including also the humans and the organization inside the company.

The main point of this document is that it does state a requirement for Safety Assurance System for the Software used in ATM operations, but there are no particular categories or assurance levels for such a purpose. Actually, the Risk Assessment for the Software shall use the ICAO matrices for Risk Assessment.

This document is important also because it states that Safety Assessment System shall apply to the changes of Software, if done during its operational use. It means that each change shall be evaluated individually and holistically, in the scope of its function in the application.

14.3.2 RTCA DO-178C DOCUMENT

Regarding the Software development in aviation, there is another fundamental framework published in RTCA document DO-178C, for defining Development Assurance Levels (DALs) which are based on the particular risk (Table 14.1). This is a document which is mostly new, comparing it with the ESAARs and, it is dealing mostly with the aviation Software in engineering systems which support operations.

In general, the higher the DAL is, the more design effort must be put into ensuring Safety because the consequences of Software failure or malfunctioning of the Complex Systems in the aviation are more severe.

Regarding the Software, there are usually three Life Cycle Processes:

1 *Planning:* The first process is to plan how to document that your Software meets the Safety objectives. There should be documents with all your proofs and evidence to demonstrate the Safety.

TABLE 14.1
The DALs for Aviation Software Based on DO-178C[11]

Failure Condition	Software Level (DAL)	Number of Objectives
Catastrophic	Level A	71
Hazardous/severe—major	Level B	69
Major	Level C	62
Minor	Level D	26
No Effect	Level E	0

[11] Refer to Table 14.1, all processes and documentation in this paragraph are presented with permission from Intland Software GmbH (Stuttgart, Germany) and they are taken from their ebook *The Basics of DO-178C&DO 254.*

2 *Development:* This includes everything which is related to the development of used Software, including requirements definitions, Software architecture development, Coding, and integration with the Complex System.

3 *Integral:* These are integral processes which need to be part of any project which deals with design of the new system. These integral processes are Verification, Configuration Management, Quality Assurance and Certification Liaison. The result of following these processes should be evidence which demonstrates that you followed the processes planning. This is actually the evidence that must be provided to certification authorities or to Regulatory Bodies.

Some of the deliverables which needs to be produced during development of the Software and the documentation to be submitted to the certification authorities or Regulatory Bodies are:

- Plan for Software Aspects of Certification
- Software Quality Assurance Plan
- Software Configuration Management Plan
- Software Development Plan,
- Software Requirements Standard, Software Design Standard, Software Coding Standard
- Software Verification Plan
- Software Requirements Data
- Requirements Document
- Software Design Description
- Source Code, Libraries
- Executable Object Code, Parameter Data Item File
- Software Verification Cases and Procedures
- Software Verification Results
- Software Configuration Index
- Software Life Cycle Environment Configuration Index
- Traceability Data
- Software Accomplishment Summary
- Tool Qualification Plan
- Tool Qualification Data
- Software Quality Assurance Records
- Software Configuration Management Records, Problem Reports

All these documents should be provided to the Regulator and to the customer, as proof that the integrity of the Software is increased (meaning the risk of fault is decreased) to the highest level for particular application.

There is another document in aviation which is more Safety related and it is document published by SAE International. This is an organization from the United States which was dedicated mostly to automotive standards,[12] but today, the range of

---

[12] Previously, SAE stood for Society of Automotive Engineers.

its activities encompasses aerospace, automotive, and commercial vehicles. This is a document ARP 4761(Guidelines and Methods for Conducting the Safety Assessment Process on Civil Aviation Airborne Systems and Equipment). As the name shows, it is document dealing with Safety of Complex Systems installed in the aircraft. It is similar to V-model explained for automotive hardware, so I will not provide details here.

14.4   SOFTWARE IN OTHER INDUSTRIES

In the year 1993, ISO and IEC established the SPICE Project with intention to support the development, validation and use of an international standard for assessment of the Software developed for critical applications in the industry. That, what is interesting, is the fact that the initiative did not come from Software Developers, but from the companies from telecommunication and defense sectors. After few years, these two organizations published (gradually) a family of six standards known as ISO/IEC 15504 (Information Technology—Process Assessment).

SPICE stands for Software Process Improvement and Capability Determination. It is actually a product suite consisting of instructions, advices and practices to assess the developed Software and its application performance, during its life cycle. The performance of the Software is assessed through the Baseline Practices Guide (BPG) where six capability levels, as the criteria for assessment (Table 14.2), are defined.

Under the "umbrella" of Baseline Practices Guide (BPG), there are 35 processes listed, each of them belonging in one of these five categories:

1 *Customer Supplier:* These are the processes which directly impact the customer and his requirements (requirements, contract, audits, reviews, etc.).
2 *Engineering:* These are the processes which directly specify how the Software requirements would be satisfied (development, implementation, testing, integration, etc.).

TABLE 14.2
SPICE Capability Levels

Level	Activity	Description
0	Not performed	The assessment in this process is not performed and it is failure
1	Performed informally	The process is only informally assessed and only outputs of the process can prove its compliance
2	Planned and tracked	The assessment of the performance was planned and there are tracking documents which can prove that the process is validated and verified
3	Well-defined	There is well-defined process to assess the software performance and as such, there is procedure and standard to be followed
4	Quantitatively controlled	The software performance is assessed through process with detailed quantitative measures which are collected and used for assessment
5	Continuously improving	The quantitative measures from previous level are embedded into business goals and quality policy of the company. As such, they are subject of continuous monitoring and improvement

3 *Project*: These are the processes regarding conducting the project for designing and manufacturing the Software (scope, costs, resources, risks, quality, Safety, etc.).
4 *Support:* These are the processes which indirectly support other processes in the project (logistics, quality assurance, documentation, payment, etc.).
5 *Organization:* These are the processes which deals with business goals of the company and deals with development, products, resources, assets, training, continuous improvement, etc.

For those who would like to get involved into SPICE more deeply, there is plenty of literature on Internet.

14.4.1 SOFTWARE FOR AUTOMOTIVE INDUSTRY

One if the areas where the Software development for critical applications is regulated is automotive industry. I will present only one of many examples how Software in the cars has caused problems. In 2013, Honda has recalled 344,000 minivans Odyssey due to bug in the Software of the cars produced between 2007 and 2008.

In particular situations, calibration of the braking system could go wrong and it would result in increasing the pressure in the braking system. This increased pressure would manifest as pressing the brake by the driver (although he will not do it) and the car would stop immediately without rear braking lights to go on. If this happens, the cars behind will not know that the car in front of them brakes and the crash is inevitable.

The new trend in automotive production are the autonomous vehicles, where the functioning of the Software makes difference between crash and safe driving. The standard Part 6 (ISO 26262:2018 Road Vehicles—Functional Safety—Part 6: Product Development at the Software Level), deals with development and testing of such a Software.

Part 6 of the standard is for vehicle SDLC (Software Development Life Cycle), which is defined as development, production, operation, service and decommissioning of Software for vehicles (cars, motorcycles, buses, trucks, etc.).

I will not continue here with ISO 26262 Part 6 because almost everything which was said in Section 11.4.1 applies also for Software. In addition, in the scope of activities of SPICE project there is particular deliverable for automotive software and it is known as Automotive SPICE (A-SPICE). Readers may find more details on the Internet, because explaining all these things in this book is too much.

14.4.2 SOFTWARE FOR NUCLEAR INDUSTRY

There are two aspects of Nuclear Industry: The military and the civil. The military aspects deals with use of nuclear ballistic missiles and civil aspect is for use of nuclear energy for producing electricity.

US Department of Defense (DoD) has produced two methods for checking Software used for ballistic missiles launch. They are known as Nuclear Safety Cross-Check Analysis (NSCCA) and Nuclear Safety Analysis and Technical Evaluation

(NSATE). Both of them (NSCCA and NSATE) have task to manage software Safety risk within intercontinental ballistic missiles and shall be provided through Independent Verification and Validation (IV&V).

The NSCCA method is based on many techniques with intention to prove, with a high degree of confidence, that the used Software will not produce any undesired operation. The overall analysis is based on two components: Technical (dealing with engineering systems) and Procedural (dealing with humans involved in operations).

The first component evaluates the Software by multiple analyses and associated test procedures to prove that it satisfies the Safety for nuclear missiles. The second component is more dedicated to Security than to Safety. It provides preventive measures to protect the nuclear missile system against sabotage, collusion, compromise, and cyberattacks.

The NSCCA implements two-step criticality analysis. The first one is to identify the specific requirements as minimum measures necessary to demonstrate that the Software used in nuclear missile system is safe according to the DoD standards. The second one is to analyze each particular software function which controls or influences the missile before and during the launch. The judgment used for this purpose is qualitative (high, medium, low), and there are proposed measures for choosing the best methods to measure the software functionality.

The NSATE deals with technical evaluation of equipment (Hardware) and it is can be done by any method for Safety assessment of engineering systems.

This is for military, but for civil use of nuclear power, the IAEA (International Atomic Energy Agency) has produced Safety Guide under the code No. NS-G-1.1 titled "Software for Computer Based Systems Important to Safety in Nuclear Power Plants". In this document, there are few chapters dealing (in 23 pages) with recommendations about Software Requirements, Software Design, Software Implementation and Verification and Analysis. This document (97 pages) is supported by many others documents and all of them are available on IAEA website.

But it must be mentioned that the Safety in Nuclear (civil) Industry is mostly based on providing excellent Quality.

14.5   SOFTWARE AND HUMAN FACTORS

The Software is actually the "point of contact" (interface) for the humans with the engineering Complex System. So, this interface shall "enable" humans (employees, operators, etc.) to do their job successfully. Or in other words, this interface shall be developed as "Poka-Yoke": It may not "disable" the humans!

The first thing is to provide interface which will allow smooth operation (process, activity, etc.) of the Complex System with all information about the process being available to operators. The capability to monitor and control vital functions of the Complex System shall be part of the Software. It has been shown that most of the bad things happen from unanticipated events (Black Swans), so the proper monitoring and timely control will help very much.

The second thing is capability to have clear control and monitoring of the data which need to be put in the system to provide correct operation. There shall be two-step verification (at least). Also, there shall be a particular balance of the amount, type, and structure of the information which will be presented to the operator. In the cases of normal and emergency conditions, the quantity of presented information shall differ in order to optimize the operator reactions.

The third thing is proper allocation of the tasks between the Complex System and the operator. It shall be optimized to help humans easy to operate. The best is, if the job is done by the Complex System and the operators should just monitor (supervise) and maybe slightly interfere with the system. In general, in cases of considerable uncertainty, it is better the decision to be made by humans, than by Software. It is good if this task allocation is more partnership than Master-Slave relation. This task allocation could be achieved in the cases where, if there is problem with the system, the computer will provide information about the problem and it will provide possible solutions, but the operator will decide what to do.

However, the task allocation must be made (I will repeat again) to "enable" humans to be active part in the operation. It has been noticed in aviation that the pilots, who fly highly sophisticated aircraft, are so passive during the flights that if a problem occurs, they need more time to understand what is going on. The total automation and Complexity of the aircraft actually make them unprepared for emergency situations.

15 How to Improve Safety of Complex Systems?

15.1 INTRODUCTION

As I have mentioned before in Section 3.5 (Reliability), the Reliability is a probability that the Complex System will function without fault for particular period of time. Not having a fault of Complex System in Risky Industries, usually means that there will not be a possible hazard for humans, assets or environment. For Complex Systems, providing high Reliability is good for providing Safety, but it is not enough.

The Safety problem with all these Complex Systems is that designers try to build them, but there is considerable risk about operators ability to understand all interdependencies and interactions. Sometimes, especially in the cases of Software, even the designers (developers) cannot predict all uncertainties and bugs. In general, embedded Complexity makes it difficult for the designers to design a safe system. Mostly, the hazards and risks come from the fact that all the potential system states are unknown. If this happens to designers, you may assume what will be the situation with operators to manage all normal and abnormal situations.

However, the proper knowledge can, very much, help with all these understandings of the processes and interdependencies and interactions. It will help also with Risk Assessment, as well as with providing proper Preventive and Corrective measures.

15.2 MEASURES FOR ELIMINATION OR MITIGATION OF THE RISK

Dealing with the risks happens after the Hazards are identified and quantified regarding their probability (frequency, likelihood) and their consequences. This quantification is actually transformation of Hazards into Risks. We always try to eliminate the Hazards or, eventually, to mitigate the Risks by decreasing the probabilities of happening and their consequences.

There are few measures which can be applied for such a purpose:

(a) *Redesign:* This is something which can be implemented during design of the Complex System. It is common to say that the Safety starts with the first steps of design process and it is true. That is the reason for doing a Risk Assessment on the early stages of the design. Imagine how much it will cost if it is done when the product is ready for sale? By Redesign we eliminate or mitigate all Safety issues.

(b) *Isolation:* This is very useful and the PPE (Personal Protective Equipment) is excellent (and simplest) example for that. PPE isolates human bodies from

DOI: 10.1201/9781003404811-18

particular adverse events. Isolating nuclear reactor with particular shields and barriers to protect the humans, assets and environment from deadly radiation, is also a good example for that.

(c) *Automation:* I have already mentioned this at the beginning of the book. However, you should be careful with that. Eventually, Automation helps with Human "unreliability", but it is part of the "Safety Paradox".

(d) *Absorption:* The Absorption is actually a process of "absorbing" adverse energy which is created during the faults in the Complex Systems or the failures of complex processes (activities, operations, etc.). Simple example is taking antacid pills when you have heartburn in your stomach. The pills will absorb the excess of the stomach acid and it will help with your heartburn (which can cause stomach ulcers). This measure is very much useful in chemical industry, pharmacy, and medicine.

(e) *Dilution:* This is also common in industry and in life: Dangerous acids (or chemicals) spilled on human skin or sensitive surfaces (Metal, etc.) can be diluted by water (or some other liquid), air-conditioning systems can dilute toxic gasses from the air before they produce damage, etc.

(f) Etc.

Whatever measure you chose for elimination or mitigation of the hazards and risks, you must be aware that dealing with the risks is creative job and it needs to be effective and efficient.

15.3   HOW WE SHOULD ASSESS THE SAFETY OF THE COMPLEX SYSTEMS?

There is a need to provide assurance that the Complex System is safe system and it will not endanger the humans, assets, environment, etc. It means there shall be a verification and validation (through argumentative and documented Safety Case) that following things could happen in the case of fault or failure:

1 A fault (failure) reasonably (during normal operation) could occur very rare.
2 If a fault (failure) occurs, it will not endanger anything (the system will "Fail Safe").

The first and most important thing about any Complex System is to gather extensive knowledge how the system is built and how it will function. This is important for the Safety Assessment of the Complex Systems which are designed, purchased, and installed by some company in any industry. Of course, it must be done for any system (process, operation, activity) in any Risky Industry.

It is assumed that the design of the Complex System is done with a due Safety diligence from the designer and now, the user of the system must take care for the operation. The accidents rarely happened during designs, although the errors and wrong decision-making are more present there than during the operations.

I wrote above "to gather extensive knowledge" and it means, the operators (users and maintenance employees) taking care for the operation and maintenance of the

Complex System, must know (in details!) how the system work and they must know any function and any limitations of each Subsystem. Going further, it is important for them to be familiar with underlying scientific and engineering laws and principles which are used to build and to operate the Complex System under consideration. They must get familiar with all control circuits and all (human and engineering) constraints, especially how the data is gathered, measured, and processed, to provide control over the system. That is the reason for this book: In my 30-years engineering experience and 16 years Quality and Safety experience, I have noticed huge ignorance of many engineers about "forgotten aspects of equipment Complexity" which are considered in this book.

So, now let's go and see how the Complex System should be analyzed. I would like to propose a procedure how it can be done. Of course, this is just proposal and you may decide to use some other steps.

15.4 CHANGE OF THE PERCEPTION TOWARD COMPLEX SYSTEMS

Engineering Complex Systems are full with Complexity and uncertainty which comes inherently from their non-linearity and constraints imposed by laws and rules to keep them operational and safe. As such, they cannot be assessed by traditional methods and, in addition they need a shift in the attitude and an expanded set of scientific tools and engineering techniques for their assessment.

In this section, I would like to emphasize some of the changes of attitude toward Complex Systems which we need to acknowledge and embrace to deal with their Complexity. The written principles below should encourage the Safety Professionals to think differently and to apply particular "paradigm shift" about how to assess the Safety of Complex Systems:

1 Designer's Team should be open-minded and ready to recognize and acknowledge Complexity, control and variety which would influence the system. Installation of the designed and tested system on the site, still needs to be considered as "unmapped territory" for its operation. You need to reconsider the uncertainty of operation in this "unmapped territory" and to be skeptical about knowledge gathered through modeling, simulation and testing. The operational and maintenance employees should be open to learn from theirs and system's failures.

2 Designers should be open to provide adaptivity necessary to cope with Complexity by identifying and creating the best version of the Complex System. This means, for example, to think about the "influences" and "interventions" on the Complex Systems by the external factors rather than to rely on the "controls" and "resilience design". Designing adaptable Complex System requires a good knowledge, creativity, open mind and bravery to investigate and accept good and strange ideas.

3 During the design, try to provide resilience, self-organization, realistic models and do not be limited only to known events. Do not forget that the system must be "tough" to deal with known risks and to be "elastic" to deal with unknown risks.

4 During the testing, pay attention to the patterns in the behavior of the Complex System. These which have been observed, must be thoroughly understood. Maybe they are a key mechanism in the emerging Complexity. These patterns could be used later to deal specifically with unconsidered side effects or even, to be a mean for avoiding undesired and uncovered side effects.

5 Do not forget that the Subsystems inside the Complex Systems are connected not only "horizontally" but also "vertically"! The hierarchy must be understood at a single scale and at many different scales by iteratively zooming in and zooming out. This should determine the external and internal borders of the system and it can help with the problems which must be addressed at a higher or lower hierarchical level. The Complex Systems must be understood from the point of interactions and interdependencies which can arise from the bottom-up through external influence factors. Relying only to control hierarchy (top-down approach) is not wise in the case of Complex Systems.

6 You may not miss the Complexity of the environment and its influence on the Complex System installed on the site. All validation and verification should be done after installation during SAT (Site Acceptance Test) by thinking about evolving an operation and maintenance of the Complex System in its operational environment. The FAT (Factory Acceptance Test) is simply not enough.

7 A Complex System often looks very different from the perspectives of the designers, manufacturers, user and maintenance staff. There are multiple perspectives and a Safety Professionals must strive to find creative ways to spread a common understanding and to eliminate or mitigate the risks.

8 Collaboration during design, manufacturing, operation and maintenance of different Teams regarding the Complex System is a necessity. At least, in the first few years of exploitation, it should be present. It should include accurate information sharing, active listening and involvement, candid dialogue based on facts, good reasoning, and making decisions based on reality.

9 Optimization on site could be often tantalizing task with Complex System. Very often, the system as whole, is sub-optimized or some of the Subsystems inside are sub-optimized. Optimization could lead toward a rigidness in the operation of the system which become unable to adapt with changing parameters. Sometimes it is wise to achieve a balance in operation instead to push blindly toward optimization. Do not forget: The goal is to have a Complex System that would continue to meet the requirements based on operation, even if a number of current influence factors change.

10 After installation and during the operation, the operating and maintenance staff must understand that, the problems and opportunities with the system, will continually emerge. Many of these emerging problems will show up in surprising and unpredicted ways. A traditional approach to risk eliminations and mitigations should take care of system Complexity using opportunity to learn from these situations.

11 Have in mind that, used Control Systems inside are mostly based on feedback mechanism and computer algorithms. Over time, sensors and actuators for operation of Feedback loops can either get out from their operational tolerances or may affect the stability. There is a must, during scheduled maintenance, to check them and to calibrate the sensors in regular intervals.

12 Pay attention to the relationships among faults and failures, rather than addressing each fault separately. In Complex System, during any faults of some of the Subsystems and components, there are also secondary faults triggered by the primary one. Looking for the Root Cause of the faults and failures will help to solve the secondary ones.

13 Relying only on the Safety Manager or Safety Department for providing Safety in the cases of operations of Complex Systems could be rectified in some areas. The nuclear reactor is, pretty much, "closed" system, protected maximally from the external influence factors, but commercial flights of aircraft are "open" system where flights are provided not only by pilots, but also by Air Traffic Controls, meteorological offices and airline centers. Here, the synergy and synchronizations of operations is important, but it is also a complex job.

The principles and approaches described above are important points of providing Safety in areas where engineering Complex Systems operate.

15.5   HOW SUCCESSFUL CAN BE A RISK MANAGEMENT?

The search for the Root Cause usually starts after the adverse event happened and the point is that this hindsight not always helps with the reasons for accident. Usually, knowing the facts after accident happened, cannot always provide good knowledge about the reasons of such a behavior of the system or about the state of the humans operating the system.

The point is that, with the Complex Systems, the uncertainty during their operation is always very high and the operators must bring their decisions in the presence of this uncertainty. In addition, the investigation and determination of possible Root Cause will provide need for some additional Preventive and Corrective Measures. It is understandable that these measures could increase the Complexity of, already highly Complex System and, as such they will bring additional uncertainty. So, due diligence must be dedicated to all these Preventive and Corrective measures.

However, in this particular case of non-linear and full with uncertainty, Complex Systems, we cannot always know all possible states of the systems. It means that there are "unknown unknowns" in our system which we need to take care to be safe for humans, assets, and the environment. Having in mind that we use probability to deal with Risks and uncertainties, maybe the total probability equations for the Complex System can be expressed by this one:

$$P\left(adverse\ event\right) = \left(\sum_{n=1}^{k} p_k\right) + P_{NA} = 1$$

Here, $P_{(adverse\ event)}$ is the probability of any incident or accident to happen; p_k in the equation are different types of incidents or accidents which may happen due to known faults and failures; and P_{NA} is the probability of adverse events triggered by the unknown states (Black Swans) in our Complex System.

The P_{NA} is actually a reason why there is wrong understanding of probability between the engineers. Many of them are not aware that, in real life, even if we have 0 (0%) probability our Complex System to be in particular state, it may happen. And even if we have 1 (100%) probability, that the Complex System may reach particular state, it may not happen.

In other words, the P_{NA} is the probability that the Hazard, caused by unknown state of our Complex System, will materialize as an adverse event. If, during analyzing the Complex System, some of the states were not assumed that can happen, there is no chance to identify the Hazards associated with them as well as to calculate the probability or severity of consequences for them.

The inferences gathered by thinking in advance, which are result of the regulations, require a process of Risk Assessment to be based on probability (likelihood, frequency) and they cannot always provide good insight. However, thinking in advance for providing Safety by implementing Risk Management have showed their values already in many Risky Industries. As support for this statement, I am providing two figures (Figures 15.1 and 15.2) with statistical data[1] from 1942 until 2021 regarding aviation accidents and fatalities. In these figures, there are gray areas showing the period where implementation of SMS become obligatory for all aviation subjects.

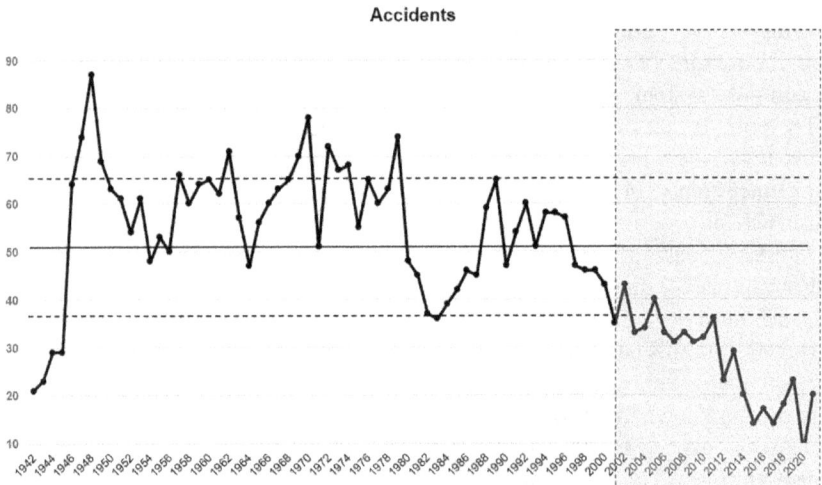

FIGURE 15.1   Statistics on all aircraft accidents from 1942 until 2021.

---

[1] Data for producing these figures is used from http://aviation-Safety.net/statistics/period/stats.php?cat=A1

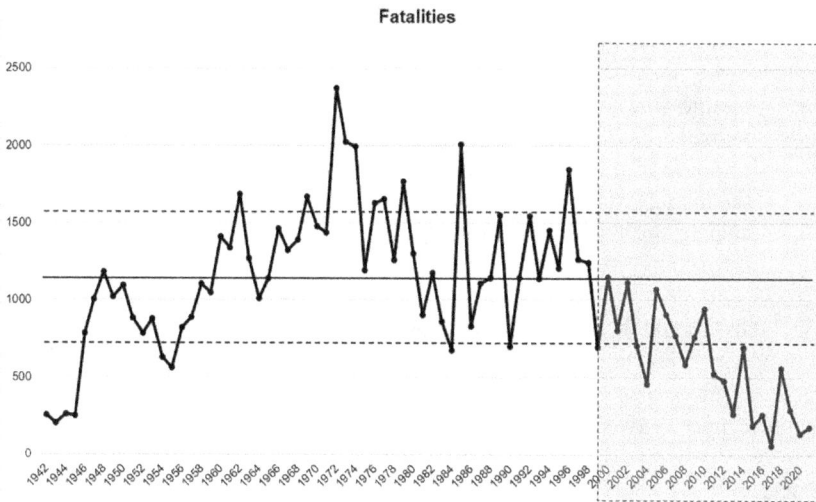

FIGURE 15.2   Statistics on all fatalities from aircraft accidents from 1942 until 2021.

The figures are self-explanatory and I will not additionally comment them. However, it can be noticed that the implementation of the systematic (proactive) way of handling incidents and accidents, gives results.

15.6   PROCEDURE FOR RISK ASSESSMENT OF INSTALLED COMPLEX SYSTEMS

This procedure for Risk Assessment of installed on site Complex Systems, applies to established Teams consisting of employees (experts) from all different departments (Safety, quality, managers, operational and maintenance) in the company—purchaser of the Complex System. These are guys who should be included in operational and maintenance activities of the operational systems. It is wise to have highly educated persons with considerable knowledge and experience in operations, processes, activities, covered by the Complex System or its operation.

It is assumed that the "mother" industry, where the company (purchaser of the Complex System) operate, already have devised clear Safety criteria and the company has accepted them or has produced stronger criteria than these of the "mother" industry. Also, it is highly recommended to do this assessment before the system is operational. It can be done later, but it will be more problematic and less effective and efficient.

The Risk Assessment of the Complex System should be executed by following these steps:

1 Define the system in detail and pay attention to
   (a) Underlining scientific and engineering laws used to build and operate the system;

(b) Define (without any doubt) the boundaries of the operation and the limitation of the system;

(c) Try to understand the number and type of operational and Safety constraints and how they are imposed (by device, by environment or by humans). Try to clarify yourself, why exactly the human constraints are imposed in particular phase of the operation of the system. Usually, if the manufacturer realize that Control System is too expensive, they will change the control engineering constrain with the human constraint (procedure).

These items should help you to diagnose, is the system under consideration a Complex System and how to proceed with assessment. Maybe, there will be a chance to remove Complexity (considering Safety) from the system and/or transfer it on some other more convenient place.

2 Decompose your system, in many ways by producing diagrams/schematics/charts about the system parts:

(a) Subsystems;

(b) Flow diagrams for interchange of information (for operation, for data and for control);

(c) Process flows of energy, data, materials, etc.;

(d) Operations supported by the system. After particular period of time (few months), if there are more than one operational system of the same type, compare the gathered data and analyze particular discrepancies between the systems, if they exist;

(e) Systems and human tasks. This relies to the operational tasks which system will support/execute and human involvement in these tasks;

(f) Procedures (operational, maintenance and management). If some process, operation or activity is not covered by procedure, then the procedure shall be produced;

(g) Links and interfaces;

(h) Interactions and interdependence between the parts of the systems. This should determine the hierarchy of controls and should be used as process model (depending on Complexity of the system). As such, the interactions and interdependencies must be considered horizontally and vertically, based on their hierarchical structure; and

(i) Interaction and interdependencies of the system with environment and with humans (operators, monitoring staff and maintenance staff).

3 If the system is not operational yet, try to perturb the system. But be cautious: Do not damage something!

It is wise if this "playing" with the system should be done in the manufacturing company during staff operations and maintenance training. Check the actions, decisions and inactions of the system in different phases of the operation. Have in mind that in the manufacturing company, it is "controlled environment" and there are experts which will help you to understand the operational balances and the limitation of the system.

4 Focus on Hazard Identification process, which needs to be done through few brainstorming sessions until there is no more hazards, ideas or concerns expressed. The point is to focus firstly on possible faults or failures of the Subsystems and components and later determine how they will affect interdependencies and interactions of the Subsystems, in the scope of the Complex System and with humans and environment.

Eventually, pay attention to the constraints imposed by the Control Systems and their possible deficiencies and flaws. Do not forget the unknown states which could show up during system operation and they could be a result of non-linear dynamics of the operation (process, activity, etc.). Although the Complex Systems have many layers (Subsystems) for protection and control, be careful because always it may exist some Single Point Failure (SPF), usually expressed as unknown system state. This SPF must be eliminated and/or mitigated, because this is the last phase where it may happen.

5 Pay attention to the processes and their interdependencies and interactions between themselves and with humans and with environment. It is important to understand how the particular hazard could materialize and there is need to make a list providing descriptions how the adverse sequences could develop.

6 Calculate the associated risks for each hazard by using criteria established by industry or by the company.

7 Maybe there will be a need to calculate time of Fault Propagation in some of the Subsystems through the system which will result with failure of Complex System operation. If necessary, try to calculate this for any hazard and associated risks. Try to eliminate or to delay this time with additional engineering controls. If not achievable try to introduce administrative controls which, as last resort, will stop operation and/or switch off the Complex System. This could save the system from bigger damage.

8 Design controls which will eliminate or mitigate the determined hazards/risks. Have in mind that there should be engineering or/and administrative controls. The engineering controls are introduced by the Control Systems, but there are places where you may use redundancy (duplication) of the Subsystems to improve Reliability. Be careful with redundancies and do not exaggerate: They increase the cost and the complexity of the system, so maybe the operation will be too expensive or the system will not be sustainable.

9 The Administrative controls should be produced as rules of behavior for humans and Operational and Safety Procedures for system and for employees. The very much important point is to spread the draft-rules and draft-procedures to other employees involved in operation and in maintenance and ask them for comments. Give them one week for comments and analyze the received comments with due diligence. If necessary, speak with the employees who proposed changes through their comments. Maybe you have missed something which is important.

10 It is obligatory to provide training for the rules and for each procedure to all operational and maintenance staff. Also, during the training, clarify what were the reasons to produce such a rules and procedures. It is obligatory to mention adverse events which could happen, if the rules and procedures are not followed. In addition, the boundaries of safe performance must be thoroughly explained.

11 Check again how each of the hazards/risks are controlled by the established controls. If there are weak controls or some of the controls are missing, make list of each of them.

12 Produce scenarios for all possible bad situations or adverse events which could happen. Strengthen the reactions or Preventive and Corrective Measures for these adverse situations by producing backup plans and emergency plans for all possible scenarios.

13 Establish operational and Safety performance indicators which will help you to assess the ongoing operations. Clarify which of them are normal and which of them are proactive.[2]

14 Based on these performance indicators, provide rules and procedures how the monitoring, operation, maintenance, control and emergency actions shall be executed. The procedures can be Operational and Safety, but the rules must establish proper hierarchy of responsibility and decision-making in any case. Do not be confused with the step 9! This step is actually a follow-up from the step 8. Do not forget that the rules and procedures are life documents and, as such, they are prone to change to adapt on new situations! The companies and their operations are dynamic entities and as such, the timely adaptation is very important.

15 Provide comprehensive training for operation, maintenance, Safety, back-up and emergency plans and all operational and Safety procedures for all staff, depending on their involvement in the Complex System's operation and/or their roles in the company. Of course, the Safety Training must be provided to every employee! Maybe it looks strange, but have in mind that the training about Complex System is actually inherent part of operation of the Complex System. However, be careful: Not every employee will gather same level of knowledge from the training. It means that, different employees even in the same area of operations, will have different expertise in maintaining operation and Complex System. Experience in similar systems and operations is important value and having experienced person in any shift is necessity.

16 As I have mentioned in Section 2.4 (Simulations), in aviation the training of the pilots and Air Traffic Controllers (ATCOs) is done on simulators. In aviation, there is also recurrent training for them.

You must be clear is this a regulatory requirement for your industry also. However, having a training by simulation of real case scenarios, is the best way to train the humans!

---

[2] Normal indicators will show you how the operation is executed and proactive indicators are limits (tolerances) beyond normal indicators which would indicate that the adverse sequence can start and as such due attention must be paid to the engineering and Safety employees.

17 Do not forget: This is a book about knowledge of Complex Systems where one reason of Complexity is its (mostly non-linear) dynamics. As such, a profound monitoring of operation of the system must be provided non-stop and the data gathered from the monitoring must be used for continuously validation and verification of the operational and Safety performance of the system. Two things should be assessed as results of the monitoring: The normal operation and Safety concerns (events). It must be done through updated continuous analysis, by addressing system, human and environment dynamics, checking the implemented changes in the materials and in the operations and possible modifications.

18 If something is wrong during monitoring, it must be thoroughly investigated. If necessary, additional measurements and testing shall be done.

19 After particular period of time (few months), if there are more than one operational system of the same type, compare the gathered operational, Safety and maintenance data and analyze particular discrepancies, if exist.

20 Continuously monitor and assess the determined risk from step 6 and continuously try to improve it.

21 Any change in the system, procedure, operation, or employee could be a warning that the process of Risk Assessment must be done again. Simply, these changes could also produce changes in already identified hazards and determined risks, but, in addition, they may produce even new hazards and new risks. Mostly changes are done due some benefit and, in such cases, the bad things which can be brought by changes are very often neglected.

IMPORTANT THINGS:

(a) At any time, during previous steps, assure yourself that everything is clear and there is no uncertainty how the system works and how it is controlled!

(b) If there is any upgrade, change or modification of the Complex System or its environment, you maybe should repeat the all procedure or, at least, some of the steps!

(c) All these steps must be documented! Documentation will be used to provide Safety Case and to get license that the Complex System can be used in operation (after installation) in particular Risky Industry.

Those, who are familiar with Safety, they can recognize that steps 4, 5, 6, 7, 8, 9 and 11 are part of the classical Risk Management. For those who like more Safety-III (Leveson!) they can recognize the steps which are dealing with the STPA and STAMP. Anyway, the procedure shown above is my view how the things should go on with the Safety assessment in the company which will operate and maintain the Complex System. The similar procedure should be applied to the services-providing companies.

16 Final Words

Here we are . . .

At the end of the book our intention was to present the knowledge forgotten by Safety Professionals . . .

This knowledge should help them to better understand the structure and operation of Complex Systems. This better understanding could help with their decision-making process, but let's be honest: All these decisions would be made in the presence of particular uncertainty which is highly associated with our lives, as well as with our Complex (and other) Systems.

At the end of the day, if the decision which has been made was OK, no one will take care. But if this decision was wrong, then it will be a problem. After such an outcome, the reason for decision will be part of the investigation and having particular knowledge could help with rectification why this was made. But do not forget: The reason for having knowledge is to stop bad decisions, not to rectify them!

The science and the engineering have produced a lot of assets (methods, methodologies, tools, instruments, etc.), which could help humans in their activities. However, knowledge achieved by education and training for using these assets is nothing without proper knowledge about functioning and operation of the Complex Systems.

As it has been stated in the book, the Safety begins with the design process. The point is to design the Complex Systems which will accommodate the humans. Totally safe system cannot be designed and manufactured and systems (from time to time) will fail. They will sometimes fail inadvertently, sometimes by system error and sometimes due to some other reason. But let's make them to Fail Safe.

Humans will fail also, but let's create a Complex System, which besides its Complexity, will help humans with monitoring and control of this system. The Complex System should be designed with capability to enable early detection of the irregularities, their fast isolation or correction, and, as much as it is possible, recovery of the normal operation.

However, it is not enough. Sometimes these Complex Systems will be designed with due attention, knowledge, skills, attitude, experience, and efforts, even in accordance with all regulations and standards, but still, they can fail. In such cases, the knowledge, skills, experience, and attitude of the operators and maintenance staff could make a difference between life and death.

In everything which we do, there is particular hazard and associated risks. In Risky Industries, there are regulations which require to identify the hazards and quantify the risks. They should be verified and validated, and sometimes there is a need to be done compromise in risk acceptance of the adverse sequences not contained by the Risks Controls (Control Systems).

DOI: 10.1201/9781003404811-19

Maybe this is a good time to get back to the beginning of the book and read again what I have said about the doctors: For them, to be better in finding what is wrong with our bodies, it is of primary importance to know what is normal with our bodies!

I hope that many of the Safety Professionals will understand that and will apply it in their professional lives!

Index

For Product Safety Concerns and Information please contact our EU
representative GPSR@taylorandfrancis.com
Taylor & Francis Verlag GmbH, Kaufingerstraße 24, 80331 München, Germany

www.ingramcontent.com/pod-product-compliance
Lightning Source LLC
Chambersburg PA
CBHW060330220326
41598CB00023B/2665